Tae-Soon Park 2/10/98

Plane Geometry

Revised Printing

Ernestine Camp

Laney College

KENDALL/HUNT PUBLISHING COMPANY
4050 Westmark Drive Dubuque, Iowa 52002

This Book Is Dedicated to My Mother
Ethel Geneva Watson

Revised Printing 1997

ISBN 0-7872-3417-6

Printed in the United States of America
10 9 8 7 6 5 4 3

Contents

Preface

This book differs from a traditional geometry text in that formal deductive proofs are not introduced at the beginning of the course. The book has been divided into two parts. Chapters One through Four provide an opportunity for students to become familiar with the vocabulary of geometry and to obtain practice in making conjectures using both inductive and deductive reasoning. Only after students have had sufficient time to build up a working background of vocabulary and geometric relationships, are formal deductive proofs introduced in Chapters Five through Ten.

I have found that when many students are forced to cope with unfamiliar vocabulary and an unfamiliar way of thinking at the same time, they often find the task overwhelming. It is not unusual for large numbers of community college students to lack an introductory background in geometry. In far too many elementary and secondary school math classes, geometry still remains in the back of the text, and often is not covered during the school year. This text attempts to address the void resulting from this problem.

When given the time to develop necessary basic skills in geometry, students are more likely to finish the course with not only the geometry they need to know for other math courses but with an appreciation for the beauty in the structure of mathematics as well.

Other features of the text:

1. The text integrates geometry with algebra. Students can use their algebra skills in many of the exercises. This strategy lessens the concern that students' algebraic skills will become rusty while taking geometry between beginning algebra and intermediate algebra. Because the algebra has been included primarily in the exercises and not to a great extent in the main body of the text, the instructor has the option of determining the amount of algebra to be emphasized in the course.
2. Some topics are spiraled throughout the book. This strategy eliminates the need for cumulative reviews. Review is a natural consequence of the organization of the text.
3. The exercises provide numerous opportunities for students to explore and discover theorems that generally are proved in other texts. Exercises with asterisk should be assigned because the information contained in them may be needed for further work in the text.
4. The text is flexible. It can be used in a lecture/discussion or small group format. Chapters One through Four can be adapted easily to a small group format. Students can read the text at home and discuss answers to exercises in small groups in class, or small groups can present the lessons and lead discussions, with the instructor clarifying topics when necessary. Chapters Five through Ten may require more active instructor participation than the beginning chapters.
5. The text is concise and is designed to be covered in a traditional geometry class that meets three hours a week. It contains the important theorems that are generally covered in such a course. Appendices contain introductions to Solid Geometry and Transformations. These can be covered in the regular course or assigned as group or individual projects.

Ernestine Watson Camp
Laney College
Oakland, California

Acknowledgements

Special thanks to my family, Bill, Billy, Daryl and Tracy Camp; my colleague, Calvin Rouse; and my students, Tracy Sorrell Turner, Kamala Parks, Stewart Watson, Marsha Jordan, Simone Patton, Guadalupe Miner and Soroya Wood, for support and encouragement during this project. Also, special thanks to Suzanne Cancilla.

To the Student

Welcome to an Exploration of Your World of Geometry!

You cannot help but be aware of the use of geometry in the world about you. When you travel about in your city, you see geometry in the shapes of traffic signs. When you buy carpet for your home, you use geometry to determine the amount of carpet required. Geometry was used by the architect who designed your home, and by the cabinet maker who made much of the furniture that you use in your home. When you cross a bridge, you can see triangular shapes used in the construction of the bridge. Every can of food that you buy, has some geometric shape. Even the planets in your world take a geometric path as they move around the sun. So you see, the world of geometry is already your world. In this course, we simply seek to explore this world in an organized manner. By doing so, we hope that your appreciation for geometry, in particular, and mathematics, in general, will be enhanced.

As you begin your study, keep in mind that you must be actively involved to get the most out of the course. This means that you should always have a pencil in your hand and paper available when you study. With these tools at hand, you will be prepared to try activities when they are suggested in the text. **This text is a working text.** It has been written so that much of what you learn will come from the exercises that you will do.

Of course there is some memorization that you will be required to do, but this memorization will not be difficult if you understand the concepts involved, and stay current with your classwork and assignments. By all means, remember to keep a positive attitude as you work, and do not be discouraged if there are problems that you cannot solve. Geometry is probably the only lower level course in mathematics where you will have an opportunity to experience the life of a real mathematician. You may be surprised to know that a mathematician's delight is a problem that cannot be solved easily.

Regardless of the grade that you receive in this course, your experience will be a worthwhile one if you have learned some geometric relationships that will be useful in your other math courses, and if you have learned why the structure of mathematics is appreciated and valued in our modern world.

Euclid, the mathematician for whom this course is named, said to one of his royal students, "There is no royal road to geometry." Keep this quote in mind as you study, and you will find that the road of involvement and diligent work is by far better than any royal road.

Enjoy your geometry!

Ernestine Watson Camp
Laney College
Oakland, California 94605

Historical Notes

Much of the information about the use of geometry in ancient cultures is taken from interpretations of written records that have survived from those cultures. A large amount of the information that is available comes from the Babylonian and Egyptian cultures, perhaps because these cultures recorded their works on durable types of writing materials such as papyri (plural of papyrus) and baked clay tablets. Of course other ancient cultures used geometry to some extent, but there is little information that has survived about these cultures.

We know from the records of the Babylonians and Egyptians that a substantial part of the geometry taught in introductory plane geometry courses today was known in some form by these cultures. Both groups used geometry to solve practical problems in measurement, surveying and construction.

It was Thales of Miletus (about 640–546 BC), a wealthy Greek merchant, philosopher and mathematician, who is credited with moving geometry from the world of concrete practical considerations to the world of abstractions. During his business trips to Egypt, Thales learned the geometry of the Egyptians and took this information back to Greece. There he not only expanded upon what he learned, but began to look at this information in a new light.

Pythagoras (about 548–495 BC), a student of Thales, also traveled to Egypt and absorbed the knowledge of the Egyptians. After spending time at the Library of Alexandria in Egypt, he founded a school in Crotona, a Greek town in the southern part of Italy. His students organized a religious society known as the Order of the Pythagoreans. This group made numerous discoveries in geometry, many of which were credited to their teacher, Pythagoras. The most famous of these discoveries was the Pythagorean Theorem, a generalization about the lengths of the sides of a right triangle.

Introductory courses in geometry take the name Euclidean geometry after Euclid (about 300 BC?), a teacher at the Library of Alexandria during the rule of the first Ptolemy. Euclid's book, The *Elements*, considered by many to be the greatest mathematics textbook ever written, contained an organized presentation of much of the geometry known in his day. This book, in tune with the philosophy of Plato, contained no practical applications of geometry. It represented the first attempt in history to organize a branch of knowledge into what we now call a deductive system. Such a system begins with assumptions about undefined terms and makes conclusions on the basis of these assumptions through a process called deductive reasoning. Many of the assumptions that Euclid used in his book are still used in varied forms as assumptions in Euclidean geometry courses today. As a result of Euclid's work, people in other fields of study still work at organizing their knowledge into a deductive system.

A French mathematician and philosopher, Rene Descartes (1596–1650) provided the stimulus for the next important development of geometry. It was Descartes' use of numbers to represent the location of points in space that gave rise to the branch of mathematics called analytic geometry. In analytic geometry, problems in geometry are studied using algebraic techniques, and algebraic properties of equations are studied by using their geometric representations.

Paralleling the development of analytic geometry was the development of projective geometry by Girard Desargues (1593–1662). Projective geometry deals with those properties of geometric figures that are not changed by projections or sections. Descriptive geometry, which is closely related to projective geometry, deals with the two-dimensional representation of three-dimensional objects by drawing different views of the objects in such a way that all the spatial properties of the object can be reconstructed from the drawing. Gaspard Monge (1746–1818) is responsible for much of the development in this field of geometry.

From the time of Euclid's *Elements,* mathematicians were bothered by his assumption about parallel lines, that is, the assumption that not more than one line can be constructed through a point not on the line, parallel to the given line. In the nineteenth century, mathematicians realized that there existed systems of geometry for which Euclid's assumption did not hold. There were several mathematicians who pioneered the development of these systems. Among them were Saccheri (1116–1733), Gauss (1777–1855), Bolyai (1802–1860), Lobachevsky (1793–1856) and Riemann (1826–1866). The geometries that they developed are called modern geometries or Non-Euclidean geometries. Although Euclidean geometry is just as applicable today as it was in the time of Euclid, the newer geometries also have been found to be useful as models especially in the field of nuclear science. Einstein found one of these modern geometries to be a more consistent model for his theory of relativity than Euclidean geometry.

Finally, we consider the development of a branch of mathematics which is somewhat related to geometry. This branch of mathematics, called topology, deals with the properties of a geometric figure that remain unchanged after the figure has been distorted by shrinking, stretching, twisting or bending. The French mathematician, Henri Poincare (1854–1912) is credited with the development of topology. Euler (1707–1783), a Swiss mathematician, is credited with making inroads into network theory, which is now considered a branch of topology.

Chapter One

Basic Concepts and Relationships

Objectives

Recognize and use basic terminology and notations for points, lines, line segments, rays, angles, lengths of segments and measures of angles

Know relationships between special angles: adjacent angles, complementary and supplementary angles, linear pair angles, vertical angles, angles formed by an angle bisector, angles formed by transversals

Know the relationship between the two segments formed by a bisector of a line segment

Know relationships between parallel lines and perpendicular lines

Measure an angle using a protractor

Convert from one unit of angle measure to another

Determine the distance: between two points on a numberline; from a point not on a line to the line; and between two parallel lines

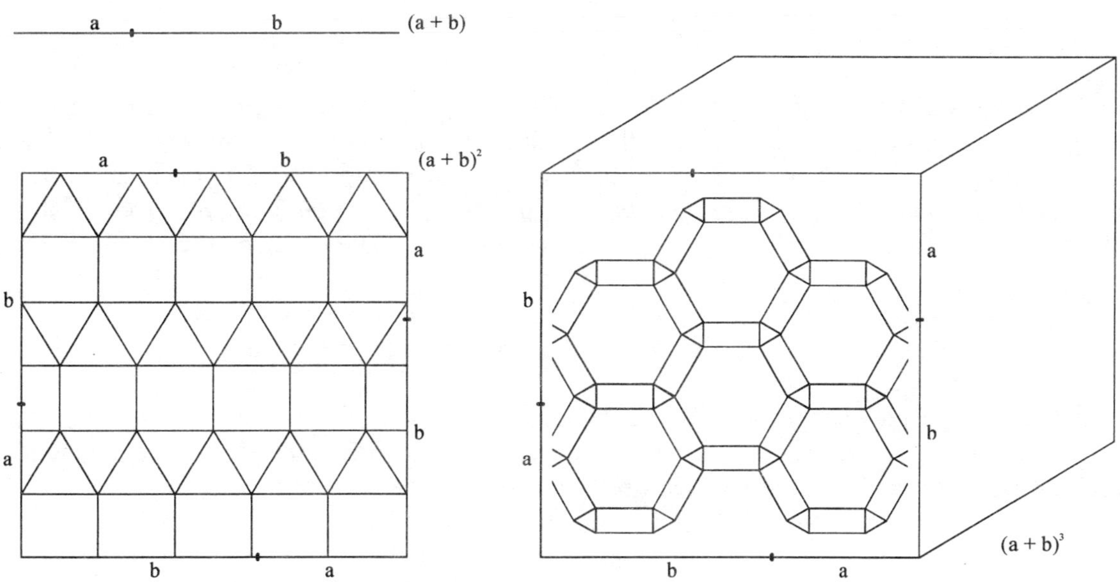

Section 1.1 Basic Concepts

The most basic of all concepts in geometry is that of a **point**. We may think of a point as a location or position in space. A geometric point has no dimension, that is, it has no length, width or thickness. It is customary to represent a point with a dot. *To name a point we use an uppercase letter of the English alphabet.* **Space may be considered as the set of all points.** In this course, we will study geometric figures that are composed of points.

We will think of a geometric **line** as a set of points that can be arranged along a straightedge. Such a line contains an infinite number of points and it extends forever in opposite directions. A line has no endpoint and it has only one dimension and that is length. In this course we will *name a line by using a lowercase letter of the English alphabet or by using the names of any two points on the line.* In the following illustration, we may refer to the given line as line ℓ, line **RS** or line **SR**

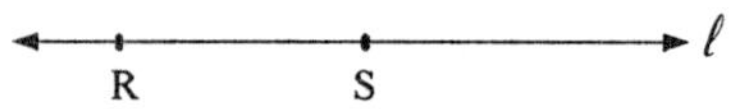

Line ℓ OR line **RS** OR line **SR**

The two arrows indicate that the line has no end. Note that points R and S are not the only points on line 1. Between any two points on a line there is always another point. When we place dots for R and S on the line, we are simply directing attention to those two specific points. Always we must keep in mind that the line contains an infinite number of points.

Points that lie on the same line are called **collinear**. In the example that follows, P, Q, and R are collinear.

When two geometric lines **intersect**, we say that they intersect in exactly one point. By that we mean they have one point in common. In the illustration that follows, lines m and n intersect in point P.

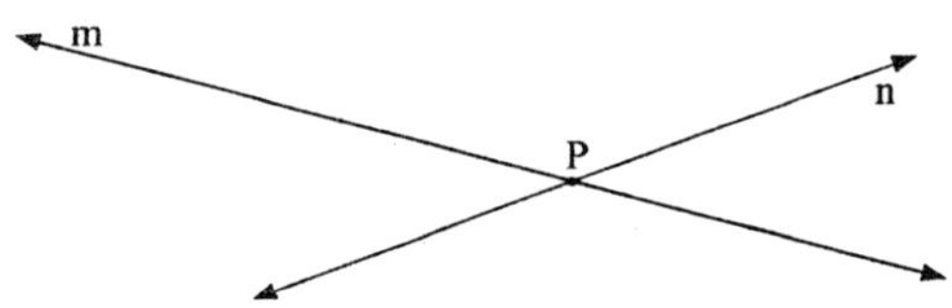

Now let us consider geometric figures that are parts of lines. A **line segment** is that part of a line which contains two points from the line together with all the points on the line that lie between those two points. A line segment differs from a line in that it has two endpoints. As you recall, a line has no endpoint. We will *name a line segment by using the names of its two endpoints.* In the illustration, we may call the line segment, line segment RS or line segment SR. The order of the letters is not important.

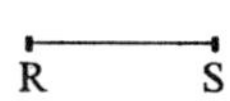

line segment **RS** OR line segment **SR**

Associated with each line segment is a positive real number that we call the **length of the line segment.** This number represents the distance between the two endpoints of the line segment. As you may recall from other math courses, we can set up a numbering system on any line so that there is a one-to-one correspondence between the points on the line and the set of real numbers. Such a line is called a numberline. When a numbering system is set up on a line, each number that is assigned to a point on the line is called the **coordinate of that point.** In the following illustration, the coordinates of points R, S and T are 6, 7 and 8, respectively. *To determine the distance between any two points on the* *line, we simply subtract their coordinates, and then take the absolute value of the difference.* The length of line segment RT is obtained by doing either of the following:

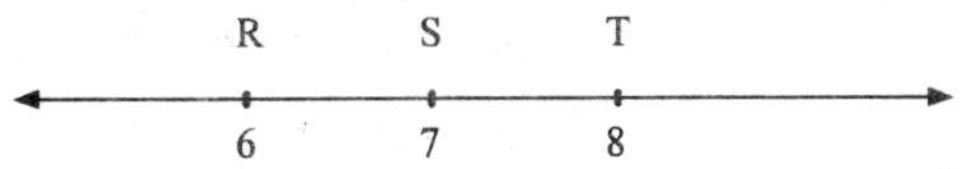

$$|8 - 6| = |2| = 2 \text{ OR } |6 - 8| = |-2| = 2$$

A **ray** is also a part of a line. It consists of a point on the line together with all points on the line that lie on the same side of that point. A ray then has one endpoint and it extends indefinitely in only one direction. We will *name a ray by first giving the name of its endpoint and then giving the name of any other point on the ray.*

In the illustration, the ray that lies to the right of R may be named ray RS or ray RT.

ray RS or **ray RT**

Notice that each point on a line separates the line into two rays that have only that point in common. These rays are called **opposite rays.** In the previous example, ray RT and ray RQ are opposite rays.

For convenience in this course, we will agree that the following *notation may be used to distinguish lines, line segments, lengths of line segments and rays:*

The line containing R and S	$\overleftrightarrow{RS}$ OR $\overleftrightarrow{SR}$
The line segment joining R and S	$\overline{RS}$ OR $\overline{SR}$
The ray containing S with endpoint R	$\overrightarrow{RS}$
The ray containing R with endpoint S	$\overrightarrow{SR}$
The length of the line segment joining R and S	RS OR SR

We agree that the use of this notation is not necessary when there is no doubt about the geometric figure being discussed. Therefore we may refer to the line through A and B as *line AB* or simply $\overleftrightarrow{AB}$.

Also, if there is no problem of confusion, we need not stress notation.

In this course, most of our work will center around **plane figures,** that is, geometric figures that lie in a flat surface called a **plane.** We will think of a geometric plane as being composed of an infinite number of points. Having two dimensions, length and width, it extends indefinitely in every direction along the flat surface. We will *name a plane by using lowercase letters of the English alphabet.*

Space, as mentioned earlier, contains all points. When we place dots and other marks on our papers to represent geometric figures, we are simply directing attention to a particular point or to a set of points that already exists in space. We must keep in mind that space contains all points whether we choose to acknowledge or to ignore their existence. For example, line segment AB is indicated by the solid line in the following illustration. The line containing the segment also exists although it is not shown by a solid line in the illustration.

Exercises 1.1

Answers to selected exercises are given in the back of the book.

For these exercises, use only those points that are named in the figures.

1. Give all possible notations for line l. Do not give notations that include the word "line."

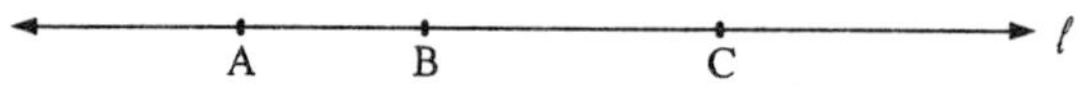

2. Give all possible notations for the line segment joining M and P. Do not give notations that include the words "line segment"

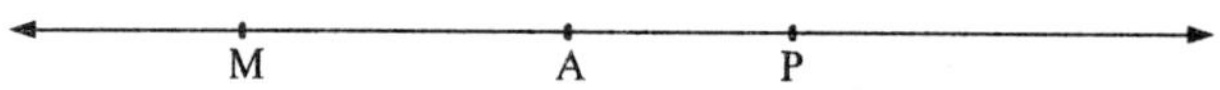

3. Give all possible notations for the ray that is shaded in the figure. Do not give notations that include the word "ray."

4. Name the two opposite rays in the drawing.

5. Explain the difference between each of the following:

   a. a ray and a line segment

   b. a line and a line segment

   c. a line segment and the length of the line segment

6. Which of the following pairs are different?

   a. $\overline{PQ}$ and $\overline{QP}$
   b. $\overrightarrow{PQ}$ and $\overrightarrow{QP}$
   c. $\overline{PQ}$ and PQ
   d. $\overleftrightarrow{PQ}$ and $\overleftrightarrow{QP}$

7. Give notation for each of the following:

a. the line containing points H and E
b. the line segment with endpoints H and E
c. the ray that lies to the right of point O with endpoint O
d. the length of the line segment joining H and E

8. Shade $\overrightarrow{KL}$ in the figure.

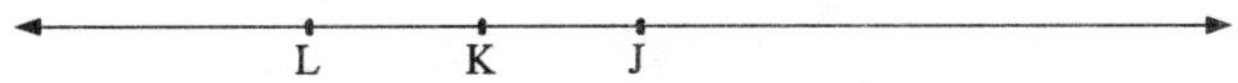

9. In the figure, $\overleftrightarrow{AC}$ and $\overleftrightarrow{BE}$ intersect in what point?

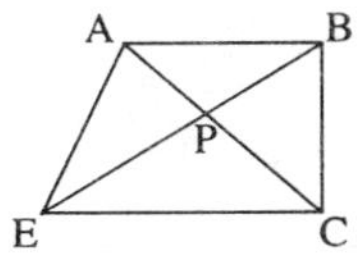

10. Name all the line segments in the drawing.

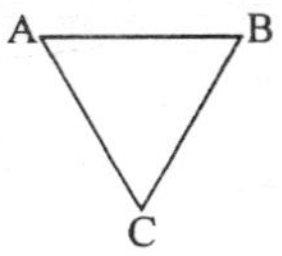

11. Find the lengths of the line segments indicated below. Do not use a ruler to measure. The figure is not drawn to scale.

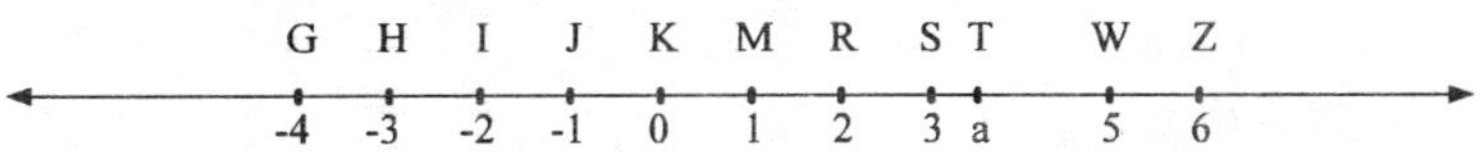

a. WR
b. RW
c. ZM
d. SW
e. KW
f. RH
g. GI
h. HJ
i. KG
j. HT

12. Make drawings that illustrate each of the following conditions:

a. line s and line t intersecting at R

b. opposite rays MN and MP

c. collinear points X, Y, and Z

13. How many points does a line contain?

14. How many lines are determined by the following number of points if no three of the points are collinear:

a. 2 points

b. 3 points

c. 4 points

d. 5 points

e. 10 points

f. n points (n is a counting number greater than 1.

15. The drawing shows line segments, AP, PB, PC, AD, AC, BC, BP, BD, CD, and DP. What other line segment may be drawn using the points shown in the drawing?

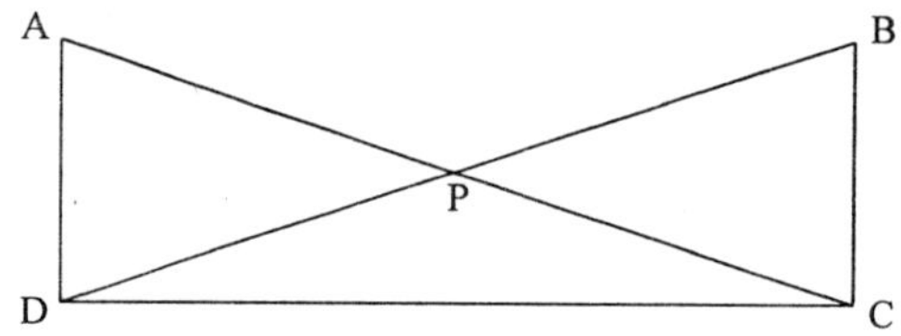

16. Plot each set of points on a rectangular coordinate system; then determine which of the sets contain all collinear points.

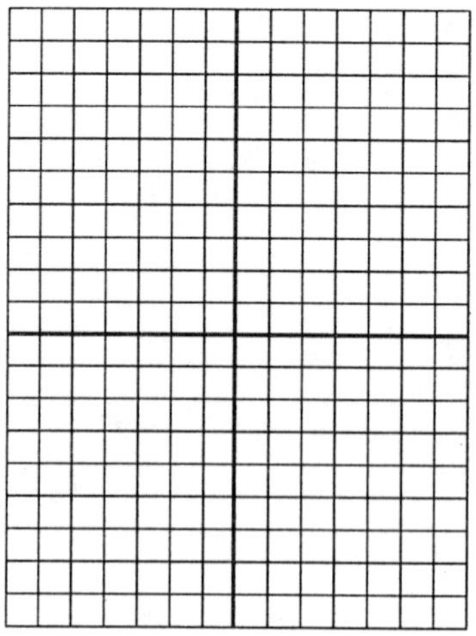

a. { (2, 3), (2, 5), (2, –4), (2, 0), (2, –3) }

b. { (5, –4), (2, –4), (4, –4), (0, –4), (6, –4) }

c. { (–3, –5), (–1, –3), (0, –2), (1, –1), (2, 0) }

d. { (1, 1), (2, 2), (3, 3), (–2, –2), (0, 0) }

17. Graph the lines represented by each pair of equations; then determine from the graph the coordinates of the point of intersection of the lines.

a. $x + y = 8$
$x - y = -2$

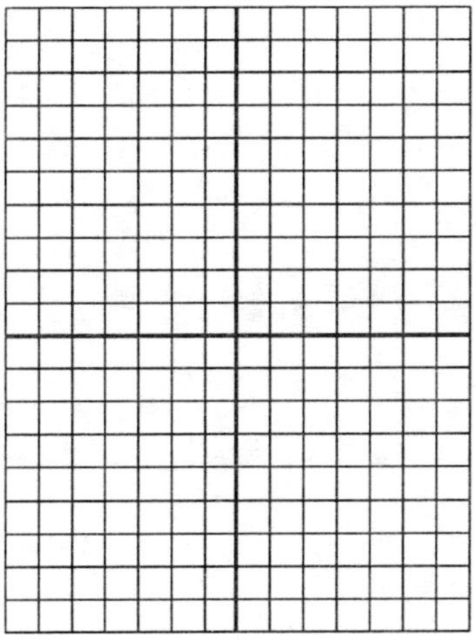

b. $y = x$
$y = -x$

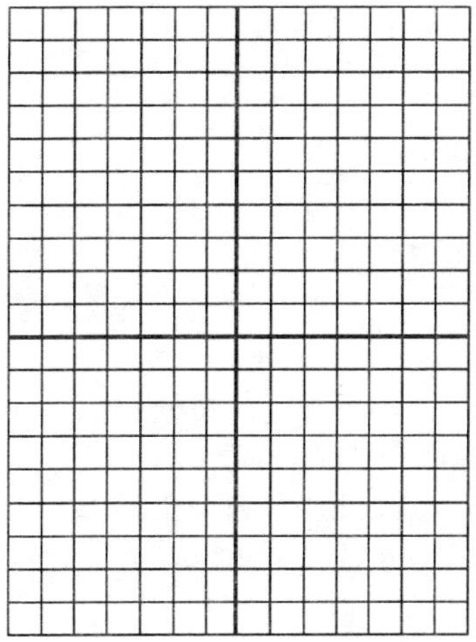

c. $y = 3x + 2$
$y = 2$

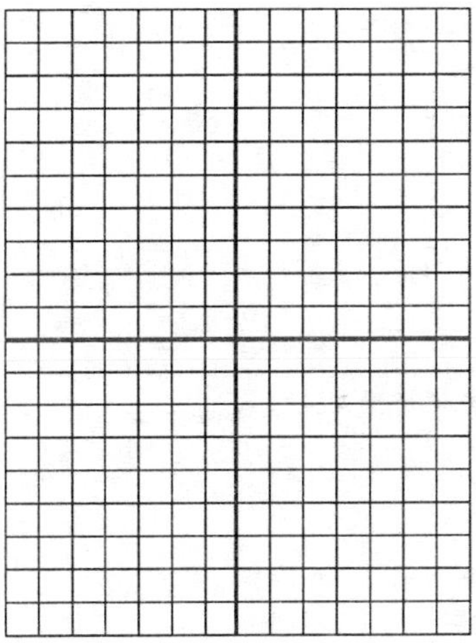

d. $2x + 3y = 6$
$x + 2y = 5$

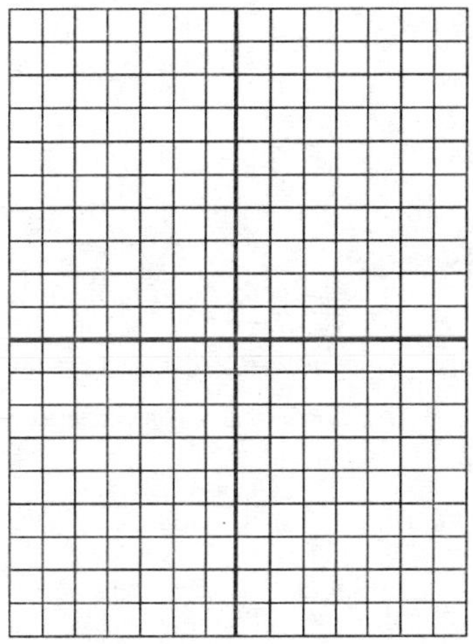

18. Without graphing the lines represented by each pair of equations, find the coordinates of the point of intersection of each set of lines.

a. $x + y = 6$
$x - y = -4$

b. $3x + 4y = 18$
$x + 2y = 8$

c. $4x - y = 5$
$2x + 3y = -1$

d. $5x + 3y = 1$
$-2x - 5y = -8$

19. Which geometric figure (point, line segment, line, plane. can be represented by each object below? Discuss the limitations of each model
   a. the sharp edge of a razor blade
   b. a sheet of notebook paper
   c. the tip of a pencil
   d. a chalkboard

20. Which of the following contain an infinite number of points?
   a. a line
   b. a line segment
   c. a plane
   d. space

Section 1.2 Angles and Angle Measure

When two rays have the same endpoint, they form an **angle**. The two rays are called the **sides of the angle**, and the common endpoint of the two rays is called the **vertex of the angle**. In the figure, A is the vertex of the angle while rays AB and AC are its sides.

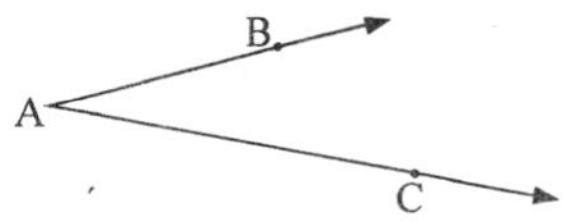

We have several ways to name an angle.

We may *use the name of its vertex.*

∠ P

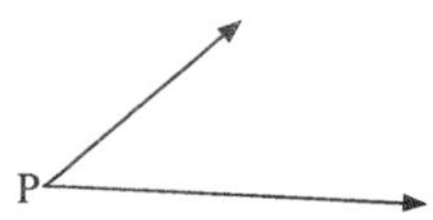

The symbol, " ∠ ," is used for the word angle.

We may *use the name of its vertex along with two other points; taking one point from each side of the angle. The name of the vertex must be the middle letter.*

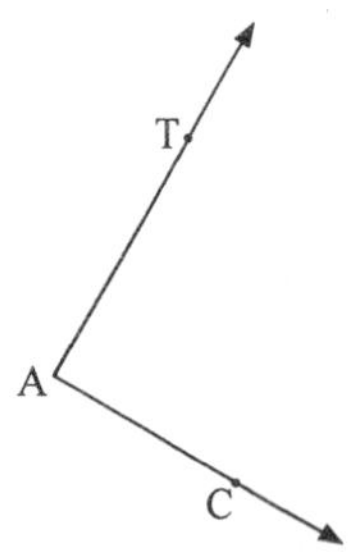

∠ CAT or ∠ TAC

We may *use a lowercase alphabet placing it between the sides of the angle.*

∠ a

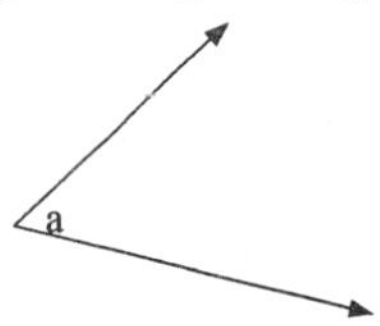

We may *use a numeral placed between the sides of the angle.*

$\angle$ 3

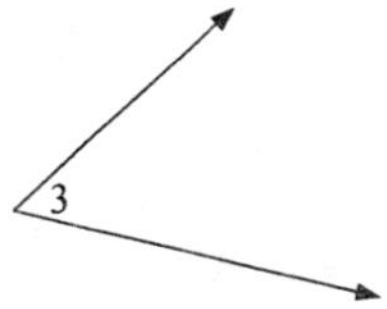

When two or more angles in a figure have the same vertex, we do not use the name of the vertex to name the angle. For example, in the illustration, $\angle$ 1, $\angle$ 2 , and $\angle$ APB all have point P as a vertex, so we do not refer to either angle as $\angle$ P. Doing so could cause confusion.

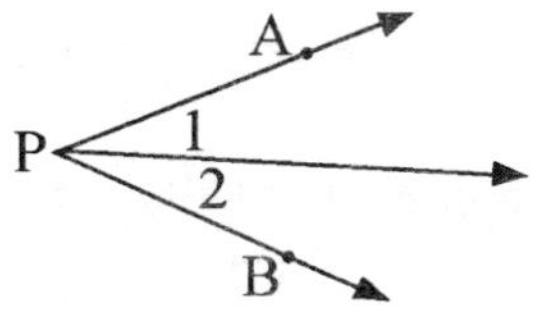

The **measurement of an angle** is determined by the amount of rotation needed to swing one side of the angle about its vertex so that it coincides with the other side of the angle. A complete revolution occurs when a side of an angle is allowed to rotate about its vertex until it returns to its original position.

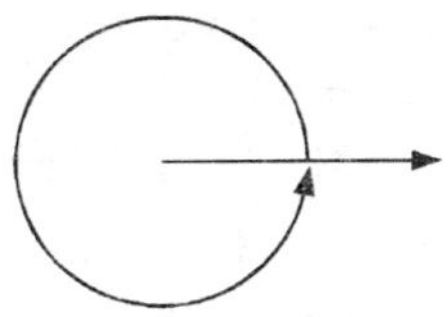

We may think of a **complete revolution** as a 360 degree rotation . The number 360 comes from the ancient Babylonians who believed that a year (a complete revolution of the earth around the sun. contained 360 days. A one **degree** unit then may be defined as 1/360 of a complete revolution. In this course we will use the degree unit to measure angles. Angles are also measured using a unit called a radian, however, this unit will not be studied in this course. *We will agree that whenever the measure of an angle is given without a unit of measure, then the unit of measure is the degree.*

We may classify angles according to the number of degrees that the angle contains.

straight angle: An angle formed by ½ of a revolution. It contains 180 degrees. Notice that the sides of a straight angle are opposite rays.

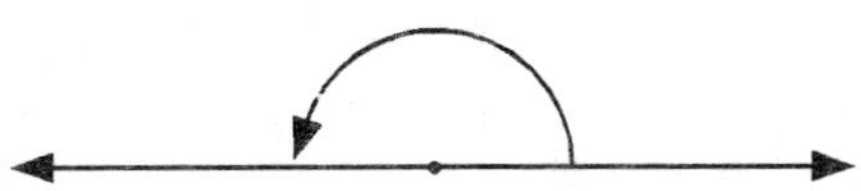

right angle: An angle formed by ¼ of a revolution. It contains 90 degrees. *A little square is often placed at the vertex of such an angle to indicate that it is a right angle.*

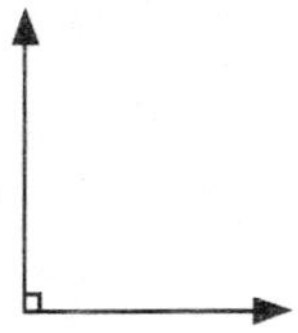

acute angle: An angle that has a rotation that is less than that of a right angle, that is, an angle that has a measure that is less than 90 degrees.

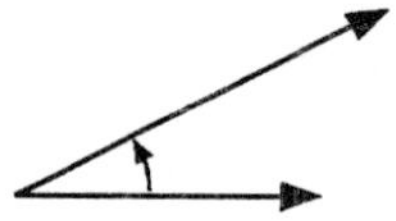

obtuse angle: An angle that has a rotation that is between that of a right angle and that of a straight angle, that is, an angle that has a measure that is between 90 and 180 degrees.

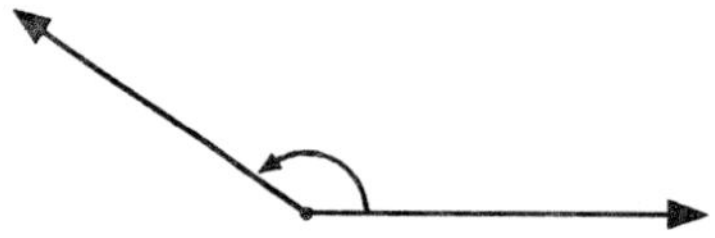

reflex angle: An angle that has a rotation that is greater than that of a straight angle and less than that of a complete revolution, that is, an angle that has a measure that is between 180 and 360 degrees.

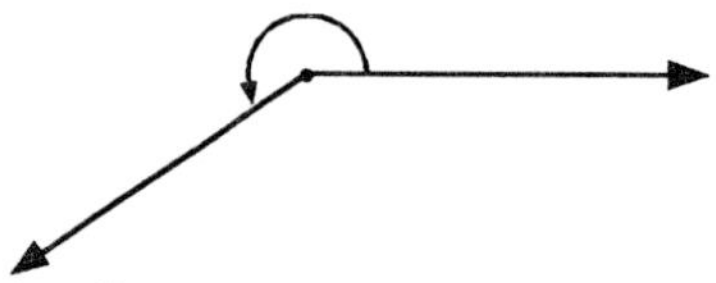

In this course, straight angles and reflex angles will not be considered because our discussions will be limited to angles that have measures between 0 and 180 degrees. Because of this restriction, when a variables is used in an expression representing the measure of an angle, we must keep in mind that the replacement for the variable is restricted to values that will keep the expression between 0 and 180.

Angles are usually measured with an instrument called a **protractor**. Some protractors are circular in shape while others are semicircular. Protractors generally have two sides so that they can be used to measure angles in either a clockwise or a counter clockwise direction.

To measure an angle with a protractor:

First place the center of the protractor on the vertex of the angle.

Next, line up the edge of the protractor with one side of the angle (either side of the angle may be used.

Read the measure of the angle by locating the place where its other side intersects the protractor.

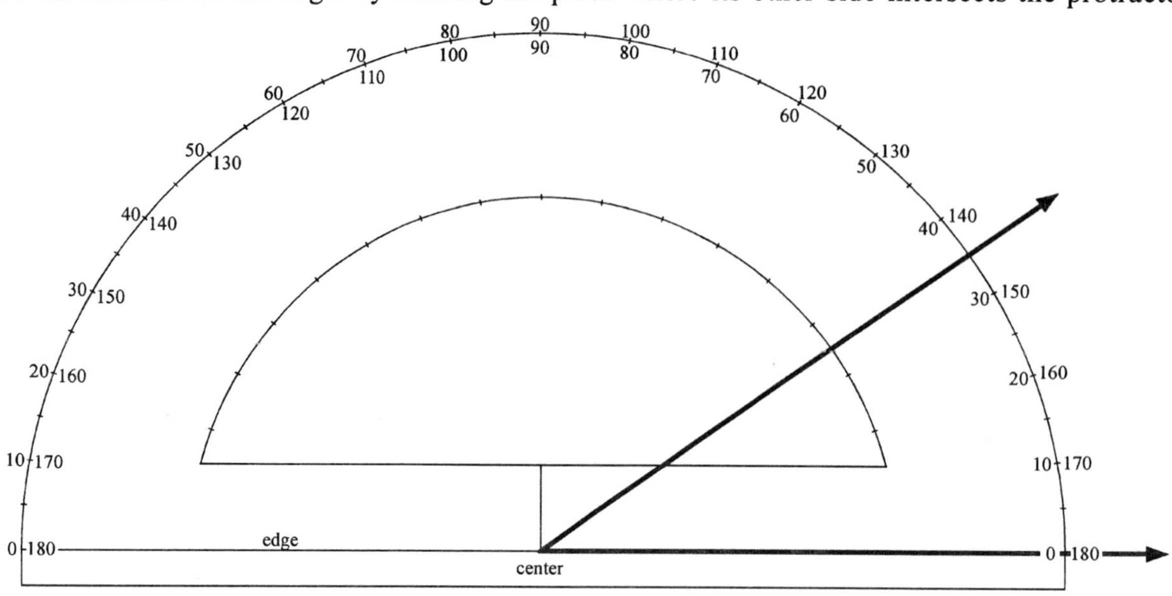

The preceding angle measures 35 degrees or 35° (***The little circle at the upper right of the number 35 represents degree measure***).

Keep in mind that measurement by its nature is always inexact. Unless indicated otherwise, when you are asked to measure an angle using a protractor, give the measure of the angle to the nearest whole degree.

Other units used to measure angles that are smaller than a degree are the **minute** and the **second.**

1 degree is equivalent to 60 minutes.
1 minute is equivalent to 60 seconds.

Therefore we may conclude that a minute is $\frac{1}{60}$ of a degree, and a second is $\frac{1}{60}$ of a minute.

We may represent 36 degrees 40 minutes and 16 seconds as $36^\circ\ 40'\ 16''$.

Notice the symbols that are used to represent the minutes and seconds.

Sometimes it is necessary to convert from one unit of measure to another. The following examples show how such conversions may be done:

EXAMPLE 1 Convert 36.482° to degrees, minutes and seconds.

Step 1 Observe that 36.482° = 36 whole degrees + 0.482 of one degree

Step 2 Change the 0.482 degrees to minutes by multiplying by 60.
$0.482 \times 60 = 28.92$ (28.92 minutes)

Step 3 Observe that 36.482° = 36 whole degrees + 28 whole minutes + 0.92 of one minute

Step 4 Change the 0.92 minutes to seconds by multiplying by 60.
$0.92 \times 60 = 55.2$ (55.2 seconds)

Step 5 Write the conversion
36.482° is equivalent to $36^\circ\ 28'\ 55.2''$

EXAMPLE 2 Convert $48^\circ\ 42'\ 16''$ to degrees.

Step 1 Convert the minutes to degrees by dividing by 60.
$42' = \frac{42}{60}$ degrees = 0.7 degrees

Step 2 Convert the seconds to degrees by dividing by 3600
$16'' = \frac{16}{3600}$ degrees ≈ 0.004 degrees

Step 3 Add the degrees and write the conversion.
48 degrees + 0.7 degrees + 0.004 degrees ≈ 48.704°
(≈ means "is approximately equal to")

In summary:

1. To change from a larger unit to a smaller unit, multiply by a power of 60.
2. To change from a smaller unit to a larger unit, divide by a power of 60.

The following line graph may help you to remember:

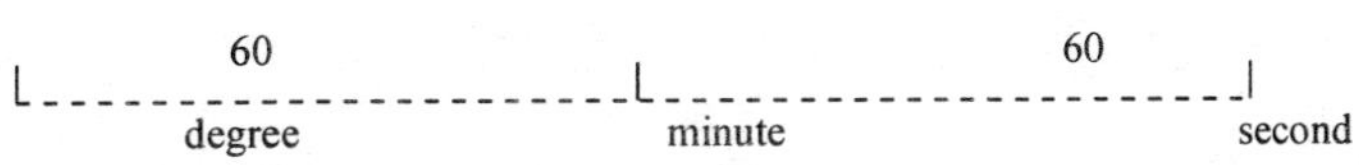

Locate the given unit of measure on the chart.

Move along the line graph toward the desired unit.

When you pass by a 60 as you move toward the desired unit, you must multiply or divide by that 60.

Multiply if you move toward the right. Divide if you move toward the left.

To convert 17 degrees to seconds, we begin on the degree mark and move toward the second mark. Since we are moving toward the right we will need to multiply. When we move from the degree mark to the minute mark, we pass through a 60, thus we must multiply by 60. When we move from the minute mark to the second mark, we pass through another 60, so we must multiply again by 60.

We then conclude that 17 degrees is equivalent to (17 × 60 × 60) seconds

Generally *we will distinguish an angle from its measure by using the symbol ∡ for the angle itself and the symbol ∠ for the measure of the angle.* However, when it is clear from context which of the two is being discussed, we will not necessarily make oral or written distinctions. Also, when the word "angle" is used, there is no need to use an angle symbol. The examples below illustrate the use of these symbols.

EXAMPLE 1 ∡ A is composed of two rays.

EXAMPLE 2 The measure of ∡ A is 30 degrees.

EXAMPLE 3 ∠ A = 30

EXAMPLE 4 ∡ A is a set of points.

EXAMPLE 5 ∠ A + ∠ B = 90

EXAMPLE 6 ∡ A has a measure of 60.

The same markings on two or more angles indicate that those angles have the same measure. Likewise the same markings on two or more line segments indicate that the segments have the same length.

Observe the following illustrations of angles and line segments:

∡ A and ∡ B have the same measure; ∡ C and ∡ D have the same measure;

∡ E and ∡ F have the same measure; $\overline{AB}$ and $\overline{CD}$ have the same length; and line segments $\overline{RS}$ and $\overline{PQ}$ have the same length.

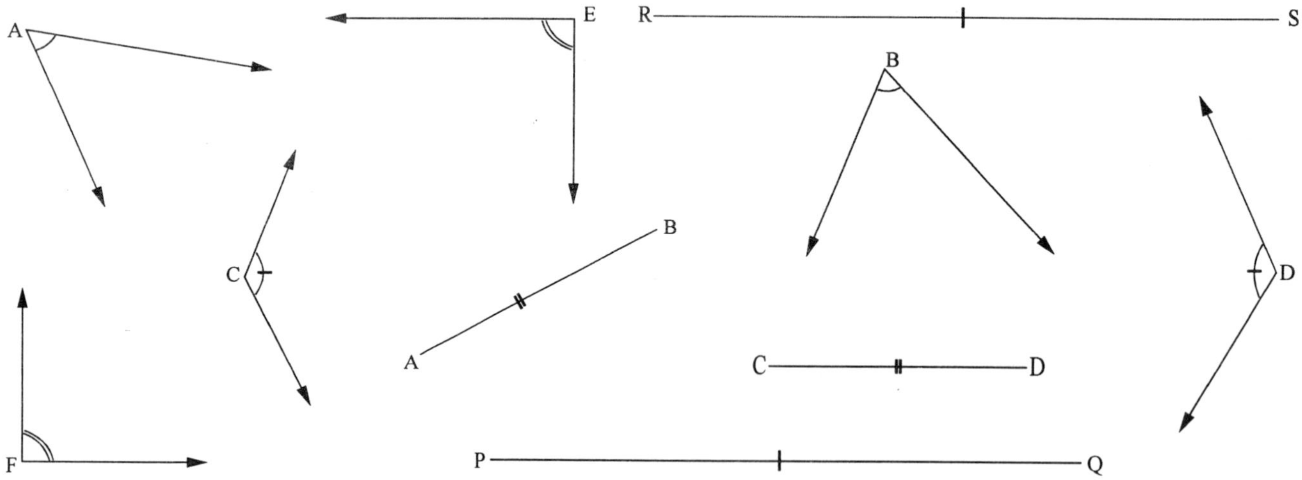

We should note that there are two ways of determining lengths of line segments and measures of angles in geometry: **direct and indirect measurement.** Direct measurement requires the use of a measuring tool such as a protractor or ruler. Direct measure is never exact, and it requires the use of approximations. On the contrary, indirect measurement does not require the use of a measuring tool. Rather, we may determine the required measurements by reasoning from information that is given. *For most of our work in geometry we will use indirect measurement. We will use direct measurement only when we are instructed to do so.*

Exercises 1.2

For these exercises consider only angles that have measures between 0 and 180 degrees. Use direct measure for exercises 5, 7, 8, 12 and 13.

1. Give 4 different names for the angle in the drawing.

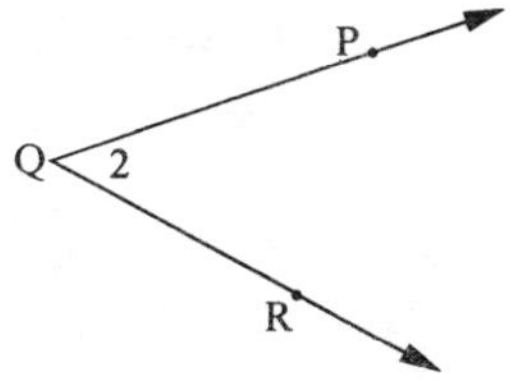

2. Name the sides of the angle in the drawing. What is its vertex?

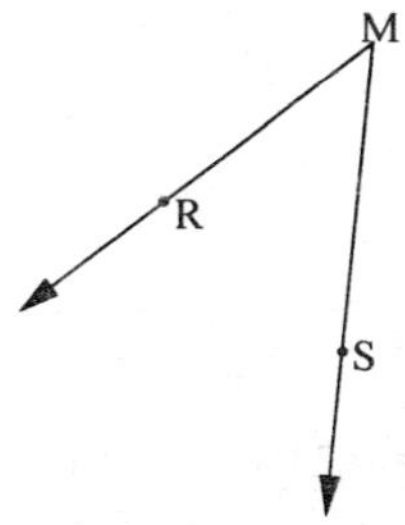

3. Name all the angles that are formed by the intersecting lines.

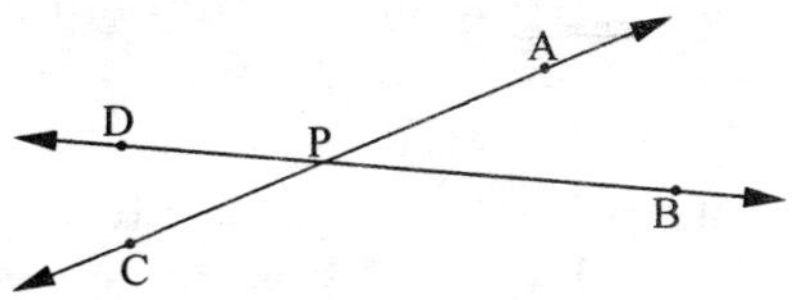

4. Three angles in the drawing have P as a vertex. Name these angles.

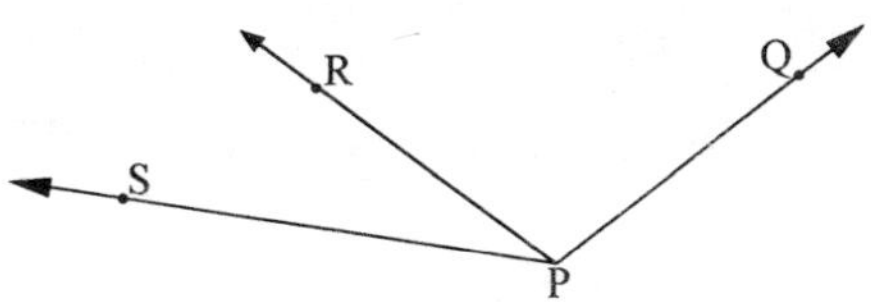

5. Identify each of the following in the drawing:

   a. an acute angle with vertex C
   b. a right angle with vertex A
   c. an obtuse angle with vertex C

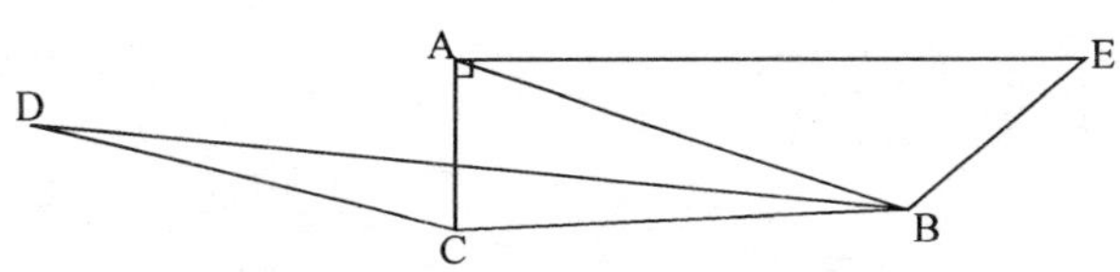

6. In the drawing, which angles have the same measure? Only consider the marked angles.

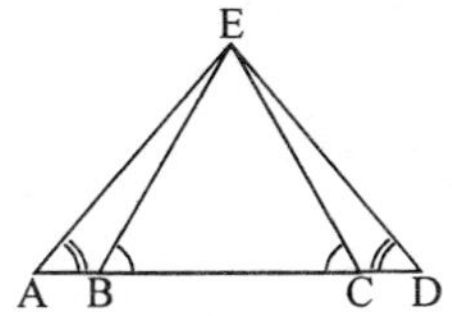

7. Find the measure of ∡ A, ∡ B, ∡ C, and ∡ D

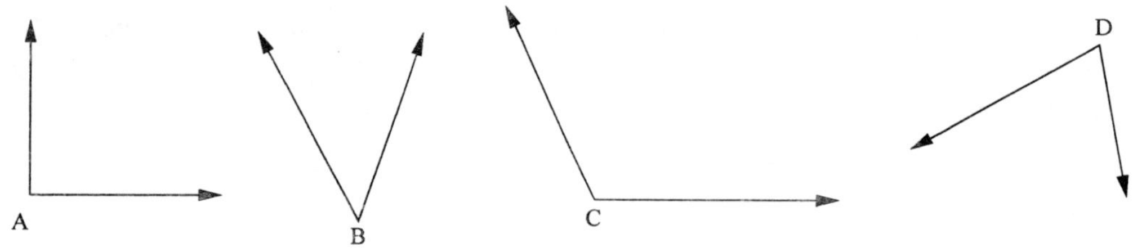

8. Find the measure of each angle in the figure.

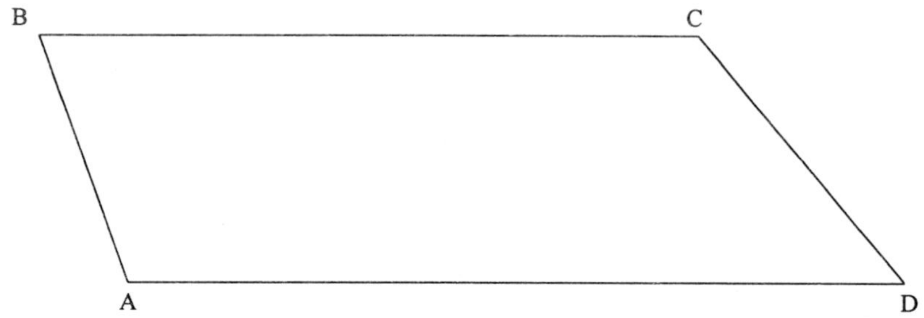

9. Convert each of the following. When necessary, round results to 4 decimal places.
   a. 150° 12′ 36″ to degrees
   b. 28° 45′ 8″ to degrees
   c. 120° 30′ 12″ to degrees
   d. 15° 57′ to degrees
   e. 42.8° to degrees, minutes and seconds
   f. ¼ degree to minutes
   g. 45 minutes to degrees
   h. 0.16 degrees to minutes
   i. ⅗ degrees to minutes
   j. 33.142 degrees to minutes
   k. 5286 seconds to degrees
   l. 57 degrees to seconds
   m. 42 degrees to minutes
   n. 37.415 degrees to degrees, minutes and seconds

10. What amount of rotation is made by:
    a. an hour hand of a clock in 2 hours
    b. a minute hand of a clock in 2 hours
    c. an hour hand of a clock in 6 hours
    d. a minute hand of a clock in 6 hours
    e. a minute hand of a clock in ½ hours
    f. a minute hand of a clock in ¼ hours

11. Find the angle formed by the hands of a clock at:
    a. 3 o'clock
    b. 10 o'clock
    c. 6:30 PM
    d. 3:15 PM
    e. 11:12 PM

12. Use your protractor to draw angles with the following measures:
    a. 30° b. 125° c. 90° d. 45°
    e. 60° f. 72° g. 120° h. 108°

13. Find the measure of each angle of the triangle.

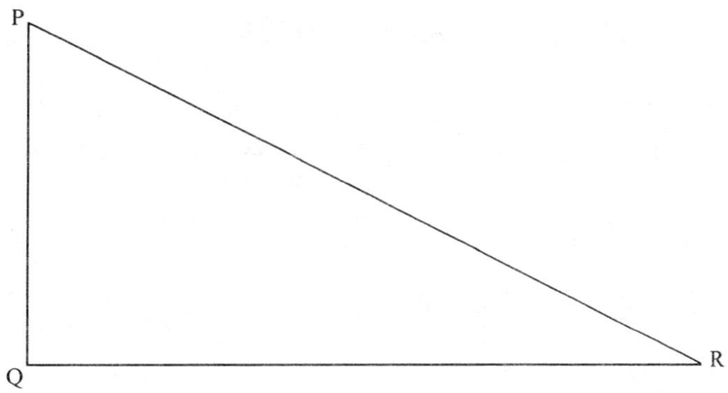

14. Find the measure of each angle:

a. The ratio of the measures of $\angle A$, $\angle B$ and $\angle C$ is 1 : 6 : 8. The sum of their measures is 120.

b. The ratio of $\angle H$ to $\angle S$ is 5 : 6. $\angle H + \angle S = 88$.

15. The sum of the measures of $\angle A$ and $\angle B$ is 47. The difference between the measures of the two angles is 17. Find the measure of each angle.

16. If $\angle K = 37$ and the measure of $\angle A$ is $(7x - 12)$, what is the value of x?

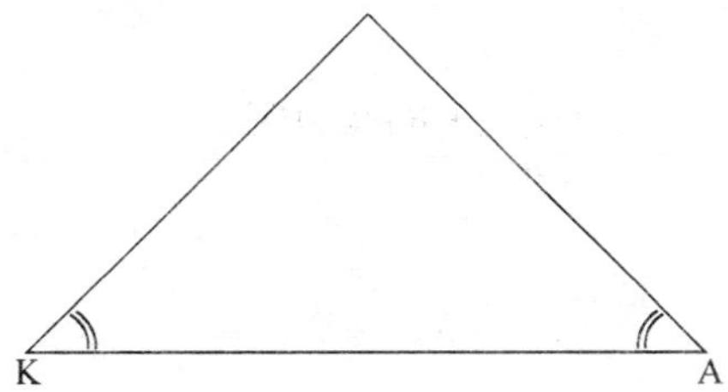

17. The measure of $\angle M$ is four more than twice the measure of $\angle N$. The sum of their measures is 157. Find the measure of each angle.

18. If $\angle R$ is a right angle with a measure of $(3x + 2)$, what must be the value of x?

19. If $\angle B$ is an obtuse angle with a measure of $(5x + 20)$. what are the possible replacements for x? Hint: $5x + 20$ must be between 90 and 180.

20. If $\angle A$ is an acute angle with a measure of $(8x + 58)$, what are the possible replacements for x?

Section 1.3 Special Relationships: Angles and Lines

When two angles have a common vertex and a common side, they are called **adjacent angles**. If two angles are adjacent, then the sum of their measures is equal to the measure of the angle that is formed by their two nonadjacent sides. In other words, the whole is equal to the sum of its parts. In Example 1 below, notice that the rotation of ∡ 1 combined with the rotation of ∡ 2 gives the rotation of ∡ PAC, that is, ∠ 1 + ∠ 2 = ∠ PAC.

EXAMPLE 1

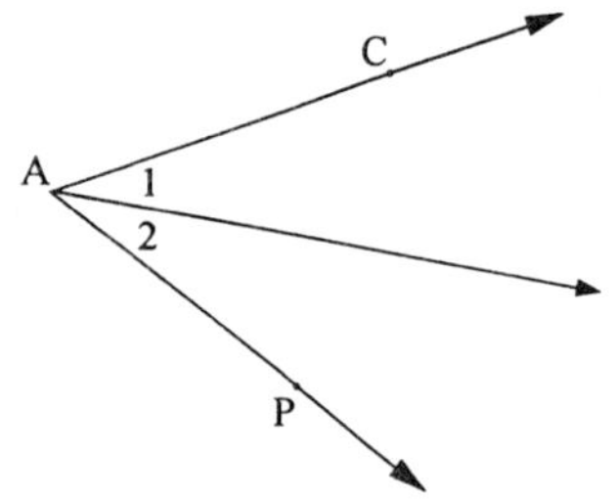

Example 2 shows that this observation may be extended to include more than two angles if each angle is adjacent to the angle that is next to it and all the angles share the same vertex. In Example 2, we conclude that ∠ 1 + ∠ 2 + < 3 = ∠ PAT.

EXAMPLE 2

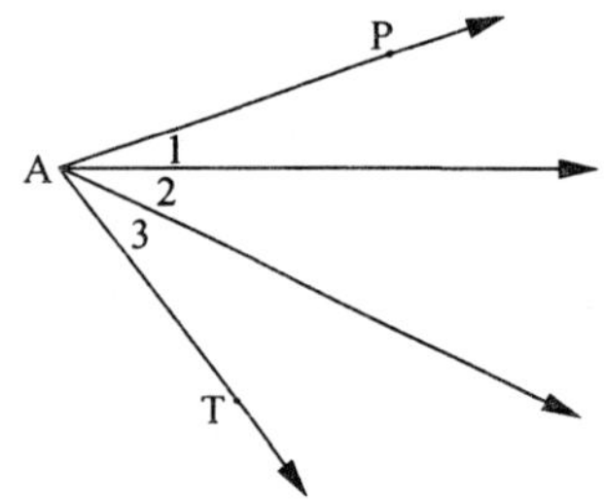

In Example 3, we conclude that ∠ TAS = 30 + 40 + 20 = 90.

EXAMPLE 3

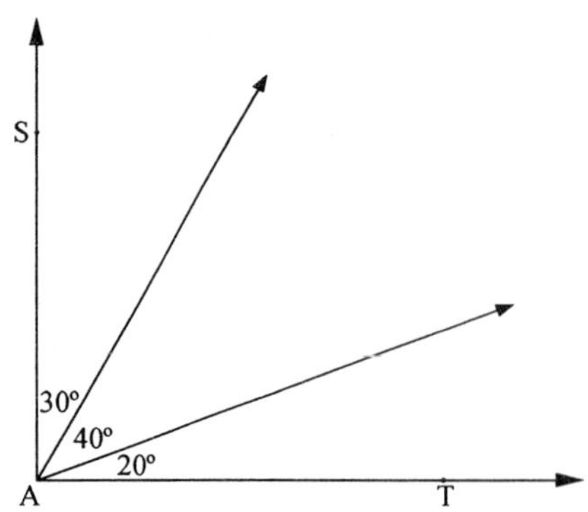

When two angles are adjacent and two of their sides are opposite rays then the angles are said to form a **linear pair**. In the illustration that follows, ∡ 1 and ∡ 2 form a linear pair.

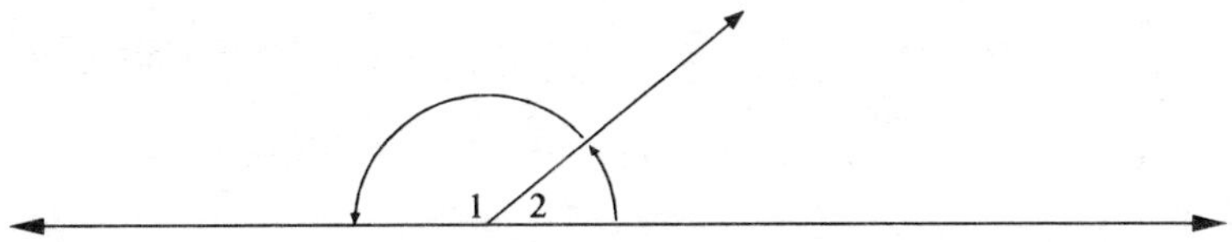

Notice that when two angles form a linear pair, their combined rotation is one–half of a complete revolution, or 180 degrees.

In the illustration, the two angles that form a linear pair have measures of 30 and 150. The sum of the measures of the two angles must be 180.

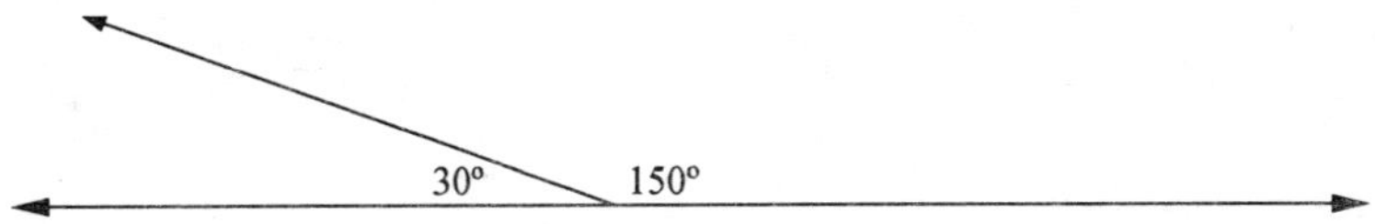

If the sum of the measures of two angles is 180 degrees, the angles are called **supplementary**. Each angle is called the supplement of the other. We may conclude then that angles that form a linear pair are always supplementary. In the illustration below ∠ A and ∠ B are supplementary. Also, ∠ m and ∠ n are supplementary since they form a linear pair.

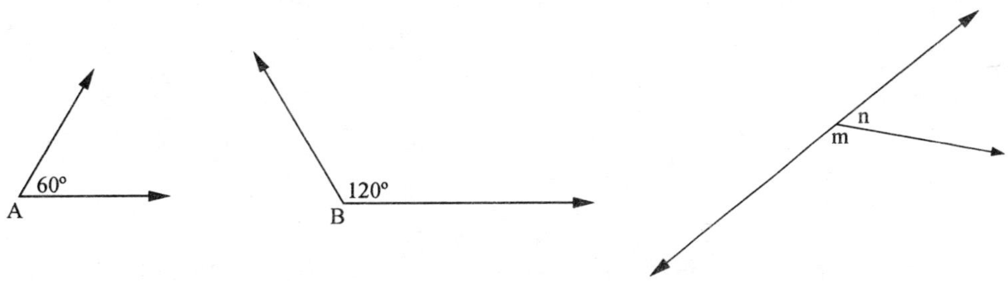

If the sum of the measures of two angles is 90°, then the angles are called **complementary,** and each angle is called a complement of the other. In the illustrations, ∠ C and ∠ D are complementary since the sum of their measures is 90. ∠ p and ∠ q are also complementary since their combined rotation is that of a right angle.

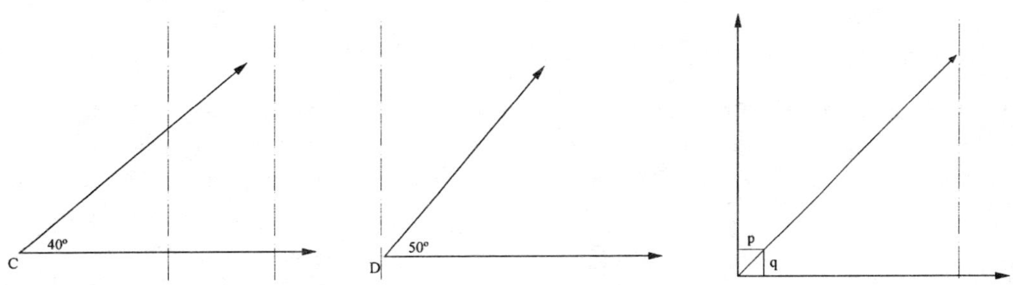

Just as we can add the measures of adjacent angles to get the measure of the "whole angle," we can add lengths of segments to get the length of a "whole segment." If points A, B, and C are collinear with B between A and C, we may add the lengths of line segments AB and BC to obtain the length of line segment AC. In the illustration we note that AB + BC = AC.

AB + BC = AC

We will use the ***notation A-B-C*** *to denote the* ***betweenness relationship of points A, B and C*** when the three points are collinear, and point B is between points A and C. Using this new notation, If A-B-C and AB = BC, then point B is called a bisector of line segment AC. In general, a **bisector of a line segment** is a point, ray, line, or line segment that separates it into two segments that are equal in length. The bisector of a line segment always contains the midpoint of the line segment. The **midpoint** is the point that separates the line segment into two segments of equal lengths. In the example that follows, point A is the midpoint of line segment MD, and also it is a bisector of the segment; while ray AT is a bisector of the same segment.

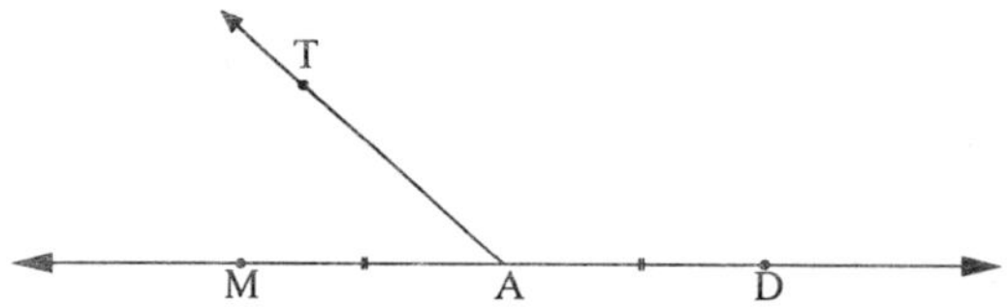

A **bisector of an angle** is a ray or line that separates the angle into two angles with the same measure. In the example that follows, ray AS is a bisector of $\measuredangle$ MAP. Line AS is also a bisector of the same angle.

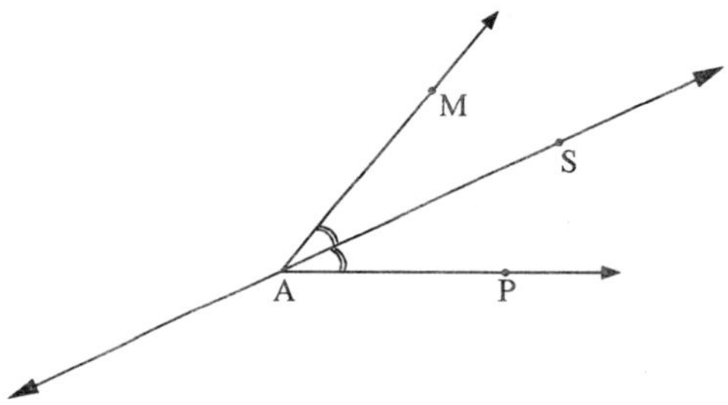

When two lines intersect in a point, notice that several linear pairs of angles are formed.

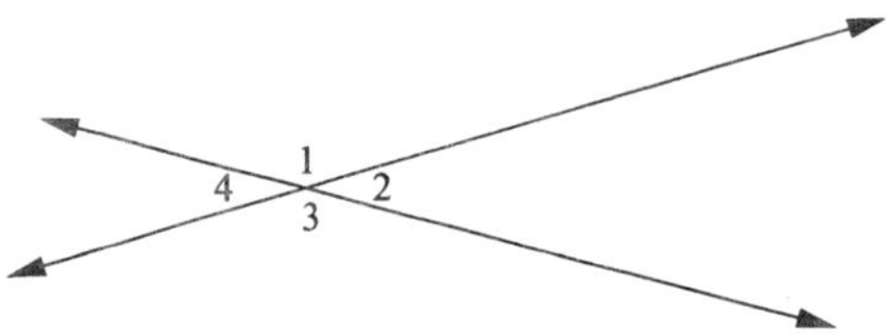

$\measuredangle$ 1 and $\measuredangle$ 2, $\measuredangle$ 2 and $\measuredangle$ 3, $\measuredangle$ 3 and $\measuredangle$ 4, $\measuredangle$ 4 and $\measuredangle$ 1

The nonadjacent angles formed by the two intersecting lines are called **vertical angles.** In the last illustration, $\measuredangle$ 1 and $\measuredangle$ 3 are vertical angles; and $\measuredangle$ 2 and $\measuredangle$ 4 are vertical angles.

Can you see why vertical angles must have equal measures? Study the following examples:

EXAMPLE 1

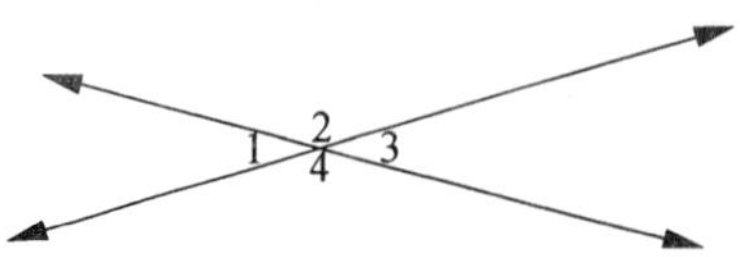

$\angle 2 + \angle 3 = 180$ since the angles form a linear pair.

Likewise, $\angle 4 + \angle 3 = 180$

$\angle 2$ then must equal $\angle 4$ since there can be only one number that will equal 180 when added to $\angle 3$.

EXAMPLE 2

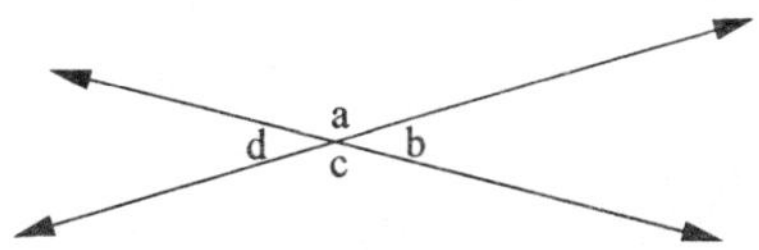

In general	$\angle a + \angle b = 180$
	$\angle c + \angle b = 180$
Thus,	$\angle a + \angle b = \angle c + \angle b$
And	$\angle a$ must equal $\angle c$
Likewise,	$\angle d$ must equal $\angle b$.

We therefore conclude that *vertical angles have the same measure.*

Now let us discuss some special relationships of lines. **There are three ways that lines can be located in space;**

1. *The two lines may intersect in a point.* In this case, the two lines must lie in the same plane. If the two lines form right angles, the lines are said to be **perpendicular**. In the following illustration, line CD is perpendicular to line AB. 1, 2, 3, and 4 are all right angles.

We *use the notation* $\overleftrightarrow{AB} \perp \overleftrightarrow{CD}$ *to indicate that the two lines are perpendicular.*

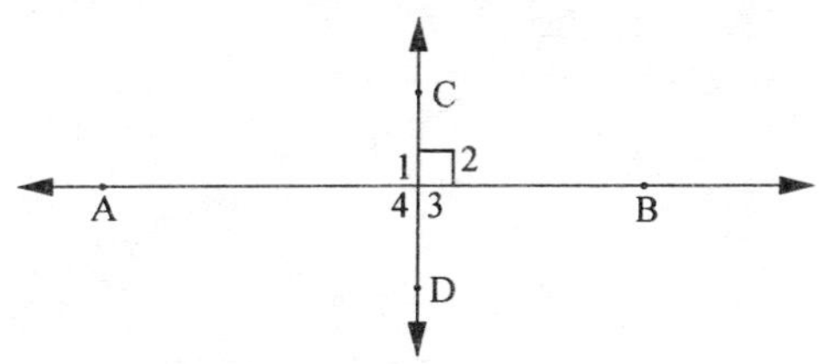

2. *The two lines may fail to intersect and also fail to be in the same plane.* In this case, they are called **skew** lines. In the figure, $\overleftrightarrow{BC}$ and $\overleftrightarrow{GF}$ are skew lines.

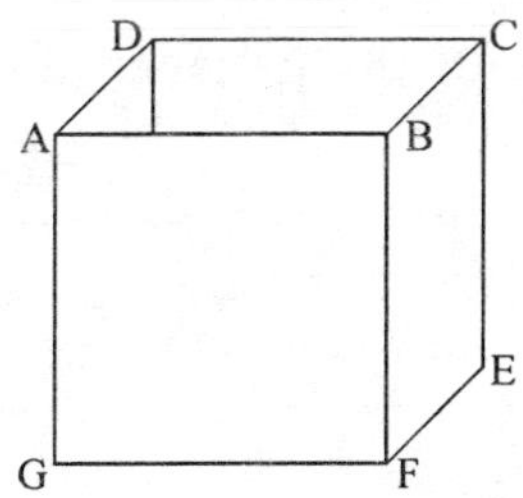

3. *The two lines may lie in the same plane without intersecting each other.* In this case we say that the two lines are **parallel.** In the illustration that follows, line n is parallel to line m, and line $\overleftrightarrow{AB}$ is parallel to line $\overleftrightarrow{DC}$.

We use the ***notation*** **n || m** *to indicate that the two lines are parallel.*

We may also *indicate that two lines are parallel by placing arrowheads pointed in the same direction on the lines.*

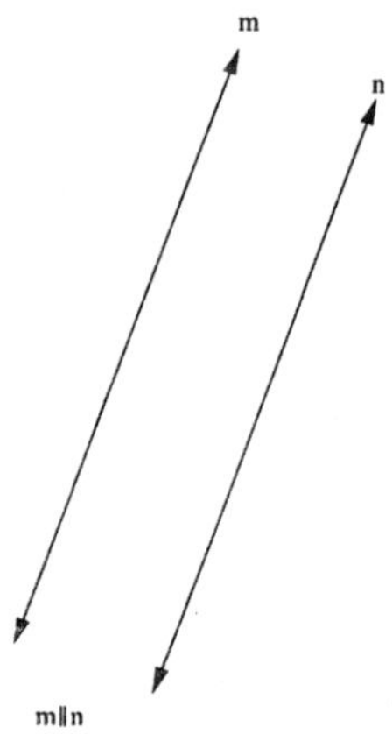

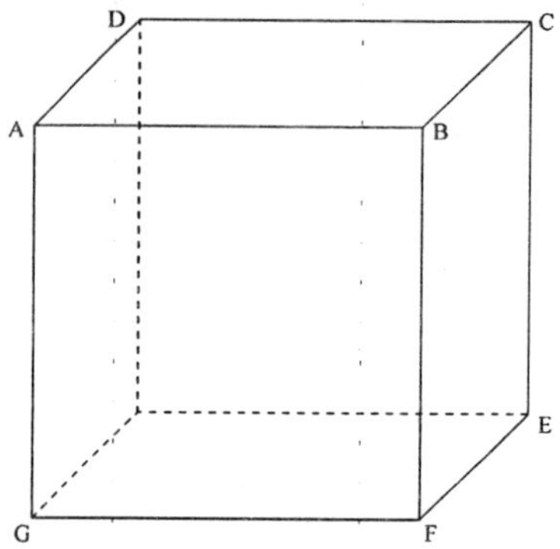

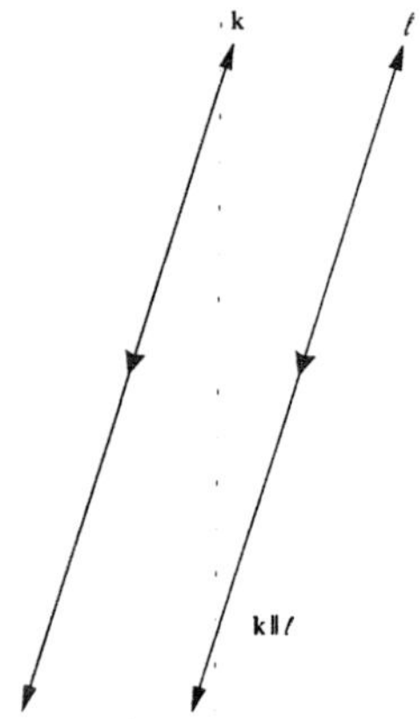

When two lines are intersected, or cut, by another line in two distinct points, special pairs of angles are formed. The intersecting line is called a **transversal** of the other two lines. We see in the illustration, the pairs of angles formed when lines m and n are cut by transversal t.

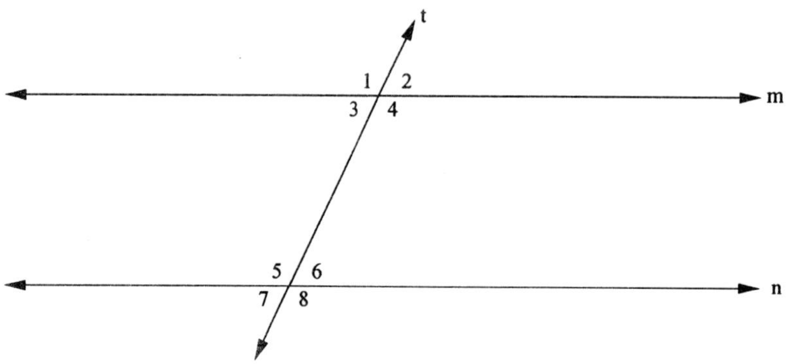

Any two angles that are on the same side of the transversal t and also above the lines m and n, respectively, are called **"corresponding angles."**

∠ 2 and ∠ 6 are corresponding angles.

∠ 1 and ∠ 5 are corresponding angles.

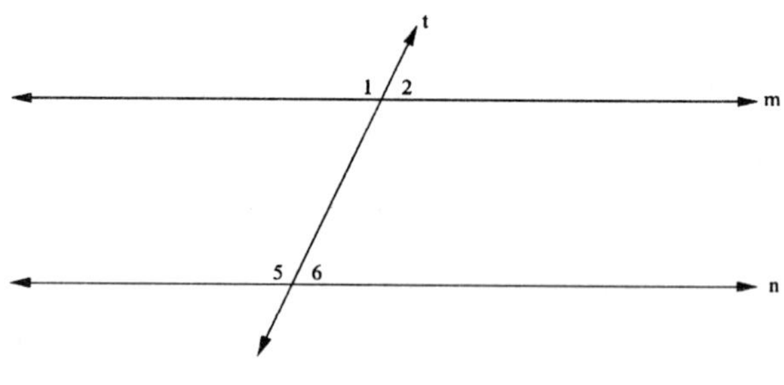

Any two angles that are on the same side of the transversal t and also below the lines m and n, respectively, are called "corresponding angles."

∠ 4 and ∠ 8 are corresponding angles.

∠ 3 and ∠ 7 are corresponding angles.

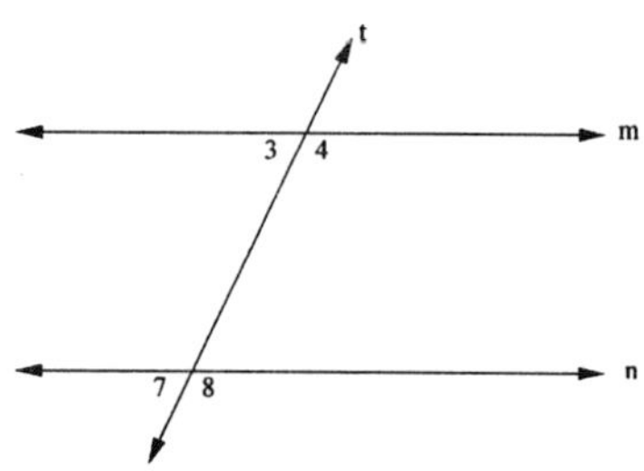

Any two angles that are between lines m and n but on the same side of the transversal t, are called **"same side interior angles."**

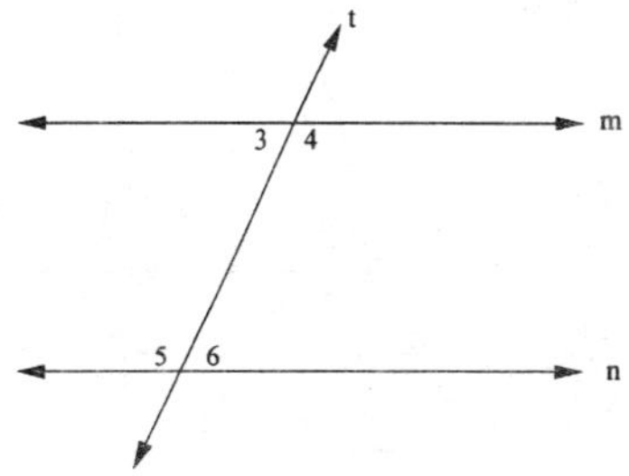

∡ 4 and ∡ 6 are same side interior angles

∡ 3 and ∡ 5 are same side interior angles

Any two angles that are between lines m and n but on different sides of the transversal t are called **alternate interior angles."**

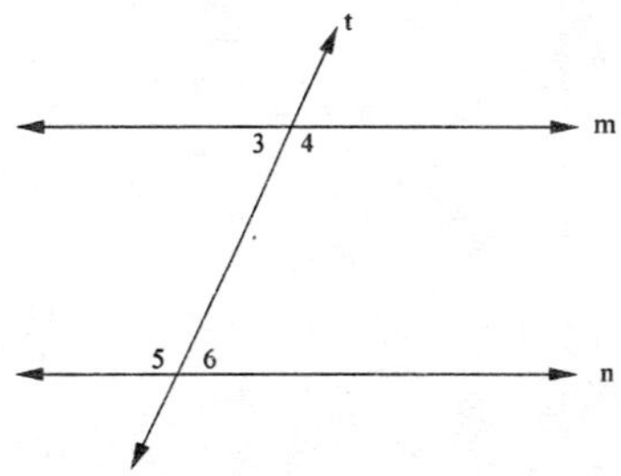

∡ 3 and ∡ 6 are alternate interior angles

∡ 4 and ∡ 5 are alternate interior angles.

Locating these special angles in drawings can be challenging. Often they may be identified by looking for the letters F, C and Z or distortions of these letters. We may identify corresponding angles by looking for the letter **F**. We may identify same side interior angles by looking for the letter **C**. We may identify alternate interior angles by looking for the letter **Z**. You can see these letters used in the following illustrations.

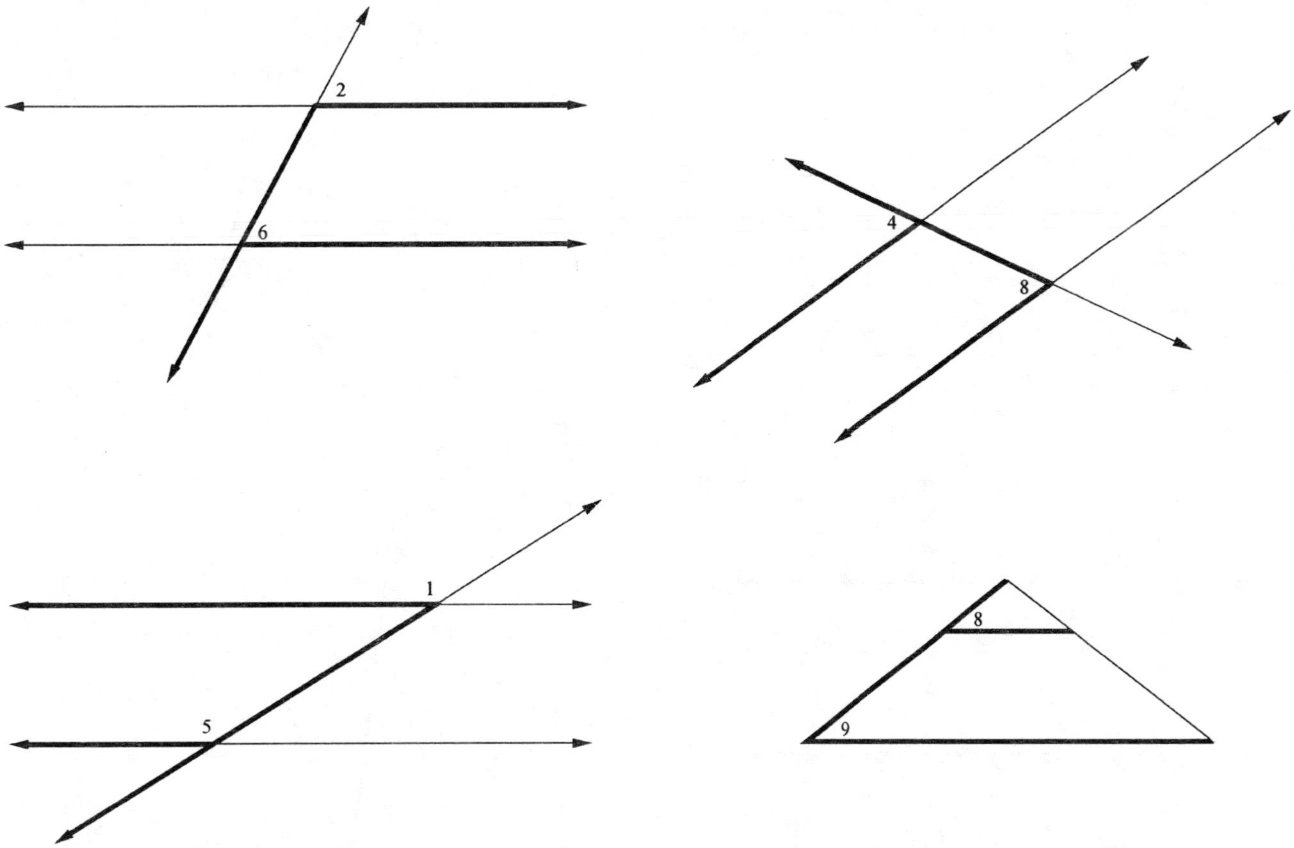

CORRESPONDING ANGLES; THE LETTER F

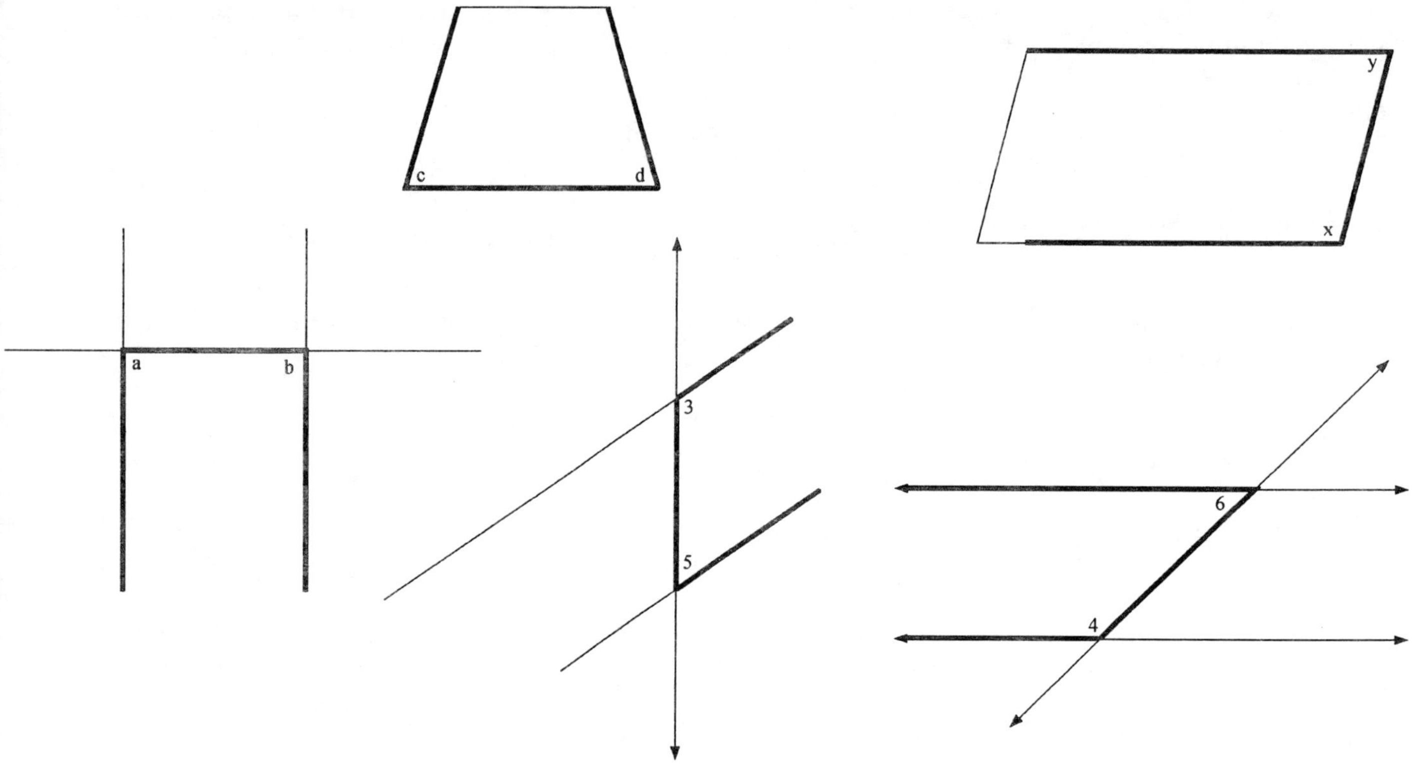

SAME SIDE INTERIOR ANGLES; THE LETTER C

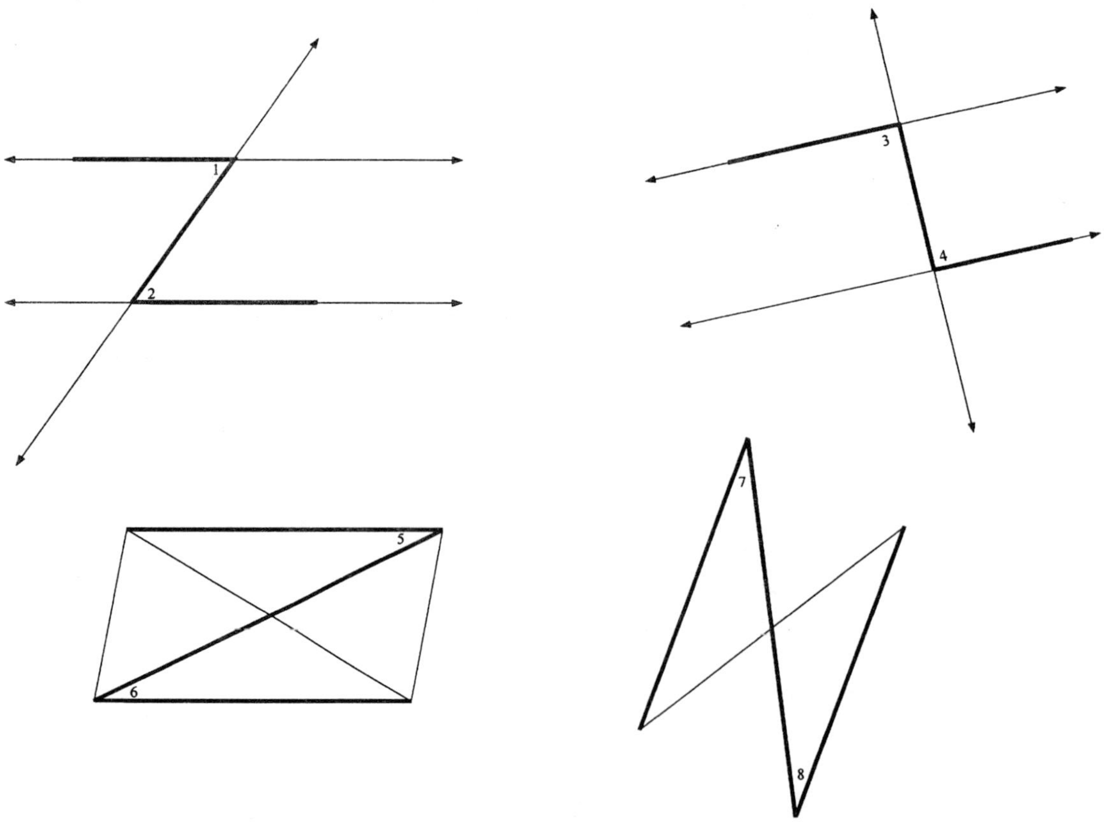

ALTERNATE INTERIOR ANGLES; THE LETTER Z

We observe that when two parallel lines are cut by a transversal, pairs of corresponding angles must have equal measures since the transversal must cut the two lines at the same slant or angle. In the illustration, corresponding angles, 1 and 2 are equal in measure.

m || n

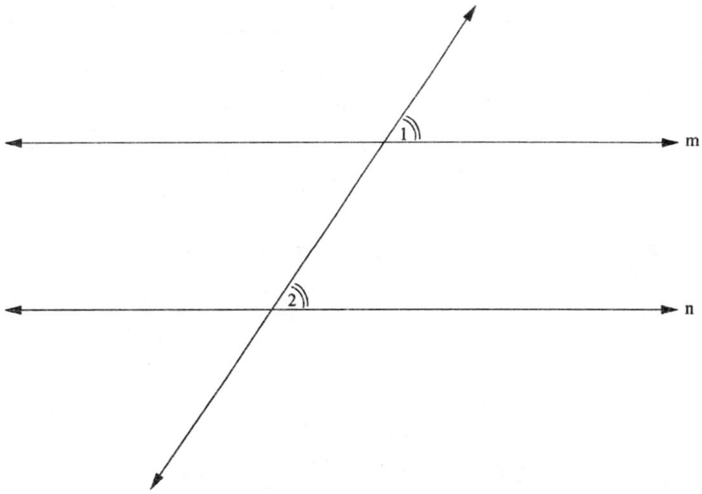

Knowing the measure of a pair of corresponding angles allows us to determine the measures of all of the other angles formed by the transversal and the parallel lines. For example, in the illustration below, $\angle 1 = 30$. Since lines m and n are parallel, $\angle 3 = \angle 1$.

$\angle$ 3 then must have a measure of 30.

$\angle 1 = 30$

m || n

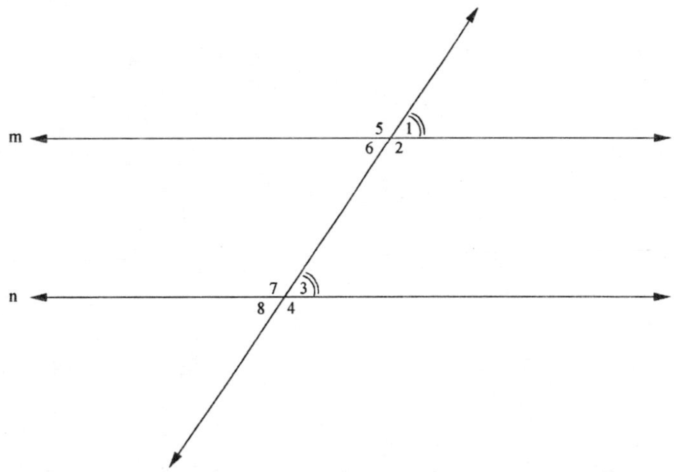

$\angle$ 2 must have a measure of 150. Why?
$\angle$ 4 also must have a measure of 150. Why?
$\angle$ 5 must have a measure of 150. Why?
$\angle$ 6 must have a measure of 30. Why?
$\angle$ 7 must have a measure of 150. Why?
$\angle$ 8 must have a measure of 30. Why?

Study the following illustration where line m is parallel to line n:

m || n

$0 < x < 180$

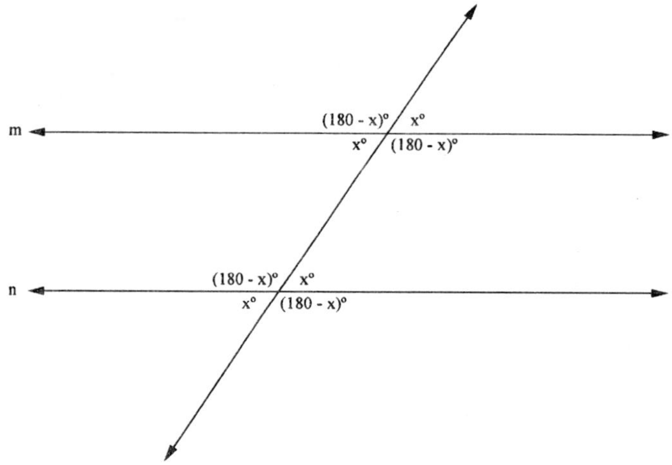

In general, *if two parallel lines are cut by a transversal,* we conclude the following:

Corresponding angles have the same measure.
Alternate interior angles have the same measure.
Interior angles on the same side of the transversal are supplementary.

Remember that these conclusions can be made only if we know that we are dealing with parallel lines!

We learned earlier that the distance between the two endpoints of a line segment is called the length of the segment. Now that we are familiar with parallel and perpendicular lines, we can learn about other distances. We will think of **the distance from a point not on a line to that line** as being the length of the perpendicular line segment joining the point to the line. In the illustration, the distance from point P to line ℓ is the length of line segment PA.

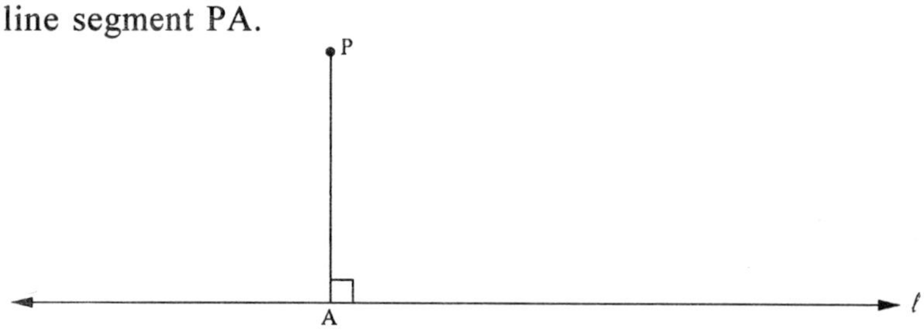

We will think of **the distance between two parallel lines** as being the length of any perpendicular line segment joining the two lines. Note that such lengths must be equal. Perhaps you have heard the saying, "parallel lines are the same distance apart." In the illustration that follows, the distance between line m and line n is the length of either line segment PA or line segment RE.

m || n

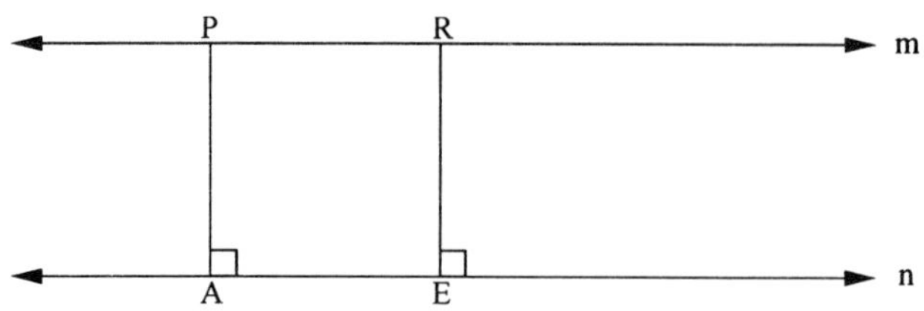

Exercises 1.3

The figures in these exercises are not drawn to scale. Use indirect measurement to complete the exercises.

1. Find ∠ MNP

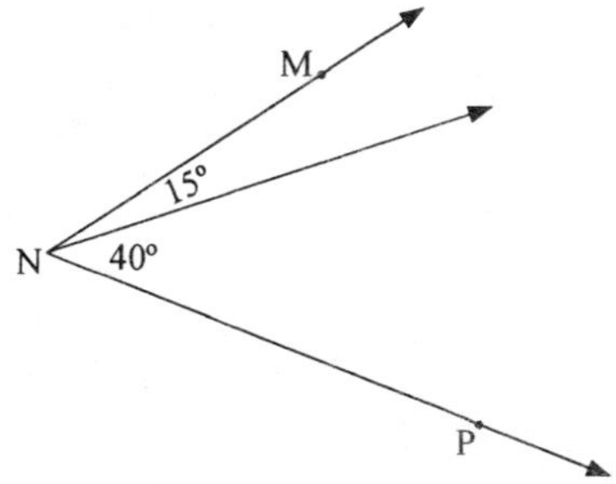

2. ∠ PQR = 50, ∠ PQS = 30, Find ∠ SQR

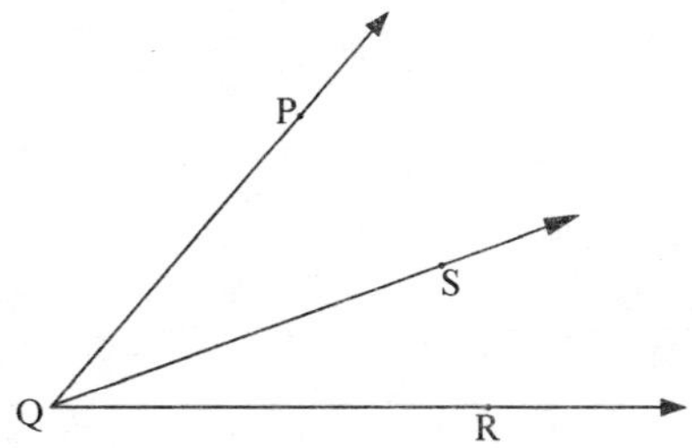

3. Fill in the blank so that the statement is true.
∠ 1 and ∠ 2 are called ______________ angles.

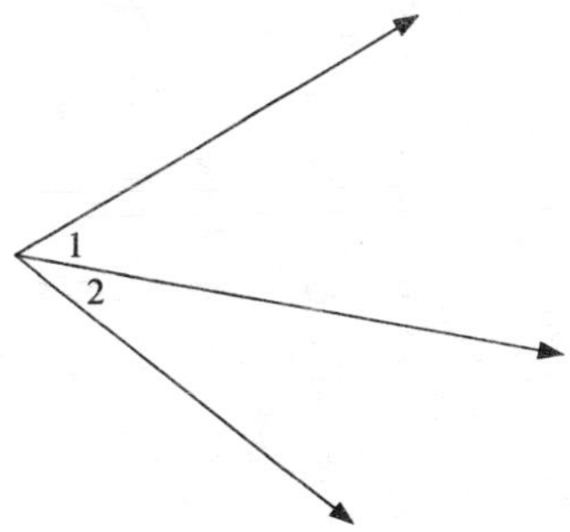

4. ∠ PAD = 62, ∠ RAP = x, Express ∠ RAD in terms of x.

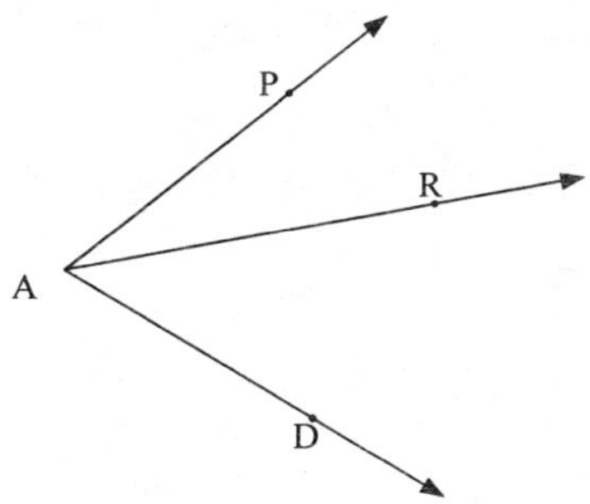

5. AB = 5 and BC = 8. Find AC.

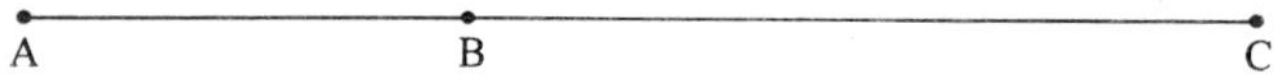

6. RT = 15 and RS = x. Express ST in terms of x.

7. Find ∠ PQS

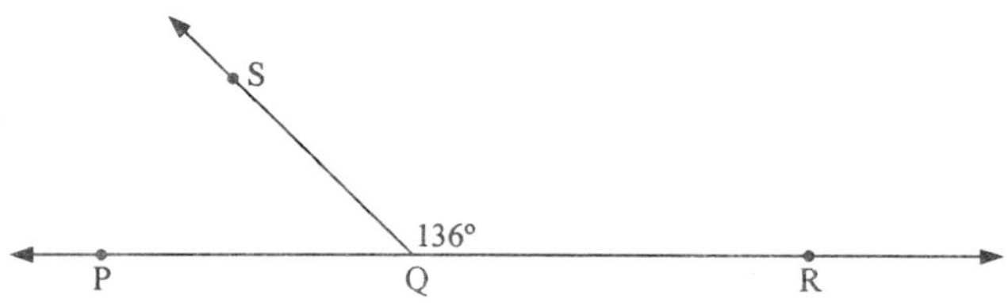

8. Fill in the blanks so that each statement is true.

a. ∠ 1 and ∠ 4 are ________________ angles.
b. ∠ 1 and ∠ 4 form a ______________ pair.
c. ∠ 1 and ∠ 3 are __________________angles.
d. ∠ 2 and ∠ 4 are __________________ angles.
e. ∠ 1 is the supplement of ______ and _______.

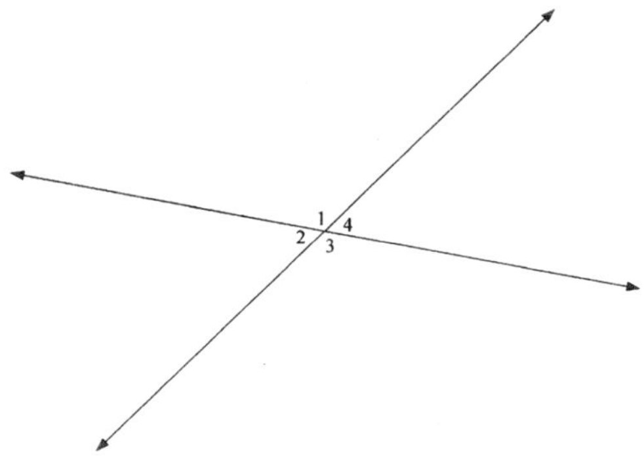

9. ∠ PAT = m. Express ∠ 2, ∠3 and ∠ 4 in terms of m.

∠ 2 = ________________
∠ 3 = ________________
∠ 4 = _________________

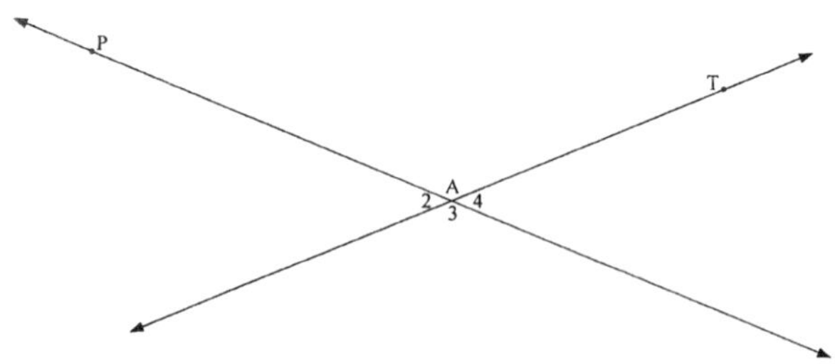

10. ∠ABC = 50, and $\overrightarrow{BD}$ is the bisector of ∠ ABC. Find ∠ ABD and ∠ DBC.

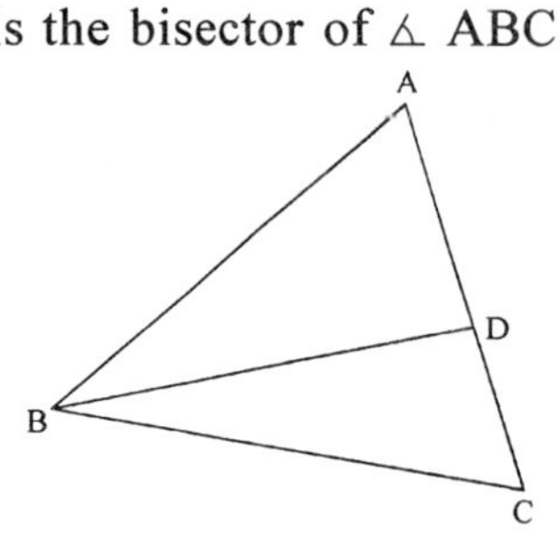

11. Fill in the blank so that each statement is true.

$\angle$ PAT = 35. $\measuredangle$ TAR has measure ____________.
$\measuredangle$ PAT and $\measuredangle$ TAR are called ______________________________ angles.

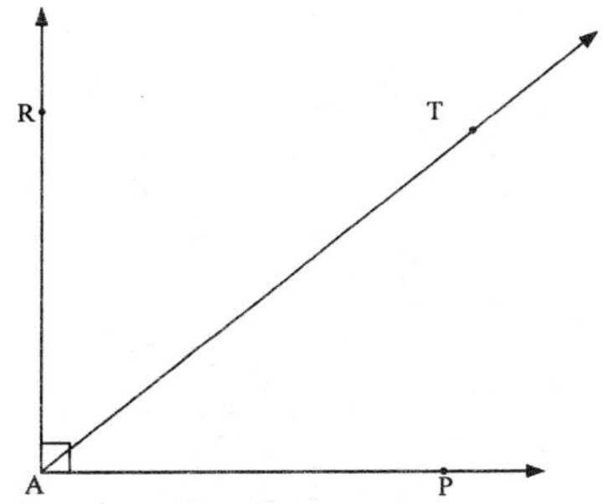

12. Use the rectangular box to identify the following:

a. a pair of parallel lines
b. a pair of intersecting lines
c. a pair of skew lines

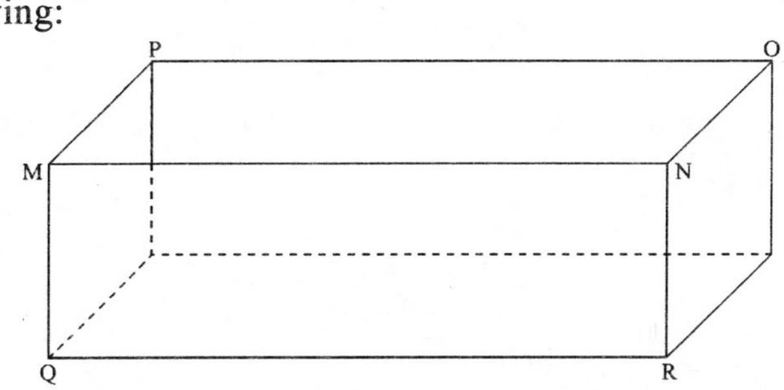

13. Identify each of the following in the drawing:

a. a transversal
b. a pair of corresponding angles
c. a pair of alternate interior angles
d. a pair of same side interior angles
e. a pair of vertical angles
f. two angles that form a linear pair

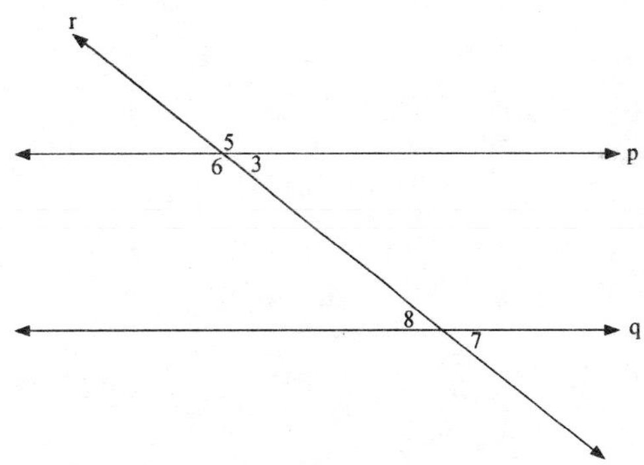

14. Find the complement and supplement of each angle when possible:

a. angle with a measure of 60° ____________________

b. angle with a measure of 87° ____________________

c. angle with a measure of 115° ____________________

d. angle with a measure of 35 ° 45′ ____________________

e. angle with a measure of 6 ° 37′ 45′′ ____________________

f. angle with a measure of a° ____________________

15. Find the measures of the angles in the following exercises:

a. The angles are complementary and the measure of the smaller angle is 40 ° less than the measure of the larger angle.

b. The angles are supplementary and the measure of the larger angle is 20 ° less than three times the measure of the smaller angle.

c. The angles are vertical and supplementary.

d. The angles are adjacent and their noncommon sides form an angle with measure 140 °. The smaller angle has a measure that is 28 ° less than the measure of the larger.

e. The angles are adjacent and their noncommon sides form an angle with measure 75 °. The difference between their measures is 21 ° .

f. The angles are supplementary. The measure of one angle is 20 ° more than four times the measure of the other.

16. Respond true or false for each statement. Be prepared to defend your response.

a. Supplementary angles always form a linear pair.

b. Perpendicular lines always form right angles.

c. Vertical angles have equal measures.

d. If two lines do not intersect, they must be parallel.

17. What is the distance from point A to line ℓ ?

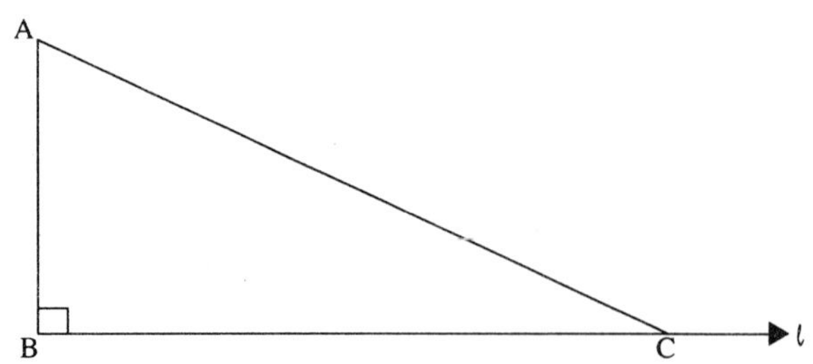

18a. Rays AB, AC, and AD are coplanar (same plane). Make a drawing for which it is true that $\angle BAC + \angle CAD + \angle DAB = 360$.

18b. Rays AB, AC, and AD are coplanar. Make a drawing for which it is true that $\angle BAC + \angle CAD + \angle DAB = 180$.

19. Find the measure of each angle.

$\angle A = ?$

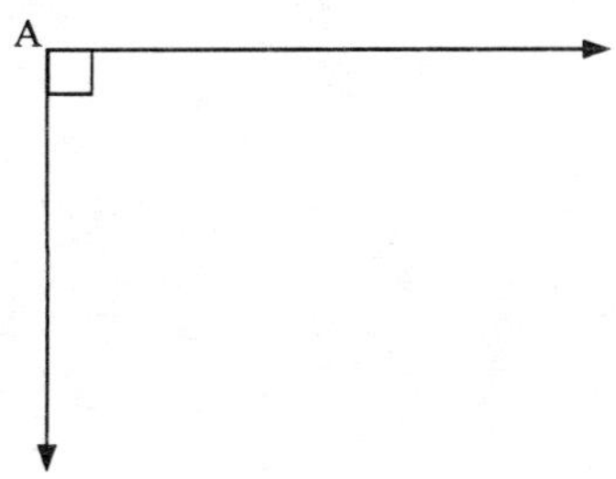

$\angle APB = ?$

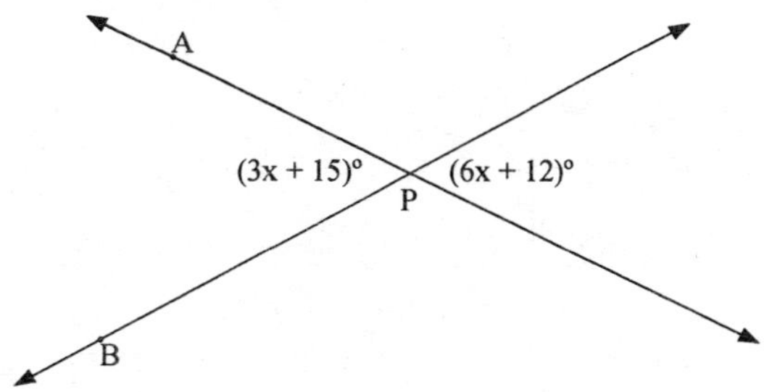

20. Find the values of the variables in each exercise:

a.

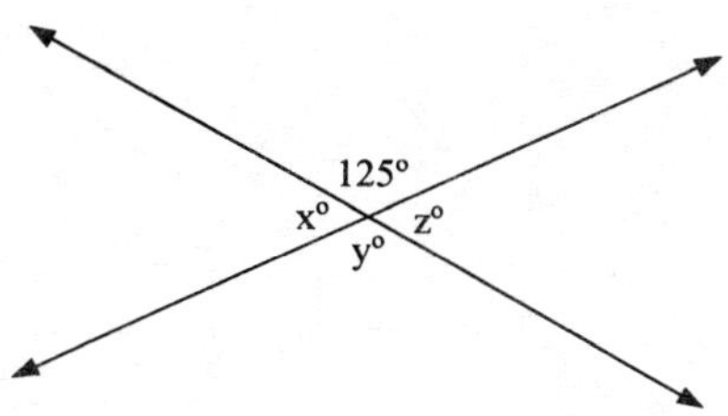

b.

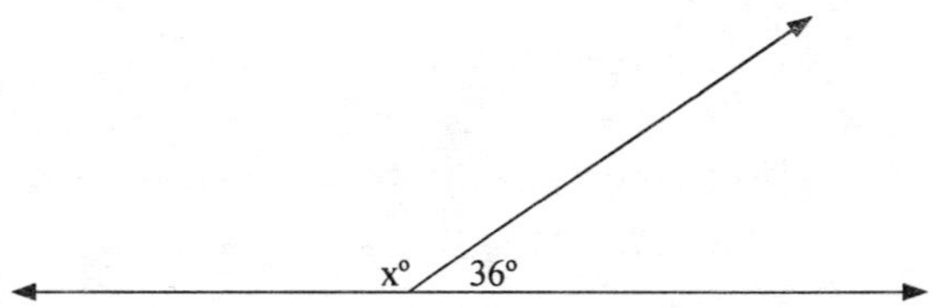

c.

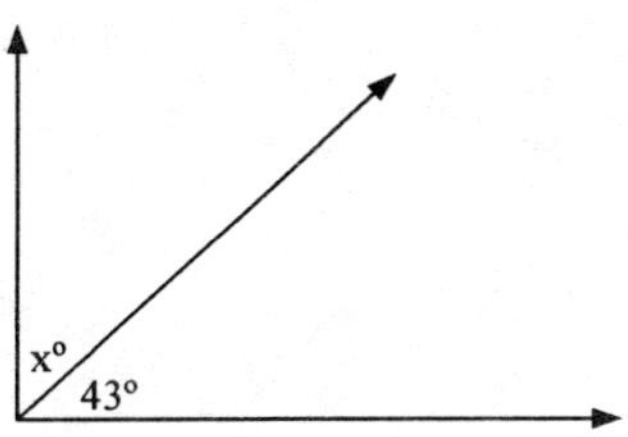

d.

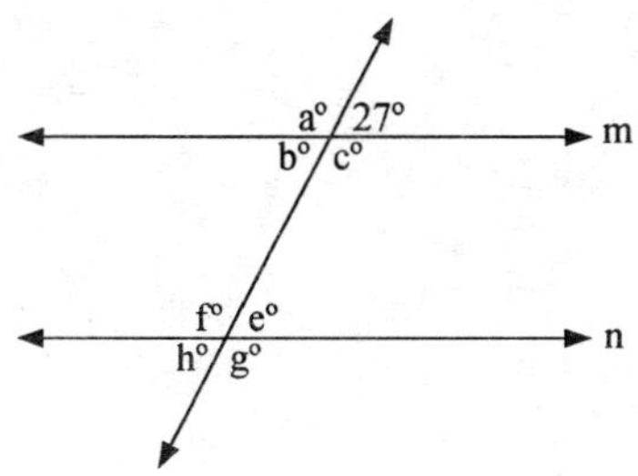

m || n

e.

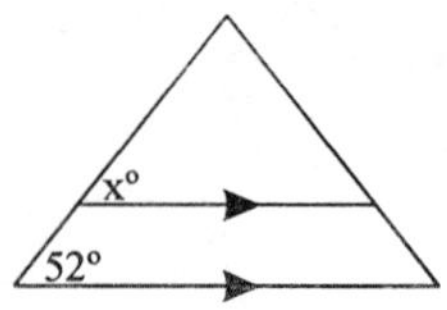

f.

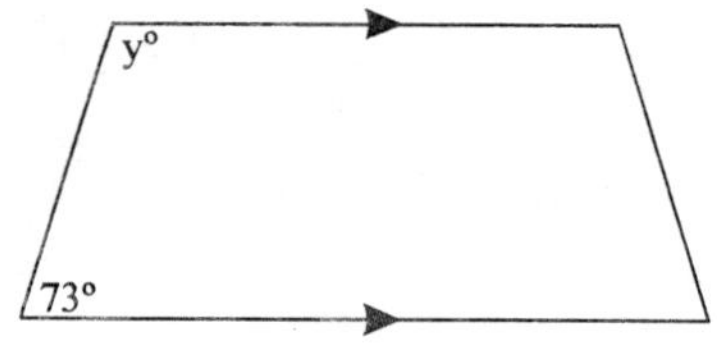

g.

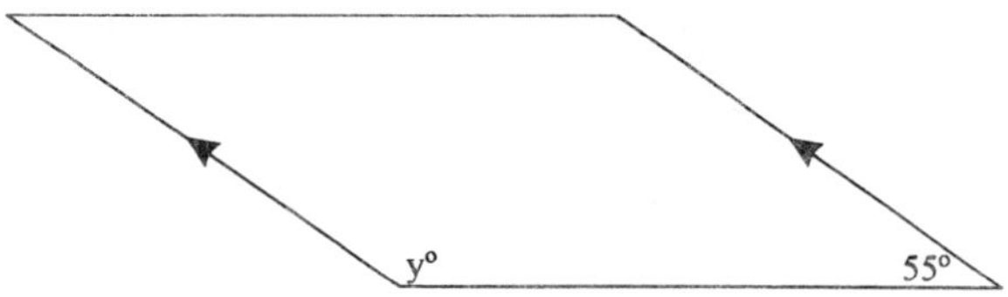

h.

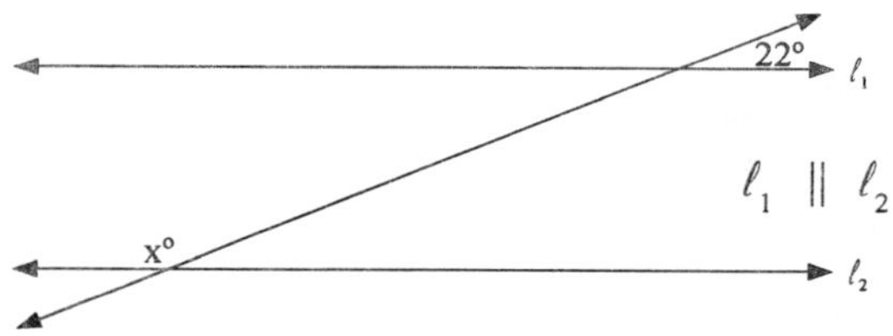

i.

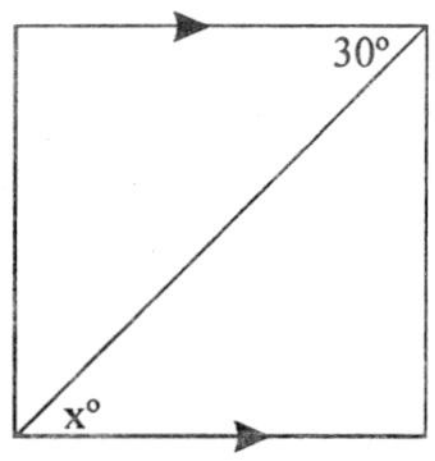

j.

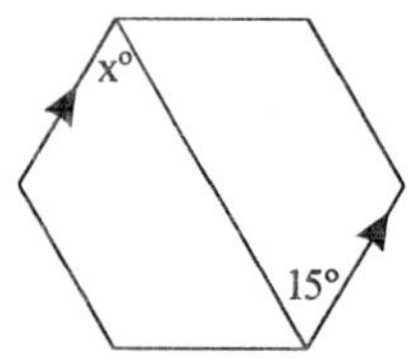

k.

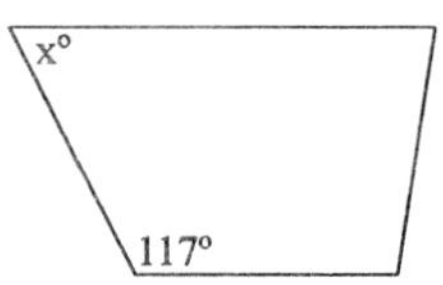

l.

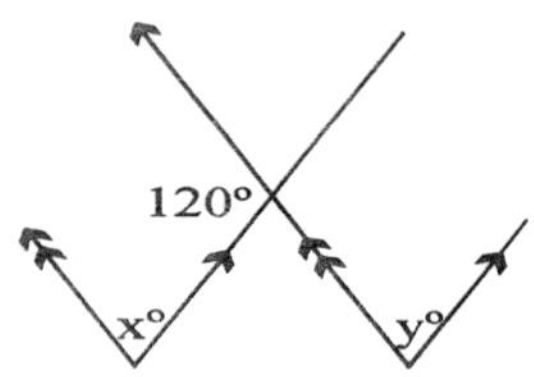

m.

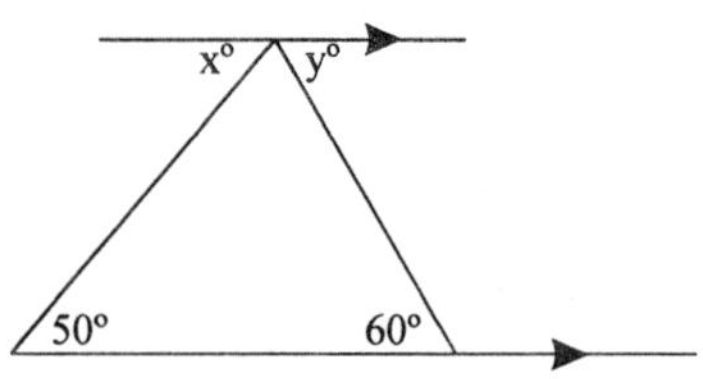

n.

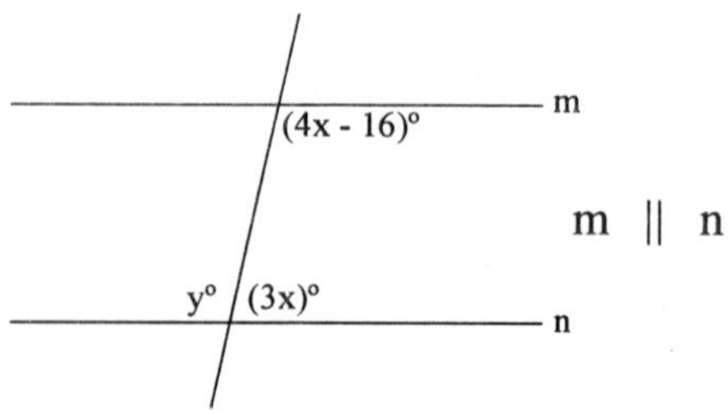

o.

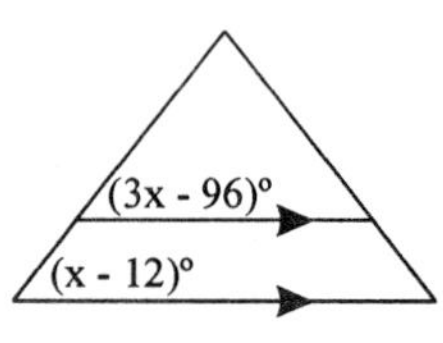

Chapter One Highlights

Vocabulary

point
line
collinear
intersect
line segment
length of a line segment
coordinate of a point
ray
opposite rays
plane
angle
vertex of an angle
side of an angle
complete revolution
degree/minute/second
straight angle
right angle
acute angle
obtuse angle
reflex angle
protractor
direct measurement
indirect measurement
adjacent angles
angles that form a linear pair
supplementary angles
complementary angles
betweenness of points
bisector of a line segment
midpoint
bisector of an angle/angle bisector
vertical angles
perpendicular lines
skew lines
parallel lines
transversal
corresponding angles
same side interior angles
alternate interior angles
distance from a point to a line
distance between two parallel lines
distance between two points

Required Skills

Find the distance between two points on a numberline when given the coordinates of the points.
Find the measure of an angle using a protractor
Convert from one unit of angle measure to another
Recognize notation for points, lines, planes, line segments, rays, angles, angle measure, lengths of line segments, angles that have equal measure, line segments that have the same length, betweenness of points, degrees, minutes, seconds, perpendicular lines, parallel lines, and right angles.

Conclusions

1. When two angles are adjacent, the sum of their measures equals the measure of the angle formed by their noncommon sides.
   Informally speaking, the whole equals the sum of its parts.

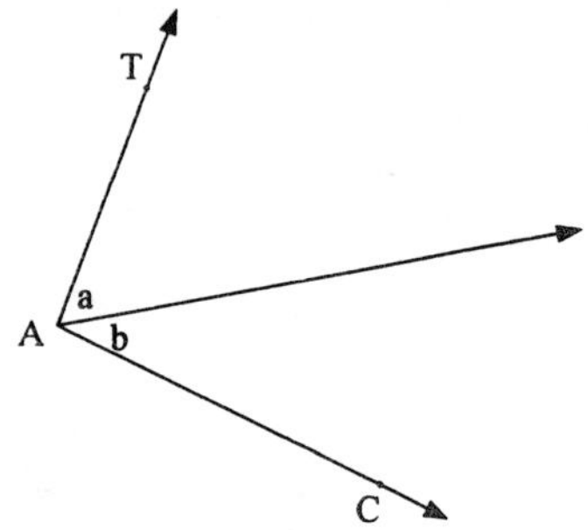

$\angle a + \angle b = \angle TAC$

2. If A, B, and C are collinear points, and B is between A and C, then AB + BC = AC.
   Informally speaking, *the whole equals the sum of its parts.*

AB + BC = AC

3. If two angles are supplementary, then the sum of their measures is 180 degrees.

4. If two angles are complementary, then the sum of their measures is 90 degrees.

5. If two angles form a linear pair, then they are supplementary.

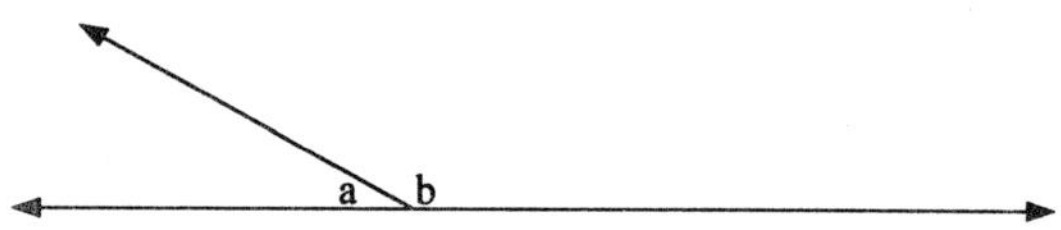

$\angle$ a and $\angle$ b are supplementary.

6. Vertical angles have equal measures.

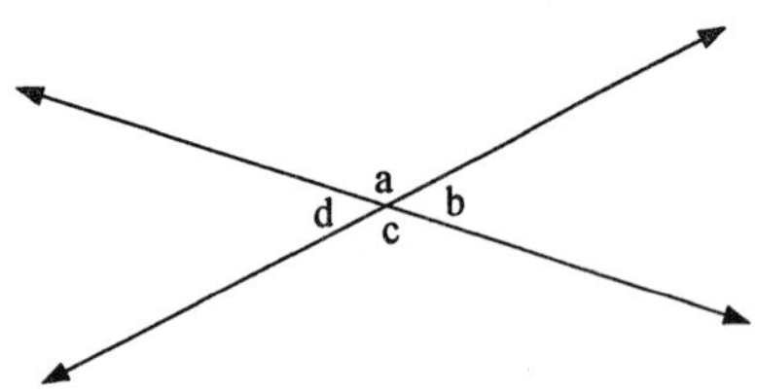

$\angle a = \angle c$

$\angle b = \angle d$

7. The midpoint or bisector of a line segment separates the segment into two segments with equal lengths.

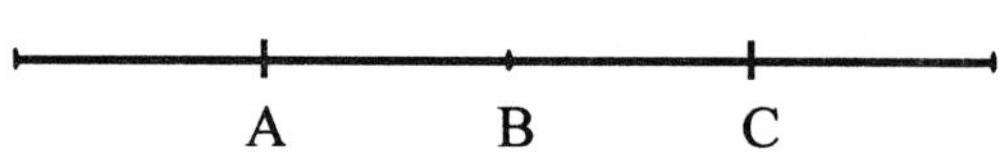

AB = BC

8. The bisector of an angle separates the angle into two angles with equal measures.

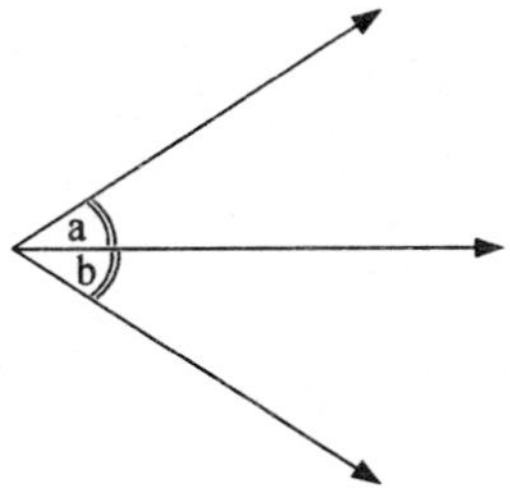

$\angle a = \angle b$

9. When two parallel lines are cut by a transversal, corresponding angles have the same measure.

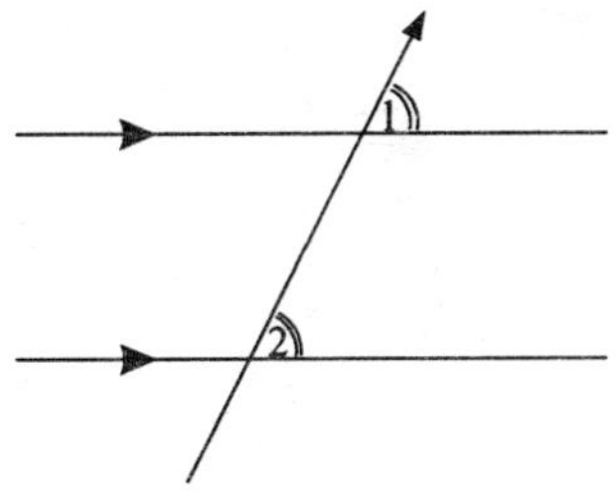

$\angle 1 = \angle 2$

10. When two parallel lines are cut by a transversal, alternate interior angles have the same measure

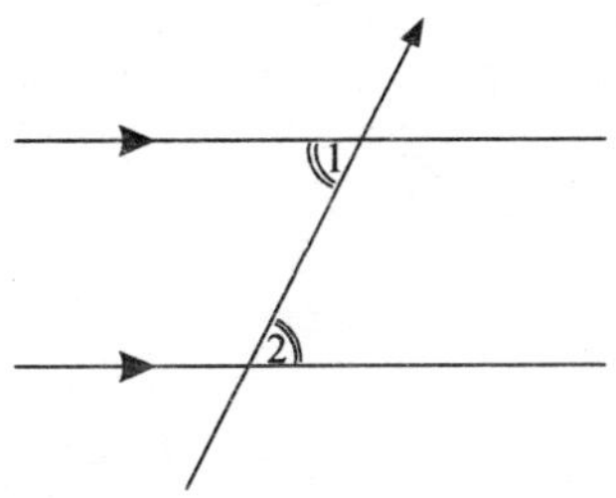

$\angle 1 = \angle 2$

11. When two parallel lines are cut by a transversal, same side interior angles are supplementary

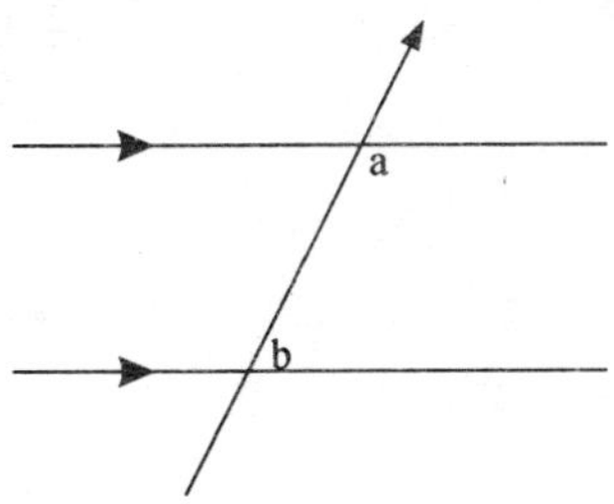

$\measuredangle$ a and $\measuredangle$ b are supplementary.

Chapter Two

Polygons

Objectives

Recognize and use basic terminology and notation for polygons
Classify polygons by number of sides
Classify triangles with respect to lengths of sides and measures of angles
Find the sum of the measures of the interior angles of a polygon
Find the measure of an angle of a regular polygon
Understand the relationships between angles and sides of isosceles, equilateral and equiangular triangles
Use the Triangular Inequality Observation to determine if a triangle can have sides of given lengths
Determine when two triangles are congruent when given certain information about the triangles
Determine when two triangles are similar when given certain information about the triangles
Find missing parts of congruent and similar triangles

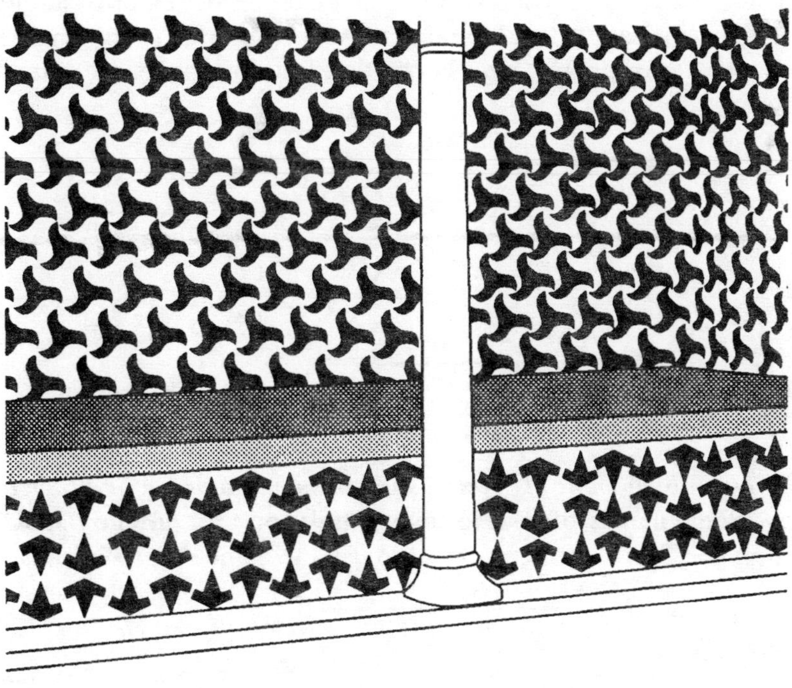

Geometric designs in the Alhambra in Granada, Spain designed by the Moors. **The Moors were forbidden by their religion from making images of living things.**

Section 2.1 Introduction to Polygons

A **polygon** is a closed geometric plane figure formed by joining line segments end to end. The line segments are called **the sides of the polygon** and the endpoints of the line segments are called the **vertices of the polygon.** Line segments that join nonadjacent vertices of the polygon are called **diagonals.** In the figure below, polygon ABCDE has sides $\overline{AB}$, $\overline{BC}$, $\overline{CD}$, $\overline{DE}$, and $\overline{EA}$; vertices A, B, C, D, and E; and diagonals $\overline{AC}$, $\overline{AD}$, $\overline{BD}$, $\overline{BE}$ and $\overline{CE}$. The polygon has angles: $\angle$EAB, $\angle$ABC, $\angle$BCD, $\angle$CDE, and $\angle$DEA.

These angles are called the **interior angles of the polygon**. Notice that a polygon must have the same number of sides and the same number of angles as it has vertices.

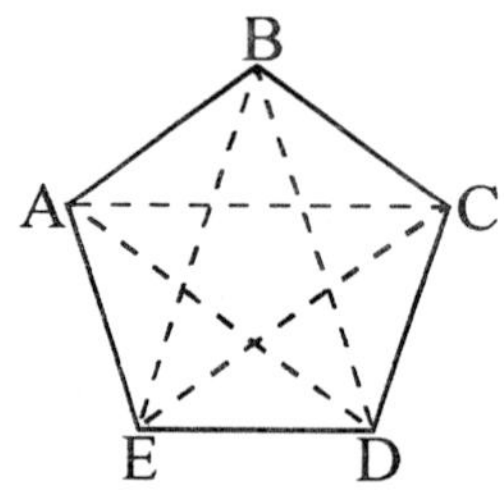

The **exterior angles of a polygon** are formed by extending the sides of the polygon. For polygon ABC, $\angle$1, $\angle$4, $\angle$2, $\angle$5, $\angle$3, and $\angle$6 are exterior angles. Notice that each exterior angle of the polygon forms a linear pair with an interior angle of the polygon. Notice also that two exterior angles are formed at each vertex of the polygon. Since the exterior angles at each vertex have the same measure, it is customary to include only one angle at each vertex when naming the exterior angles of a polygon. The exterior angles of polygon ABC are $\angle$1, $\angle$2, and $\angle$3. When we refer to the angles of a polygon without indicating interior or exterior angles, we assume that interior angles are intended.

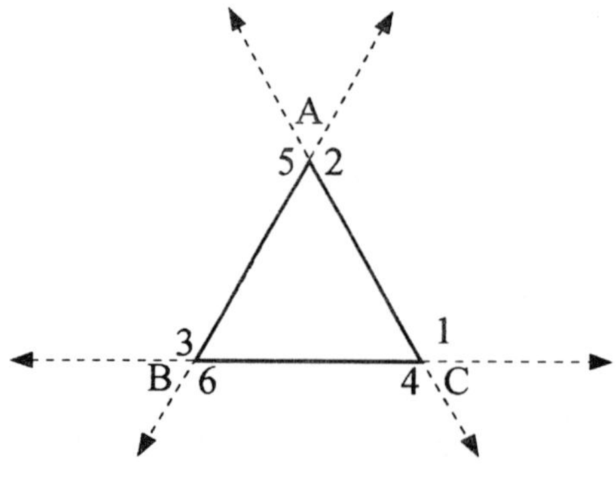

We may *name a polygon by naming its vertices in consecutive order in either a clockwise or counterclockwise direction.* Some names for the following polygon are: polygon ABCD, polygon ADCB, polygon CDAB, and polygon BADC. It is incorrect to use the name ACBD because the vertices are not in order.

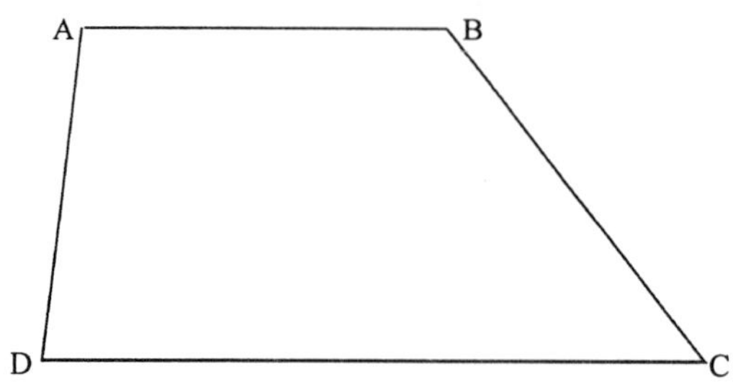

We may classify a polygon according to the number of sides that it has:

Number of sides	Name of Polygon
3	**triangle**
4	**quadrilateral**
5	**pentagon**
6	**hexagon**
7	**heptagon**
8	**octagon**
9	**nonagon**
10	**decagon**
11	**undecagon**
12	**dodecagon**
.	.
.	.
.	.
n - sides where n is a counting number and n > 2	**n-gon**

With the exception of the Latin prefix, *nona*, the underlined prefixes come from the Greek. The word *polygon* also comes from the Greek. (*poly* meaning *many* and *gonis* meaning *sides*).

We may also *classify polygons as convex or concave*. A **convex polygon** is one in which the measure of each interior angle is less than 180 degrees. A **concave polygon** has an interior angle with a measure greater than 180. In our study we will only consider convex polygons, so we agree that whenever the term polygon is used, unless stated otherwise, it will refer to a convex polygon.

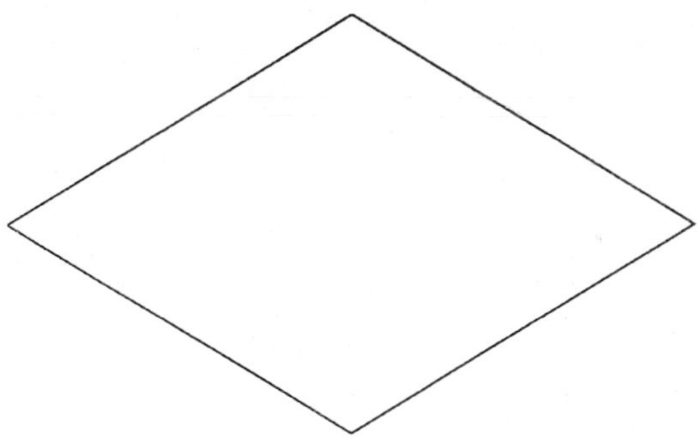

Convex

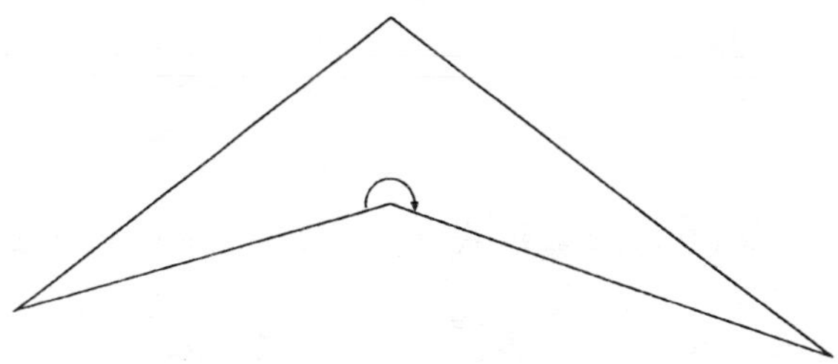

Concave

A polygon is said to be **regular** when it is both equiangular ***and*** equilateral. Note that the word "and" means that both conditions must hold if a polygon is regular. Of course **equiangular** means that all angles have the same measure, and **equilateral** means that all sides have the same length. A **square** is an example of a regular polygon. It is both equiangular and equilateral. A rectangle is equiangular, but it may not be equilateral. Therefore a rectangle is not a regular polygon.

Let us now concentrate our attention on triangles and quadrilaterals.

We have two basic ways by which we may classify a triangle. We may classify it according to the type of angles that it contains or according to the relative lengths of its sides.

Classification of Triangles by Angles

acute triangle: all angles are acute

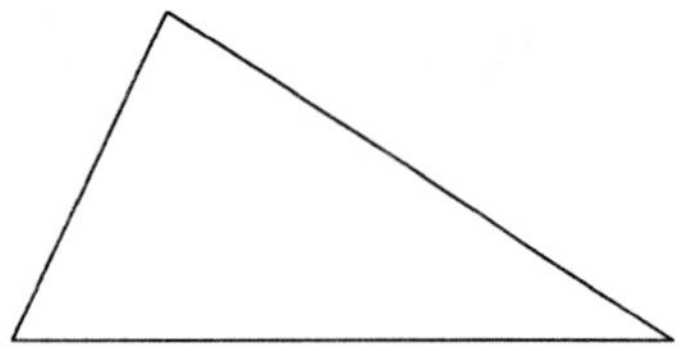

right triangle: has a right angle (the side opposite the right angle is called the **hypotenuse,** the other two sides are called **legs**).

hypotenuse $\overline{AB}$
legs $\overline{AC}$ and $\overline{BC}$

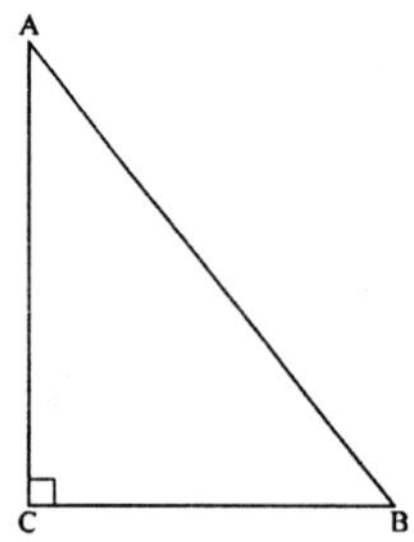

obtuse triangle : has an obtuse angle

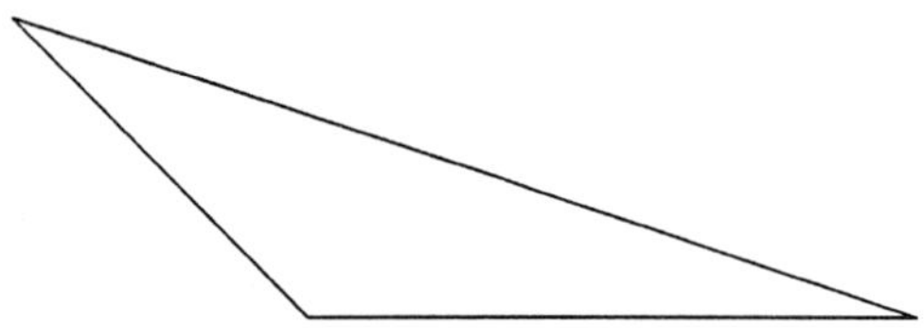

equiangular triangle: all angles have the same measure

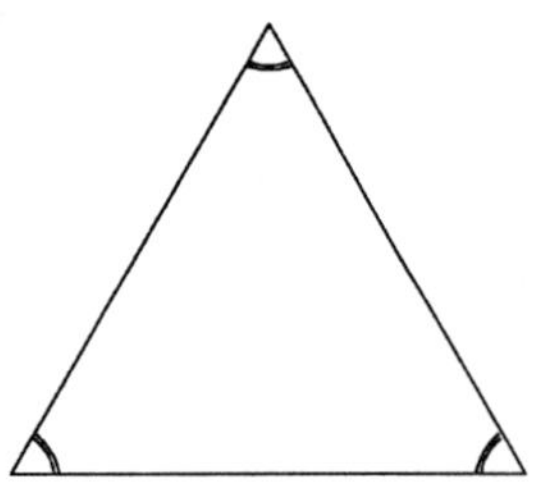

Classification of Triangles by Sides

scalene triangle: all sides have different lengths

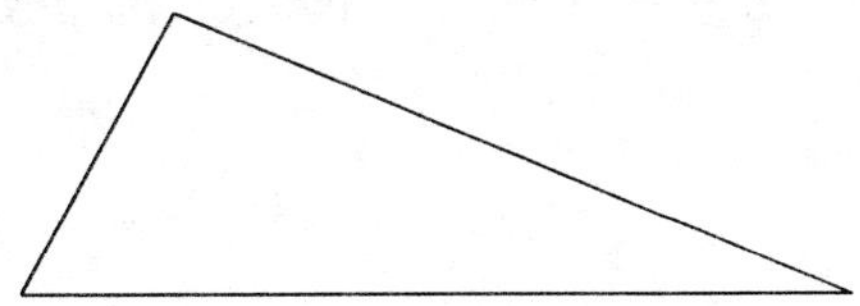

isosceles triangle: at least two sides have the same length

The two sides with the same length are called **legs.** The other side is called the **base.** The angles opposite the sides of equal length are called the **base angles.** The angle opposite the base is called the **vertex angle.**

legs	$\overline{AB}$ **and** $\overline{AC}$
base	$\overline{BC}$
vertex angle	$\angle A$
base angles	$\angle B$ **and** $\angle C$

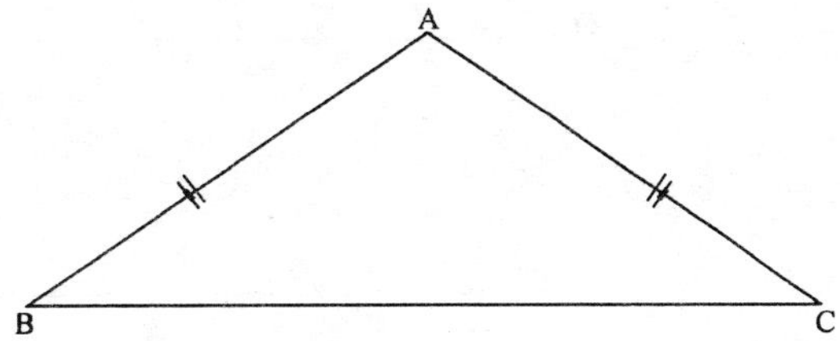

equilateral triangle: all three sides have the same length

(Note that an equilateral triangle must also be isosceles since it has at least two sides of the same length).

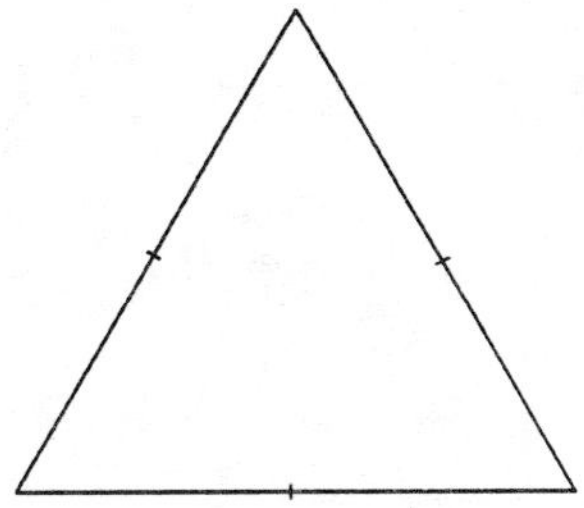

Now for quadrilaterals:

We may classify a ***quadrilateral*** *according to the number of parallel sides that it has:*

trapezium (plural—trapezia or trapeziums): a quadrilateral with no parallel sides

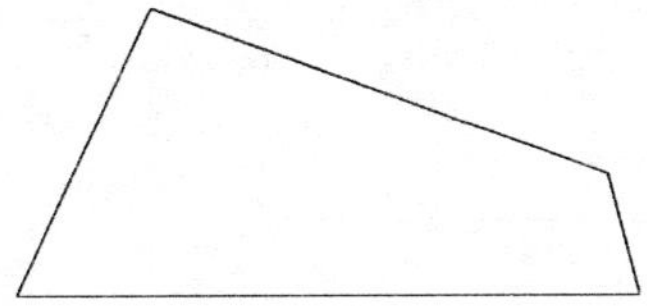

A kite is an example of a trapezium

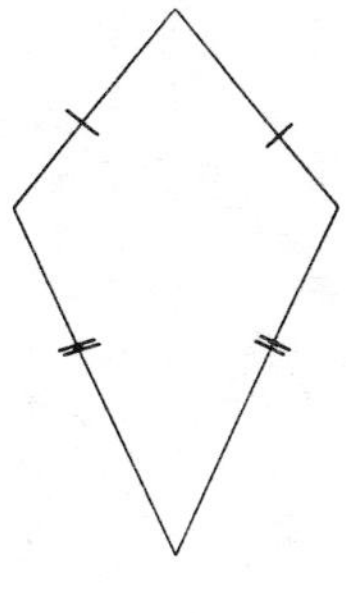

trapezoid: a quadrilateral with exactly one pair of parallel sides. The parallel sides are called **bases.** The nonparallel sides are called **legs.** The two angles opposite each base are called **base angles.** When the legs have the same length, the trapezoid is called an **isosceles trapezoid**.

bases	**$\overline{AB}$ and $\overline{DC}$**
legs	**$\overline{AD}$ and $\overline{BC}$**
base angles	**$\angle A$ and $\angle B$**
base angles	**$\angle D$ and $\angle C$**

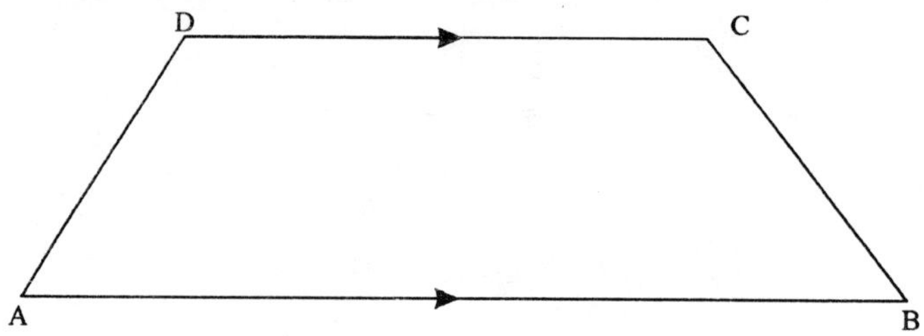

Trapezoid MNOP is an isosceles trapezoid

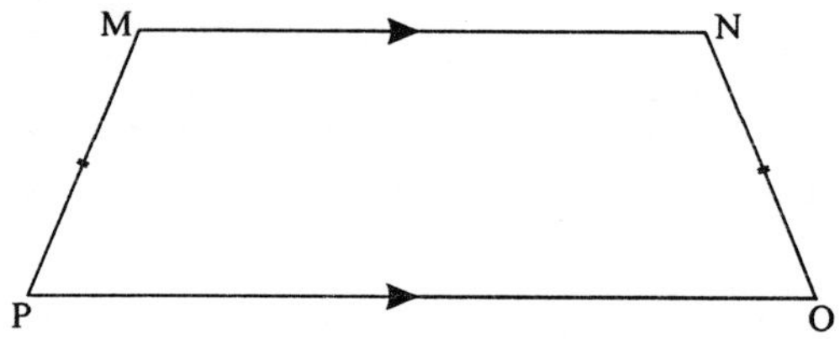

parallelogram: a quadrilateral with two pairs of parallel sides
Each pair of parallel sides may be called bases.

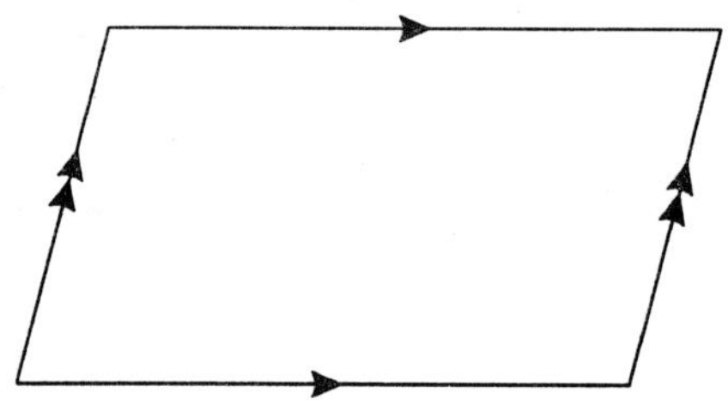

rectangle: an equiangular quadrilateral
Note that a rectangle is also parallelogram.

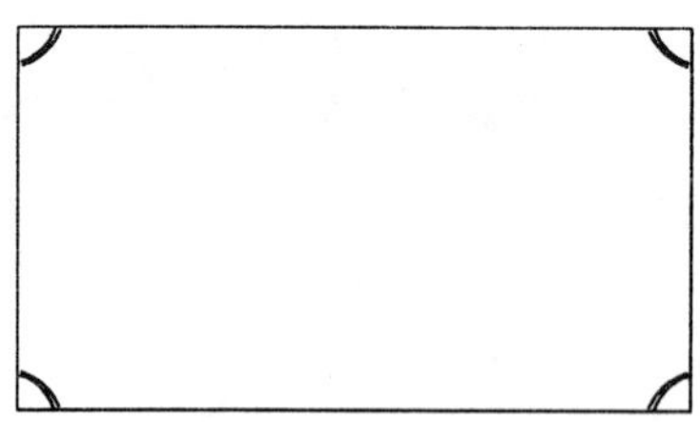

rhombus: an equilateral quadrilateral
Note that a rhombus is also a parallelogram.

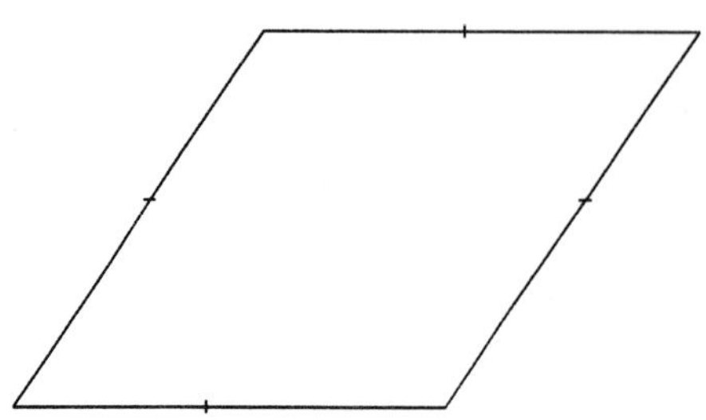

square: an equiangular and equilateral quadrilateral
Note that a square is also a rhombus, a rectangle and a parallelogram.

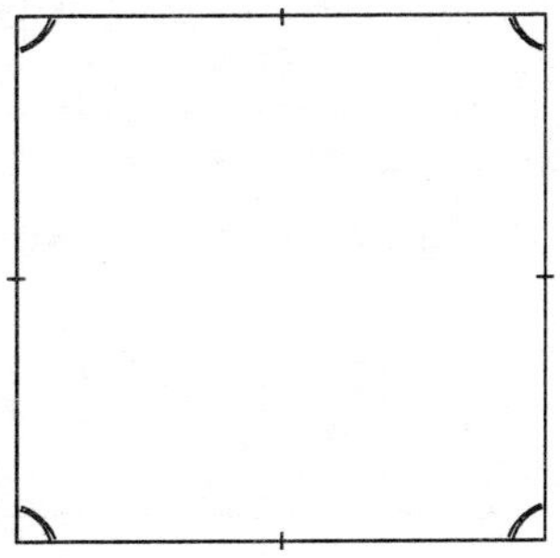

We will now investigate some properties of polygons, paying special attention to the properties of triangles.

If we cut out the angles of a triangle and place their vertices together as shown in the figure, we observe that the total rotation of the three angles is 180 degrees. We conclude that *the sum of the measures of the angles of a triangle is 180.*

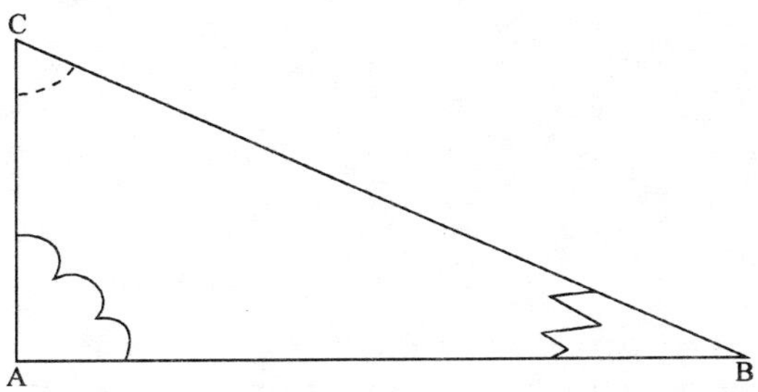

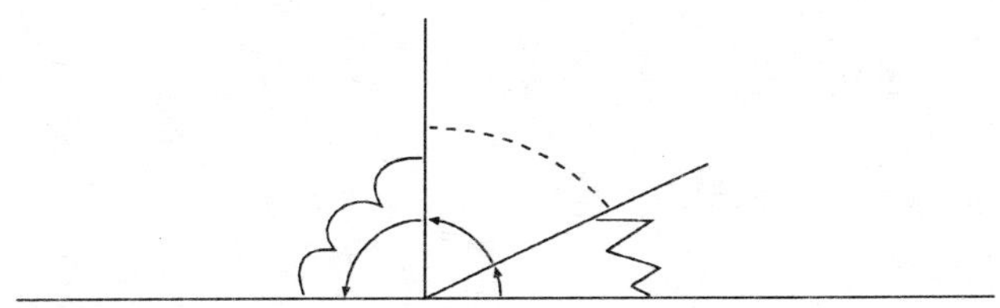

Now let us see if we can determine the total measure of the angles of a polygon with more than three sides. In the illustration, we select a vertex in each polygon, then we draw all the diagonals from the vertex. The diagonals separate each polygon into triangles. For each polygon, we observe that the sum of the measures of the angles of the triangles is the same as the sum of the measures of the angles of the polygon.

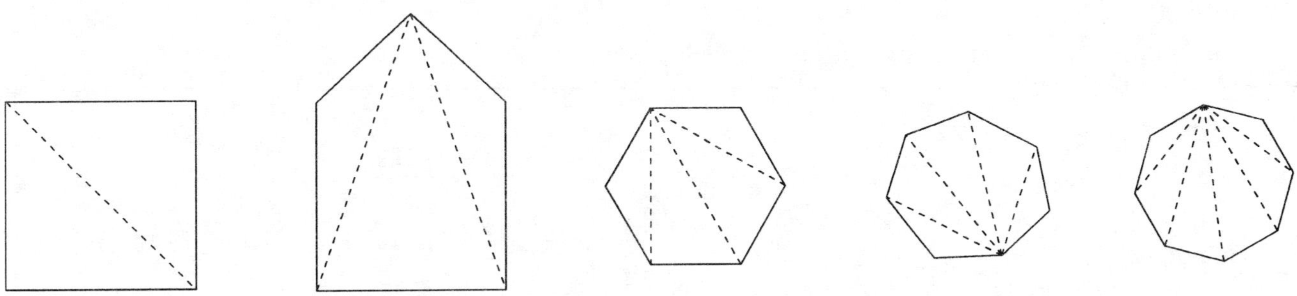

We then count the triangles that are formed in each polygon, and multiply this number by 180 to obtain the sum of the measures of the angles of the polygon.

The results are shown in the chart below. Study the chart carefully. Do you notice a pattern?

Number of sides	4	5	6	7	n, n > 2
Number of triangles	2	3	4	5	n – 2
Sum of angle measures	2(180)	3(180)	4(180)	5(180)	(n – 2)180

We conclude that the *sum of the measures of the angles of a polygon with n sides is the product of (n – 2) and 180.*

polygon with n sides, where n > 2 sum of measure of angles is (n – 2)180

EXAMPLE:

Find the sum of the measures of the angles of a hexagon.

A hexagon has 6 sides, so n = 6.	(n – 2) 180
(n – 2) = 6 – 2 = 4	(6 – 2) 180
We multiply 4 times 180 to get 720.	4(180)
The sum of the measures of the angles of a hexagon is 720.	720

If a polygon with n sides is regular, we can determine the measure of each angle of the polygon by dividing the product (n – 2) 180 by n.

regular polygon with n sides
measure of each angle $\frac{(n-2)180}{n}$

EXAMPLE:

Find the measure of each angle of a regular hexagon.

Again, n = 6	
We use the formula	$\frac{(n-2)\ 180}{n}$
We replace n with 6	$\frac{(6-2)\ 180}{6}$
The measure of each angle of a hexagon is	120

We now can see why *each angle of an equiangular triangle must have a measure of 60 degrees and why each angle of a rectangle and square must have a measure of 90 degrees.*

Here is another interesting conclusion that we can make:

If two angles of a triangle have the same measures, respectively, as two angles of another triangle, then the remaining two angles of the triangles have the same measure.

Observe the triangles below:

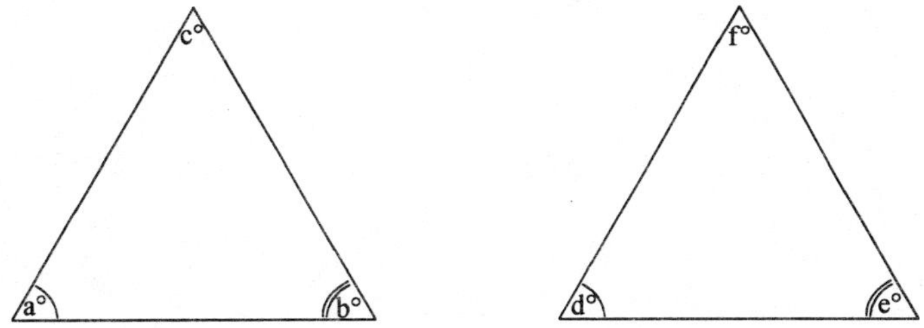

In the triangles,

$a + b + c = 180$

$d + e + f = 180$

$\mathbf{a} + \mathbf{b} + c = \mathbf{d} + \mathbf{e} + f$ since both sums equal 180

$\mathbf{a} + \mathbf{b} = \mathbf{d} + \mathbf{e}$ since $a = d$ and $b = e$

Now can you see why c and f must be equal?

Let us consider some properties of special triangles. First, we will draw an isosceles triangle with legs $\overline{AB}$ and $\overline{AC}$. If we cut out the base angles, $\angle B$ and $\angle C$, as shown in the illustration, and then place one angle on top of the other so that their vertices and sides coincide, we observe that the two angles of the triangle have the same measure.

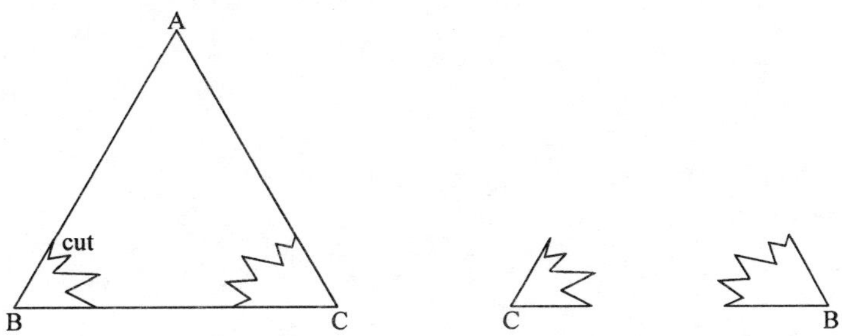

It is reasonable to assume that this result will be the same for any isosceles triangle that we may draw, so we make the following conclusion:

If a triangle has two sides with the same length, then the angles opposite those sides have the same measure

More informally we may say, *the base angles of an isosceles triangle are equal.*

If this ➔ 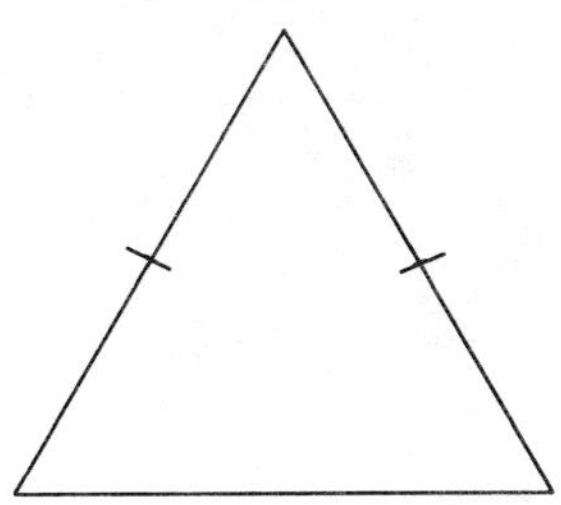 **Conclude this ➔** 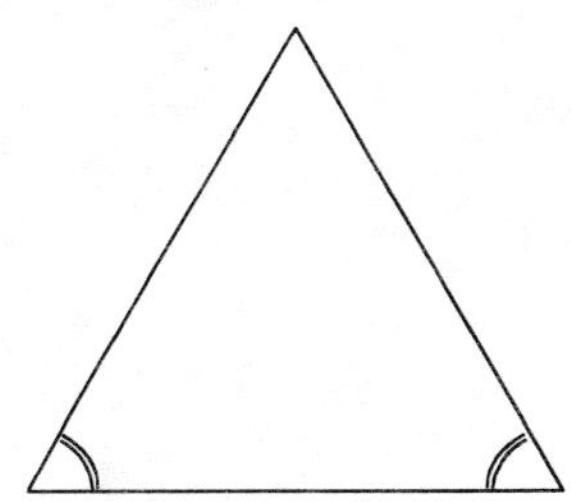

Since every equilateral triangle is also isosceles, we can conclude that *every equilateral triangle is also equiangular.* Do you see why this must be so?

If this → 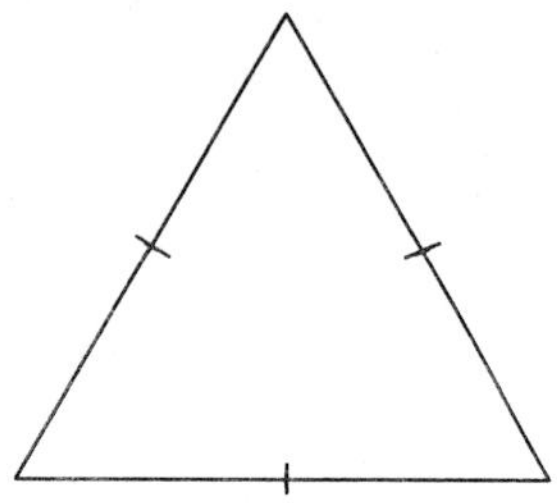

Conclude this → 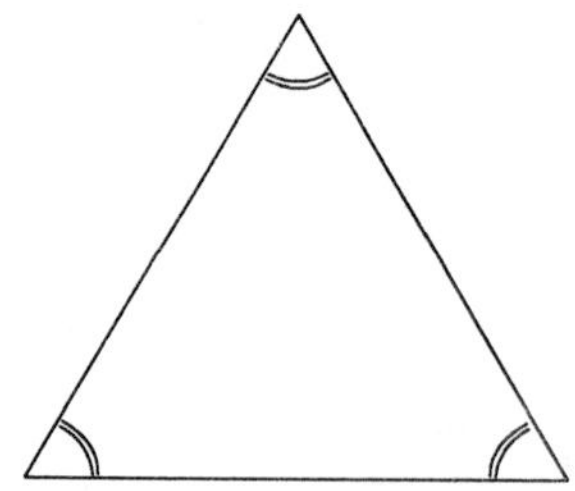

Now let us draw a triangle with two angles that have the same measure. If we cut the triangle into two parts as shown in the illustration, and place one part on top of the other so that $\angle$B and $\angle$C coincide, we observe that sides $\overline{AB}$ and $\overline{AC}$ of the triangle have the same length.

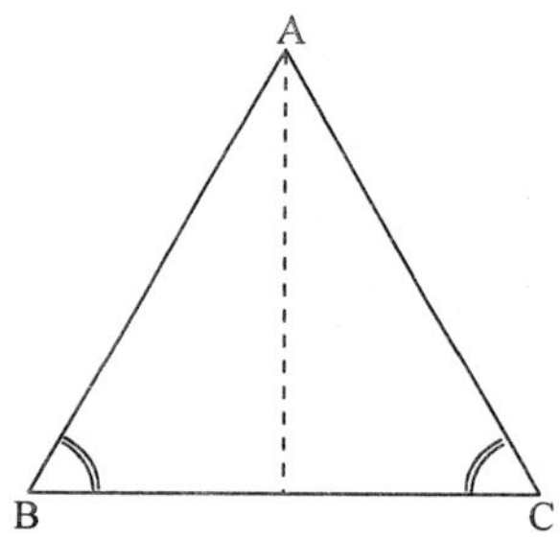

There is no reason to believe that this observation will differ if we consider any other triangle with two angles having the same measure. We therefore conclude:

If two angles of a triangle have the same measure, then the sides opposite those angles will have the same length.

If this → 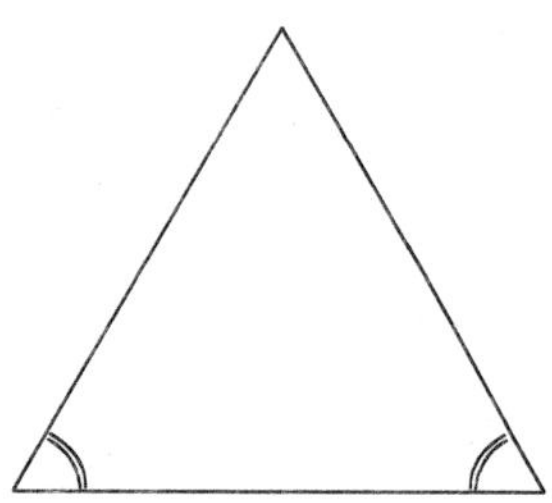

Conclude this → 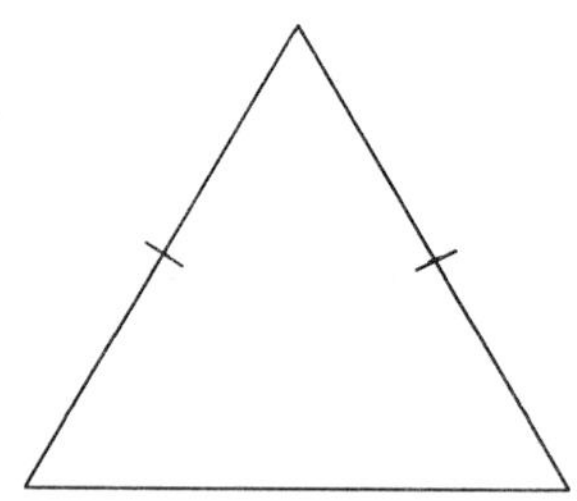

Furthermore, *we conclude that every equiangular triangle is also equilateral.*

If this → 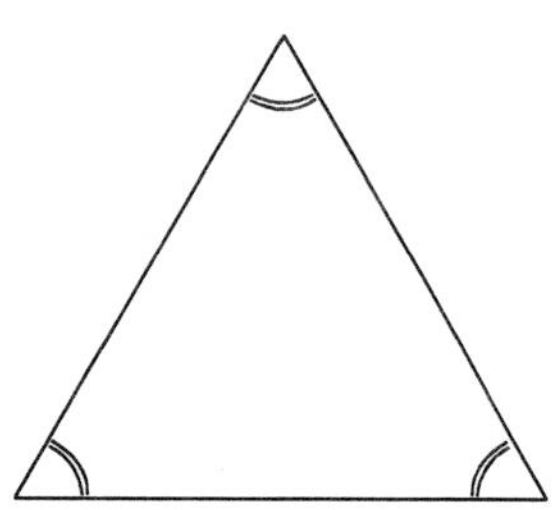

Conclude this → 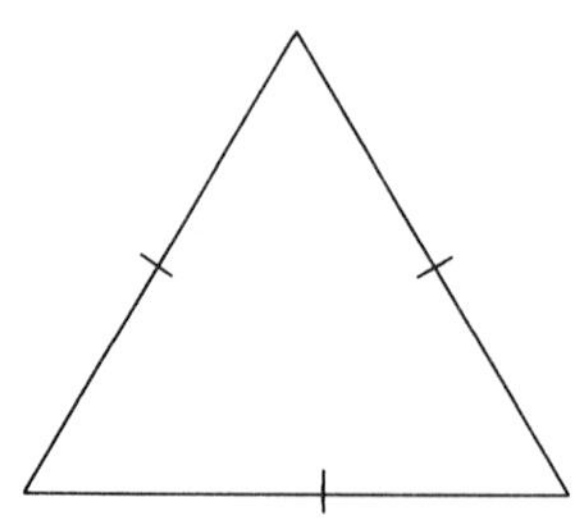

Now for a final observation about triangles:

Let us imagine taking a trip along the sides of the triangle shown below. The trip must begin at point A of the triangle, and it must end at point B of the triangle. The trip can be made in either of two ways:

1. We can travel first from A to C , and then travel from C to B.

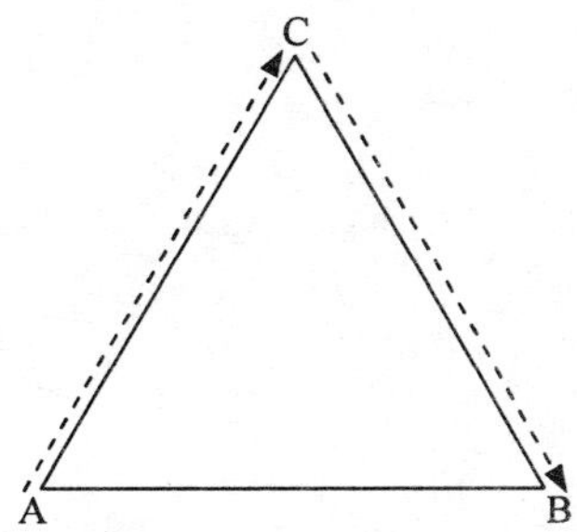

2. Or, we can travel straight from A to B.

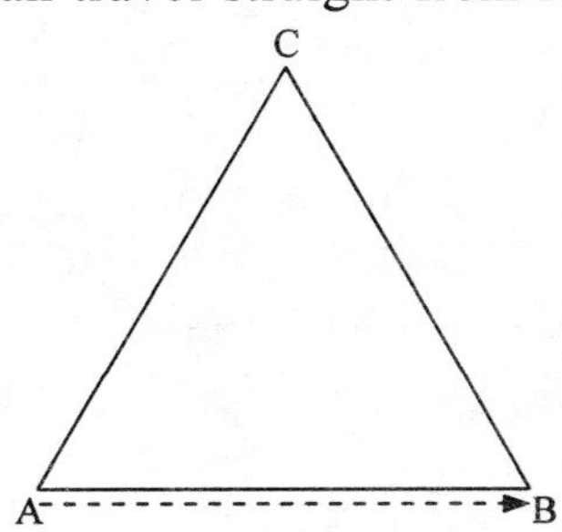

Obviously, the shorter trip will be the direct one from A to B, that is, AC + CB > AB. This rather simple conclusion is very important in the study of mathematics. For now, we will call it the **triangular inequality observation.** Let us restate this observation in the following way:

The sum of the lengths of any two sides of a triangle is always greater than the length of the remaining side of the triangle.

For triangle ABC we conclude: **AC + CB > AB**
AB + BC > AC
AC + AB > BC

The triangular inequality observation allows us to conclude that the triangle below can not exist with sides of lengths **3, 5** and **9**

3 + 5 < 9

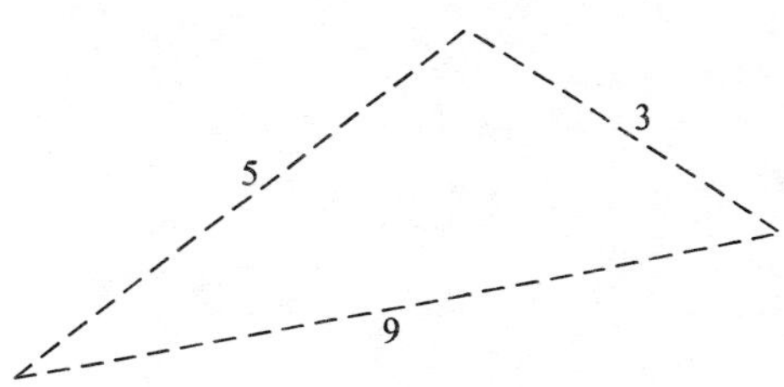

Exercises 2.1

1. Try to draw a triangle in each block of the following chart. The triangle in each block must be of the type indicated by both the column and the row in which that block is located. One triangle has been done for you. Notice that the triangle is both a right triangle (because it is in the second row) and an isosceles triangle (because it is in the second column).

	scalene	isosceles	equilateral
acute			
right		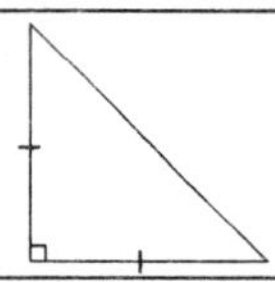	
obtuse			

2. If possible, give an example of each of the following polygons:
   a. a triangle that is isosceles but not equilateral.
   b. a triangle that is equilateral but not isosceles
   c. a rhombus that is not a square
   d. a square that is not a rhombus
   e. a polygon that is equiangular but not equilateral
   f. a rectangle that is not a square
   g. a square that is not a rectangle
   h. a trapezium
   i. an isosceles trapezoid
   j. a trapezoid that is not isosceles
   k. a parallelogram that is not a rhombus or rectangle

3. What is the common name for a regular quadrilateral?

4. Determine the sum of the measures of the angles of each polygon:

a. triangle	b. quadrilateral	c. pentagon	j. n–gon $n > 2$
d. hexagon	e. heptagon	f. octagon	
g. nonagon	h. decagon	i. 15-gon	

5. Determine the measure of each angle of the following polygons:

a. regular triangle

b. regular quadrilateral

c. regular pentagon

d. regular hexagon

e. regular heptagon

f. regular octagon

g. regular decagon

h. regular n-gon

6. Triangle MNP is a right triangle. What is its hypotenuse? _________
What are its legs? ______________________________

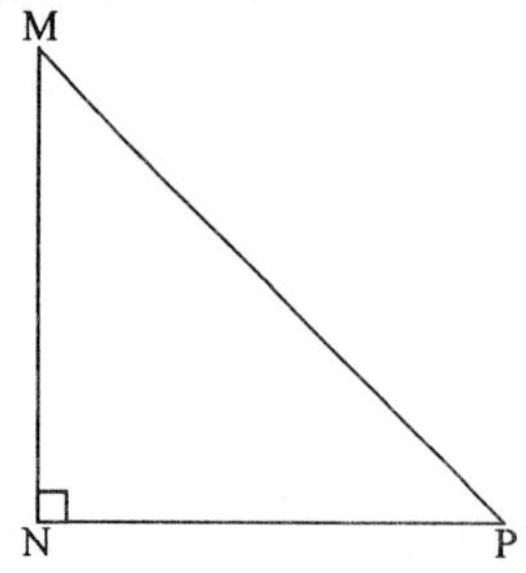

7. Identify the legs, base, base angles and vertex angle of isosceles triangle NAP

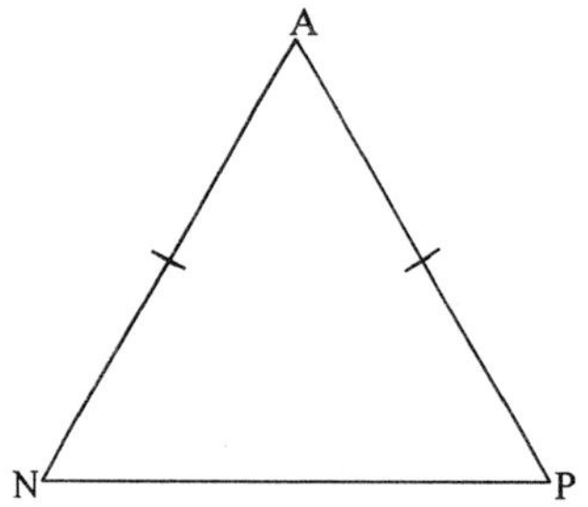

8. Identify the bases and legs in trapezoid HOME.

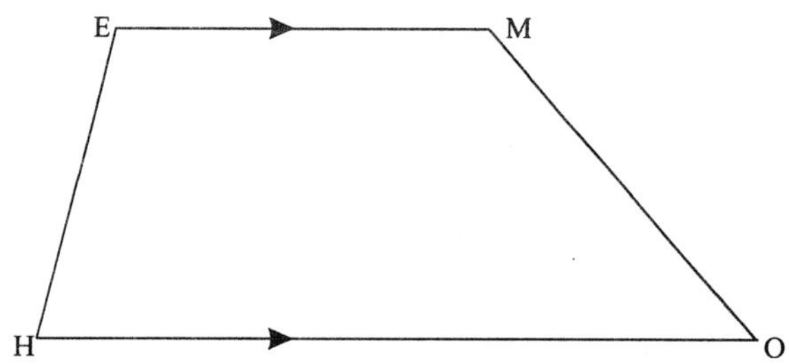

9. Identify the sides, angles, vertices and diagonals of the polygon ABCDE.

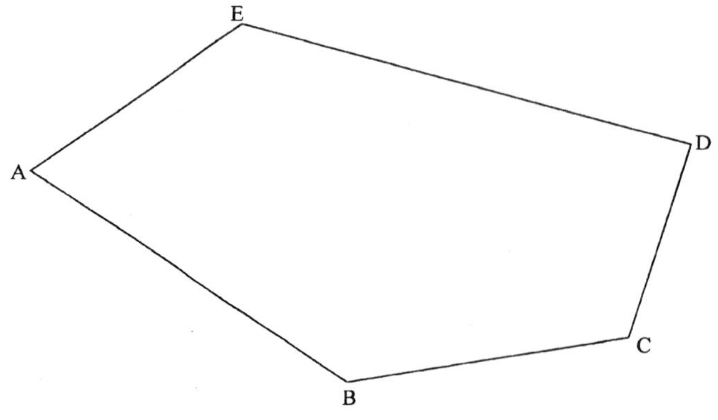

10. Find the measure of the angles or the lengths of sides for each polygon.

a. One acute angle of a right triangle has a measure that is twice that of the other acute angle. Find the measure of each angle of the right triangle.

b. The angles of a triangle have measures that are in the ratio of 2:3:5. Find the measure of each angle.

c. The measure of the largest angle of a triangle is three times the measure of its smallest angle. The measure of the remaining angle of the triangle is five more than the measure of the smallest angle. Find the measure of each angle.

d. Find the lengths of the sides of a rhombus if two of its adjacent sides have lengths (5q–3) and (2q+93), respectively.

e. Find the lengths of two sides of an isosceles triangle if its legs have lengths (13x – 4) and (5x + 8), respectively.

f. Find the measures of the angles of an isosceles triangle if its vertex angle has a measure that is twice that of its base angles.

g. Find the measures of the angles of a triangle if the sum of the measures of two of its angles is 150; and if the difference between the measures of the same two angles is 10. Hint: You may want to use two equations to solve this one.

11. Determine the number of diagonals in each polygon:

a. pentagon

b. hexagon

c. dodecagon

d. n-gon

12. Show how four isosceles triangles with the same size and the same shape can be placed together to form each of the following:

a. a rectangle that is not a square
b. two different parallelograms that are not rectangles
c. a square
d. a triangle
e. a trapezoid

13. How many squares can you count in this figure? The answer is not sixteen.

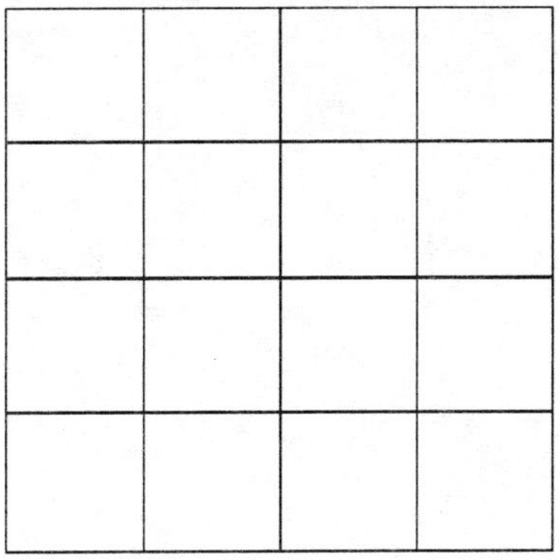

14. Look at the illustration below to see why the numbers 1, 3, 6, 10, etc. are called triangular numbers.

Can you find the 50th triangular number?
Can you find the nth triangular number where n is a counting number?

15. Look at the illustration below to see why the numbers 1, 4, 9 and 16 are called square numbers.

Can you find the 25th square number?
Can you find the nth square number where n is a counting number?

16. Look at the illustration below to see why the numbers listed are called rectangular numbers. Which counting numbers are missing?

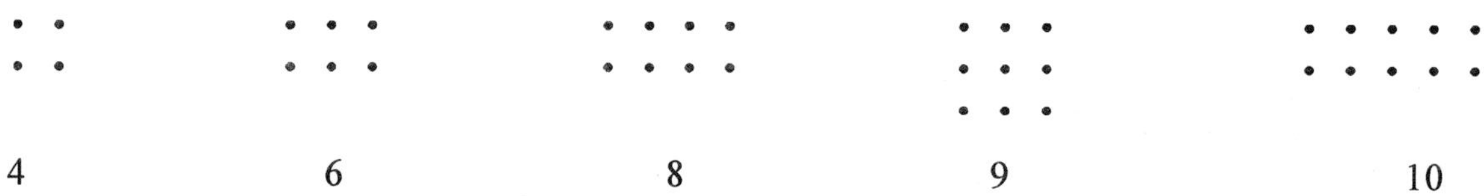

17. Plot each set of points on a rectangular coordinate system. Connect the points in order with line segments to form a polygon, that is, join A to B then B to C, etc.; then identify the polygon.

a.	A (2, 3)	B (4, 1)	C (8, 2)	D (6, 4)
b.	A (2, −3)	B (6, −3)	C (6, −1)	D (2, −1)
c.	A (−1, 1)	B (−1, 3)	C (−3, 3)	D (−3, 1)

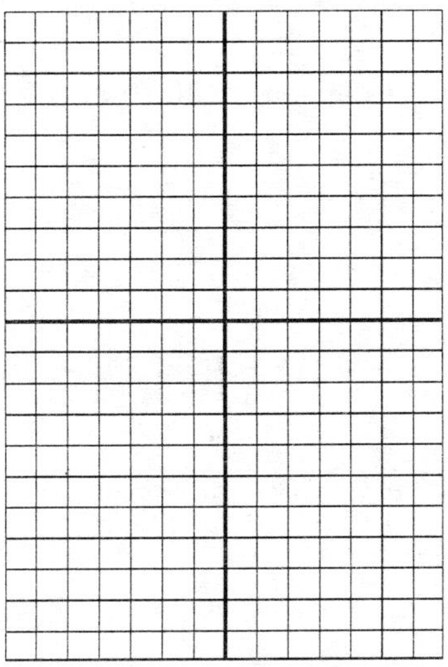

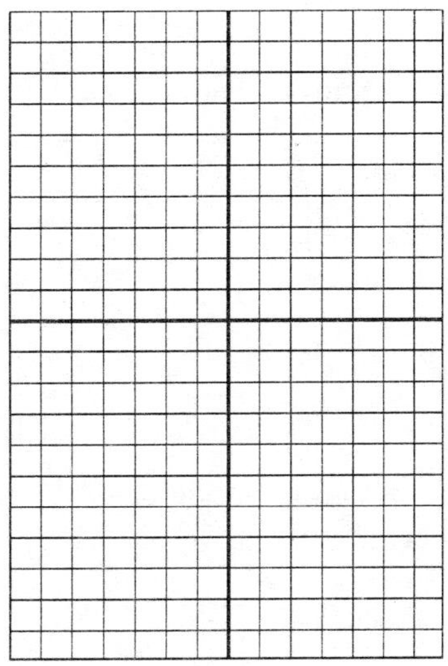

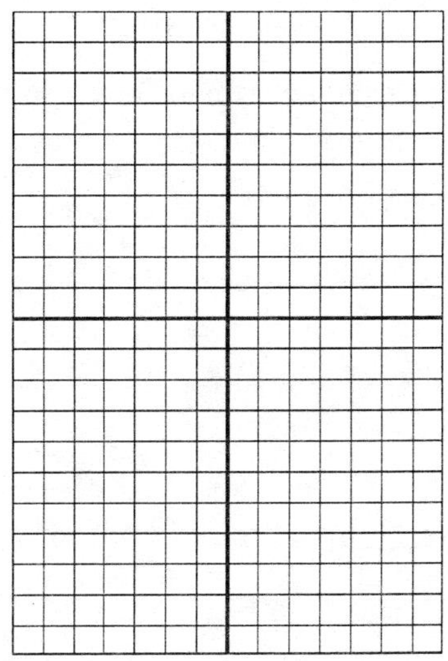

18. Which of the following sets of numbers can be the lengths of sides of a triangle?
a. 6, 6, 6 b. 3, 4, 5 c. 5, 8, 13 d. 2, 5, 8

Section 2.2 Congruent Triangles

Figures are said to be **congruent** if they have the same shape and the same size. The word "congruent" comes from two Latin words, "con" meaning "with" and "gruere" meaning "to agree." Geometric figures are congruent if they can be made to fit or coincide. The concept of congruency is important in the development of many geometric concepts. It is also an important concept in industry where it forms the basis for the mass production of parts. Observe the congruent polygons shown below. The symbol $\cong$ means "is congruent to." The symbol $\sim$ indicates that the polygons have the same shape, while the symbol $=$ indicates that they have the same size. We agree to use the symbol $\triangle$ for the word triangle.

$\triangle$ ABC $\cong$ $\triangle$ DEF

The triangles can be made to fit by mentally flipping one over so that it can be made to coincide with the other one.

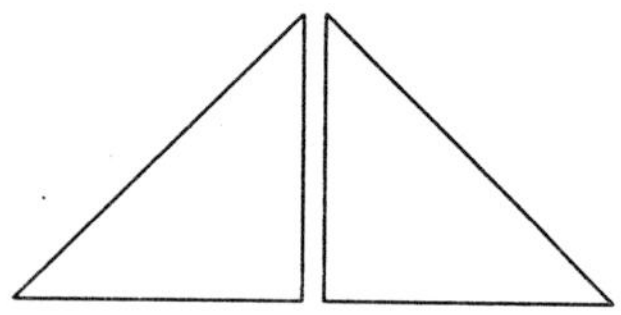

The two hexagons are congruent. They can be made to fit by mentally sliding one hexagon so that it can be made to coincide with the other one.

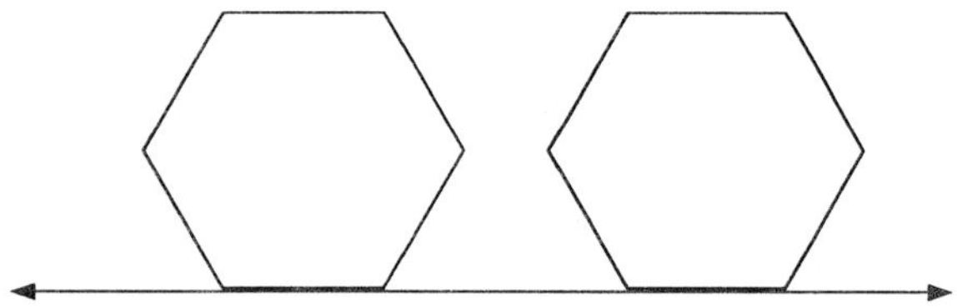

The two quadrilaterals are congruent. They can be made to fit by mentally rotating one quadrilateral so that it can be made to coincide with the other one.

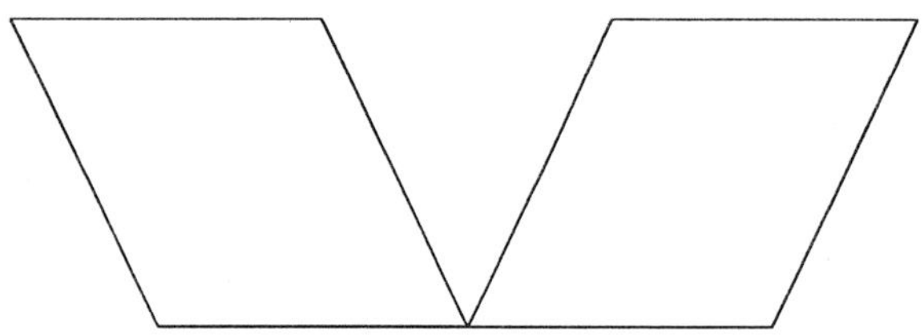

Although in some cases we may determine by observation whether two polygons are congruent by mentally moving them to make them coincide, such a task may not always be easy to accomplish. For this reason, it is often more practical to determine congruency through the use of indirect methods. Before proceeding, we will agree that polygons can be made to coincide if all of their corresponding parts are congruent, that is, if all of their corresponding angles have the same measure, and if all of their corresponding sides have the same length.

We then define **two polygons to be congruent** if a correspondence can be set up between their vertices so that all corresponding angles have the same measure and all corresponding sides have the same length.

The phrase, "set up a **correspondence between their vertices,**" simply means that we can match each vertex of one polygon with exactly one vertex of the other polygon, and vice versa.

Of course the vertices that are matched together are the ones that will coincide if the polygons are mentally made to fit one on top of the other. Similarly, we define congruent angles as angles that have the same measure and congruent segments as segments that have the same length.

In the illustration, △ ABC is congruent to △ DEF.

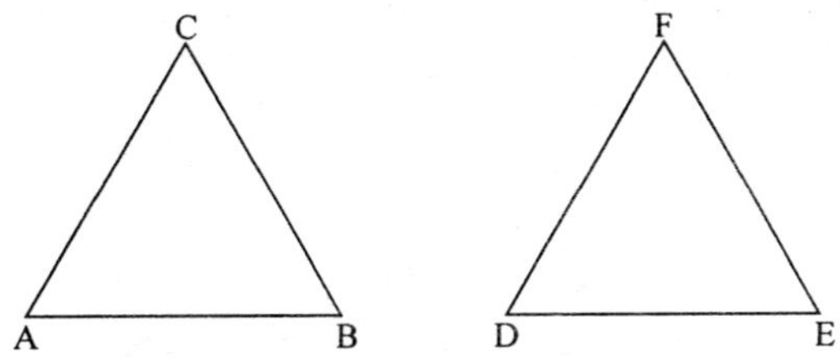

Here is the correspondence between their vertices:

A corresponds to D
B corresponds to E
C corresponds to F

To indicate this correspondence, we may use the notation: **ABC ⟷ DEF.**

For these congruent triangles, there are three pairs of corresponding angles:

∠ A = ∠ D   **∠ B = ∠ E**   **∠ C = ∠ F**

For these congruent triangles, there are three pairs of corresponding sides:

AC = DF   **AB = DE**   **BC = EF**

We write △ **ABC** ≅ △ **DEF.**

Note that when we use the symbol, ≅ , it is good practice to write the names of the corresponding vertices of the two polygons in the same relative order. When the symbol is used in this way, we can determine congruent sides and congruent angles from the notation. Mentally moving the polygons or actually seeing the polygons will not be necessary. For example, if we know that △ **PQR** ≅ △ **MNO,** we can write the correspondence immediately:

PQR ⟷ MNO

vertex P corresponds to vertex M (Both points are named first)
vertex Q corresponds to vertex N (Both points are named second)
vertex R corresponds to vertex O (Both points are named third)

From the correspondence, we obtain the following information:

∠ P = ∠ M   **∠ Q = ∠ N**   **∠ R = ∠ O**

PQ = MN   **QR = NO**   **PR = MO**

Now let us investigate the conditions necessary for drawing triangles that are congruent. As concluded earlier, triangles will be congruent if we ensure that they have *six pairs* of corresponding parts congruent.

1. The three angles of one polygon must have the same measures, respectively, as the three angles of the other polygon;
2. The three sides of one polygon must have the same lengths, respectively, as the three sides of the other polygon.

A little investigation, however, will convince us that all these conditions are not necessary to ensure congruent triangles.. We will see that if the corresponding parts are selected with care, we can draw congruent triangles provided *only three pairs of corresponding parts* have the same measures or lengths. Let us consider six situations.

1. Suppose we are asked to draw a triangle with sides having three specified lengths. More specifically, we will draw a triangle with sides of lengths 6, 8 and 10 units. We begin in this way:

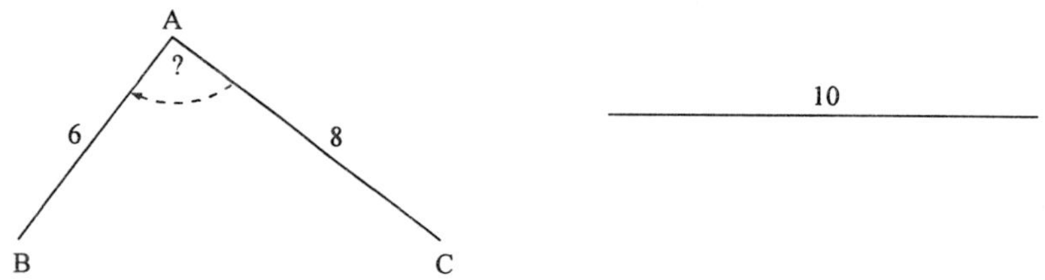

Now imagine that we must swing, or rotate, sides $\overline{AB}$ and $\overline{AC}$ apart just enough so that the third segment will fit. Notice that we have no choice concerning the amount of rotation for angle A. The three given conditions alone determine the final shape and size of the triangle formed.

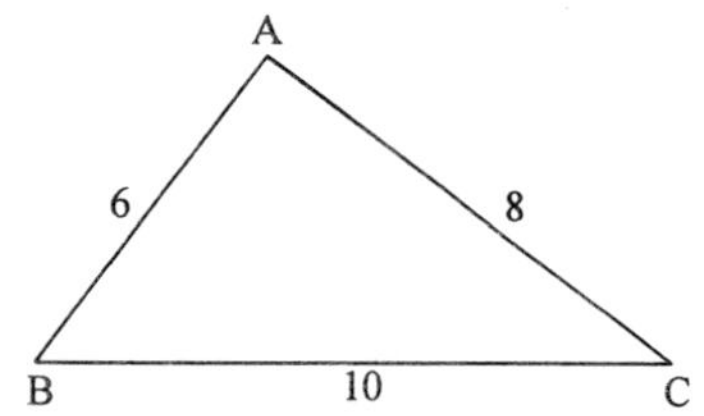

We say that the triangle is uniquely determined by the three conditions (the lengths of the three sides). If we draw another triangle with sides of the same lengths as those of △ ABC, the triangle will have the same shape and size as △ ABC. In other words, whenever a triangle is uniquely determined by certain given conditions, all triangles specified by those same conditions must be congruent.

We conclude:

Two triangles are congruent if three sides of one triangle have the same lengths, respectively, as three sides of the other triangle. We agree to use the notation **SSS** *to represent this conclusion.*

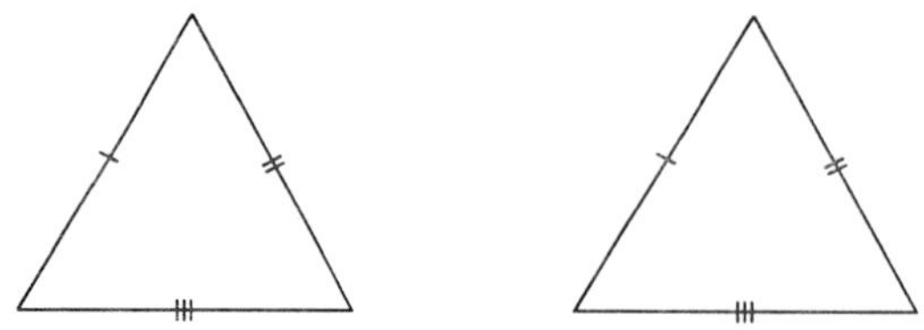

2. Now let us see what happens when we are asked to draw a triangle given the lengths of two of its sides and the measure of the angle included by those two sides. (Note: The included angle of two sides of a triangle is the angle of the triangle that includes, or contains, those two sides)

More specifically, we will draw $\triangle$ ABC with AB = 4, AC = 2 and $\angle$ A = 120.

First we draw line segment AB.

Next we draw angle A.

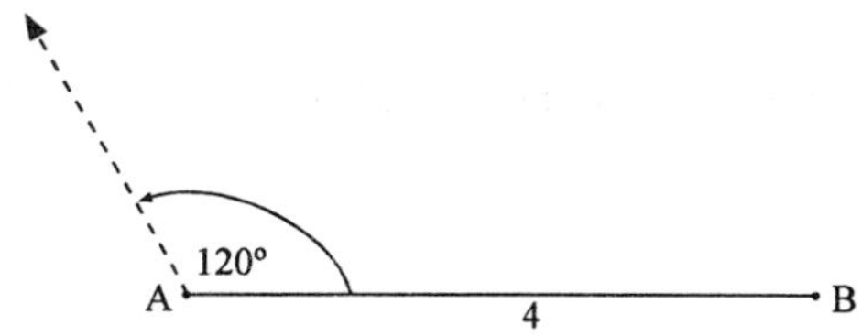

Then we draw line segment AC.

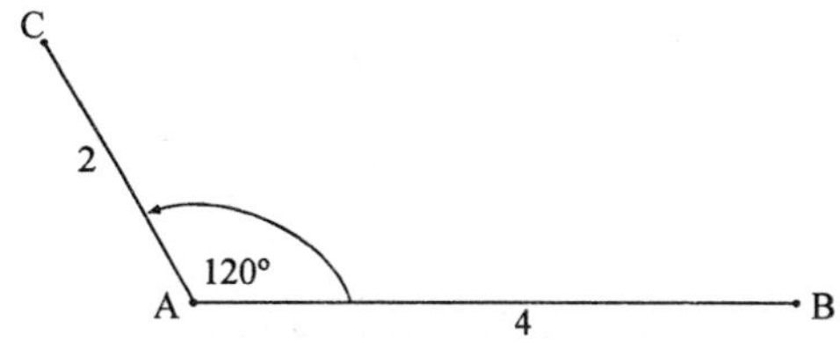

Here is the triangle:

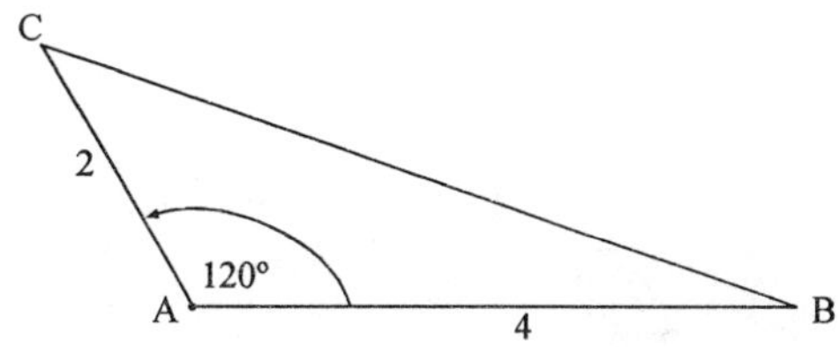

Again, the three conditions imposed on the triangle uniquely determined its size and shape. We had no choice concerning the location and length of the third side of the triangle. It had to have B and C as endpoints. Any other triangle drawn with the same three specified conditions will have the same size and shape as $\triangle$ ABC.

We conclude:

Two triangles are congruent if two sides and the included angle of one triangle have the same lengths and measure, respectively, as two sides and the included angle of the other triangle. We agree to use the notation **SAS** *to represent this conclusion.*

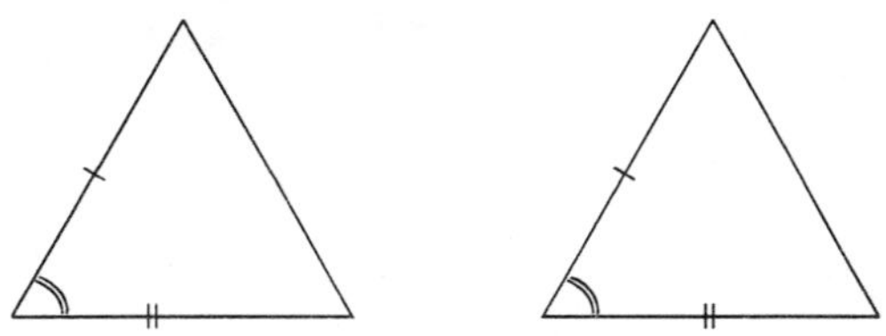

3. Now let us draw a triangle when given the measures of two of its angles and the length of the included side for those angles. (Note: The included side of two angles is the side of the triangle that has the vertices of these two angles as endpoints). More specifically, we will draw triangle ABC with AB = 2, ∠ A = 70 and ∠ B = 40.

We begin by drawing line segment AB.

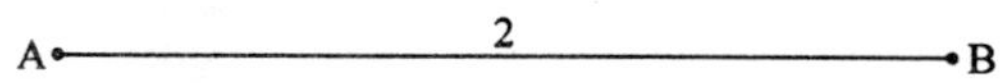

Next we draw ∠ A and ∠ B.

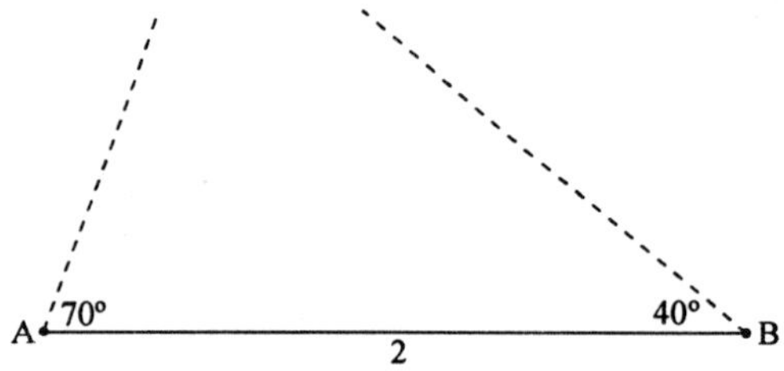

Then we extend the two sides of the two angles until they intersect to get the completed triangle.

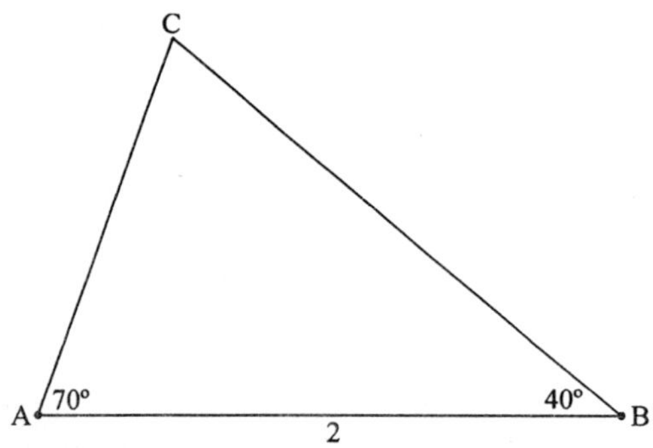

Notice that the third vertex of the triangle, vertex C, must be the intersection of the two sides of the given angles. We have no choice concerning the location of vertex C. The three conditions imposed upon the triangle uniquely determined the size and shape of the triangle. All triangles drawn with the same three specified conditions will be congruent.

We conclude:

Two triangles are congruent if two angles and the included side of one triangle have the same measures and length, respectively, as two angles and the included side of the other triangle. We agree to use the notation **ASA** *to represent this conclusion.*

4. We now consider the following triangles in which two angles and the side opposite one of them have the same measures and length, respectively, as two angles and the corresponding side of the other triangle.

Since the triangles have two pairs of angles with the same measures, the third pair of angles must have the same measure, that is, $\angle B = \angle E$.

We conclude:

Two triangles are congruent if two angles and the side opposite one of them have the same measures and length, respectively, as two angles and the corresponding side of the other triangle. We agree to use the notation **AAS** *to represent this conclusion.*

When using AAS to make conclusions about triangles, we must be sure that the given corresponding sides are *both* opposite the corresponding angles of the same measure.

We may use AAS to conclude that the triangles are congruent

We may not use AAS to conclude that the triangles are congruent

We can use similar methods to conclude that right triangles are congruent given the following conditions:

5. *Two right triangles are congruent if the hypotenuse and a leg of one triangle have the same lengths, respectively, as the hypotenuse and a leg of the other triangle. We agree to use the notation* **HL** *to represent this conclusion.*

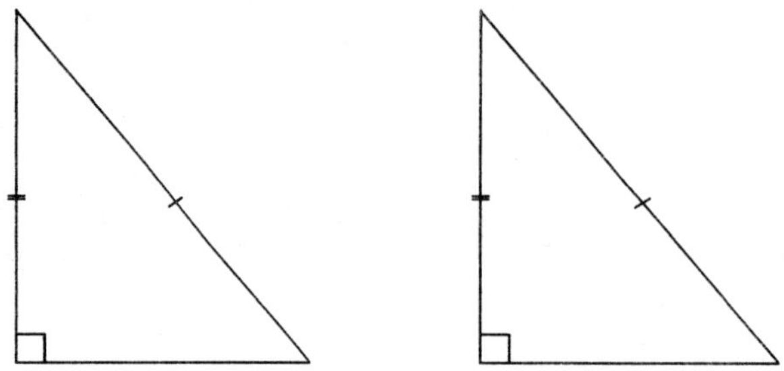

6. *Two triangles are congruent if the hypotenuse and an acute angle of one triangle have the same length and measure, respectively, as the hypotenuse and an acute angle of the other triangle. We agree to use the notation* **HA** *to represent this conclusion.*

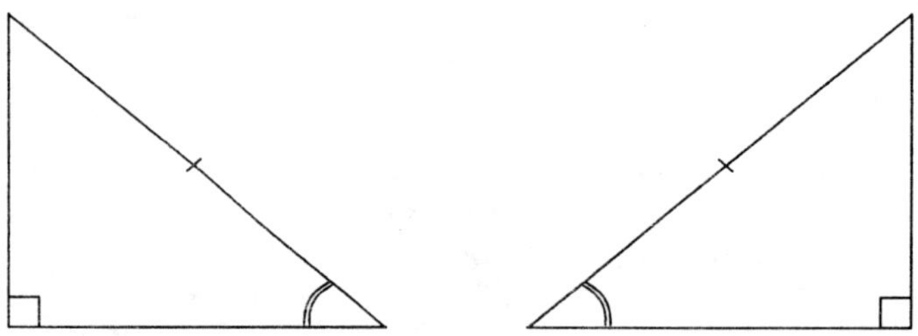

In summary, we conclude that triangles are congruent by **SSS**, **SAS**, **ASA**, **AAS**, **HL** and **HA**. Observe that AAA and SSA were excluded from this list. We can not conclude that two triangles are congruent if three angles of one triangle have the same measures, respectively, as three angles of another triangle. Neither can we conclude that triangles are congruent if an angle and two sides of one triangle have the same measure and lengths, respectively, as an angle and two sides of the other triangle. The following examples show why this is so.

EXAMPLE

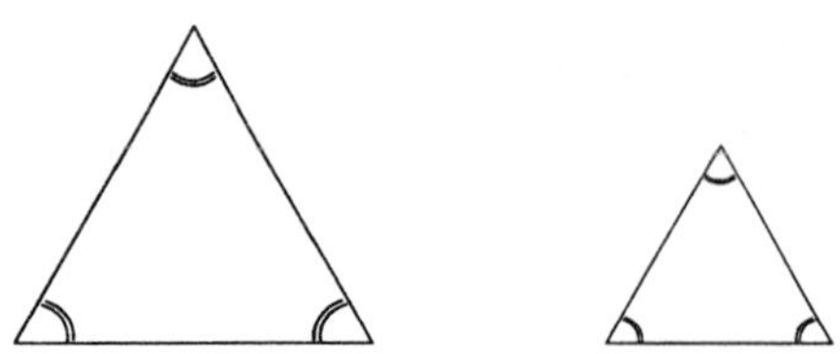

These two equiangular triangles have three pairs of angles with equal measures, but clearly the triangles are not congruent.

In the next example, we observe two triangles that have angles with the same measure and two pairs of sides with the same lengths. As you will see, the triangles are not congruent.

EXAMPLE

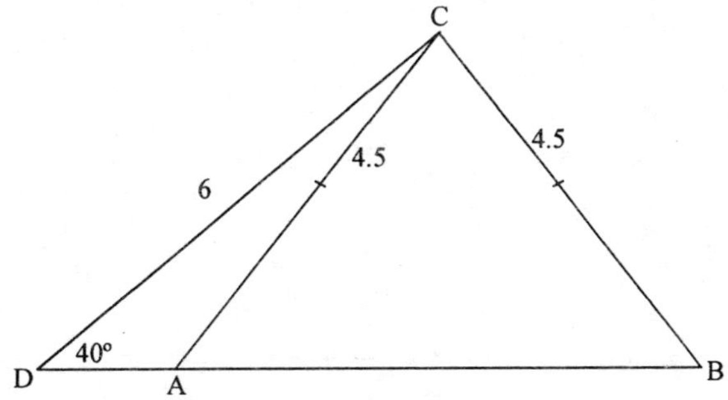

Consider △ DCB and △ DCA. If we mentally take the two triangles apart, we can see the angles that have the same measure and the sides that have the same lengths. Obviously, the triangles are not congruent.

Exercises 2.2

Figures are not drawn to scale, so do not use direct measurement or observation.

1. △ PQR ≅ △ NAD. List the conclusions that can be made about the sides and angles of the triangles.

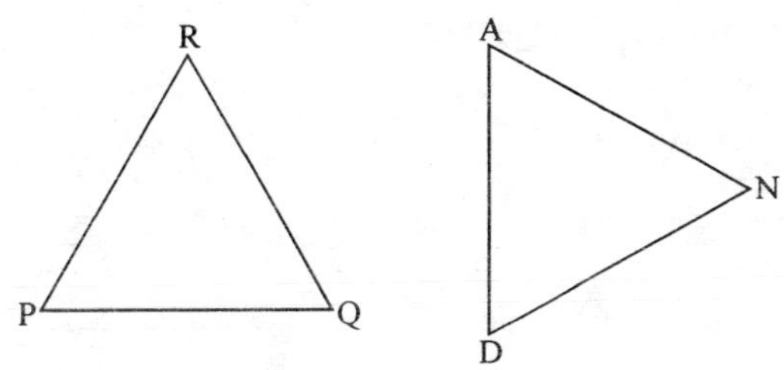

2. Fill in the blanks for the congruent triangles in the figure.

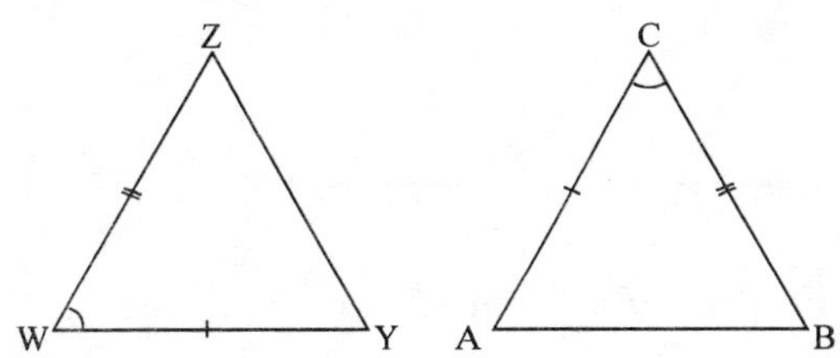

△ WYZ ≅ △ ___ ___ ___

3. Which of the following conditions will ensure congruent triangles?

a. AAA	b. AAS	c. ASA	d. SSA
e. SSS	f. SAS	g. HL	h. HA

4. Draw two equilateral triangles that are not congruent.

5. Are all equiangular triangles congruent? Support your response.

6. Draw two isosceles triangles that are not congruent.

7. Which of the following triangles are congruent?

a.

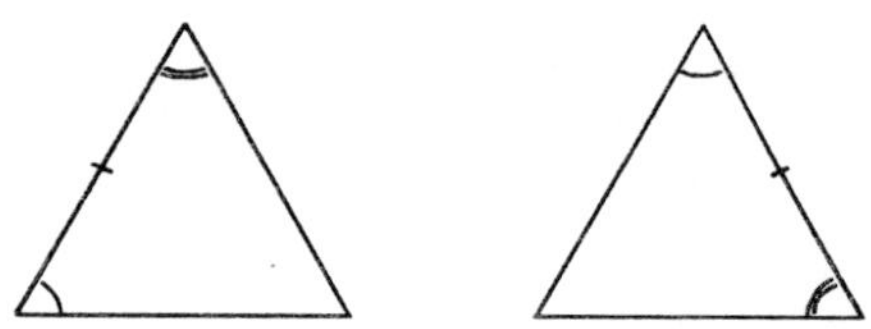

b.

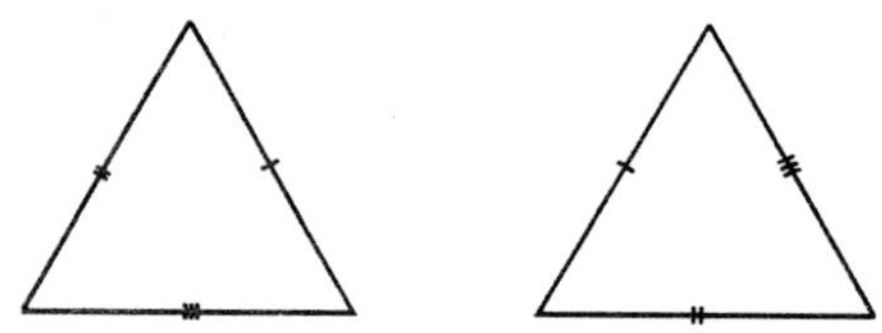

c.

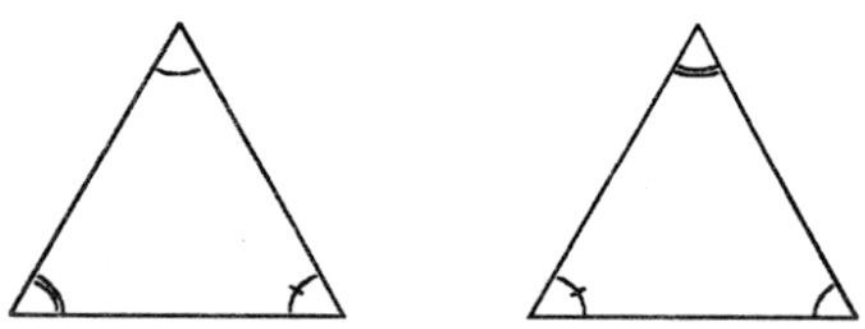

d.

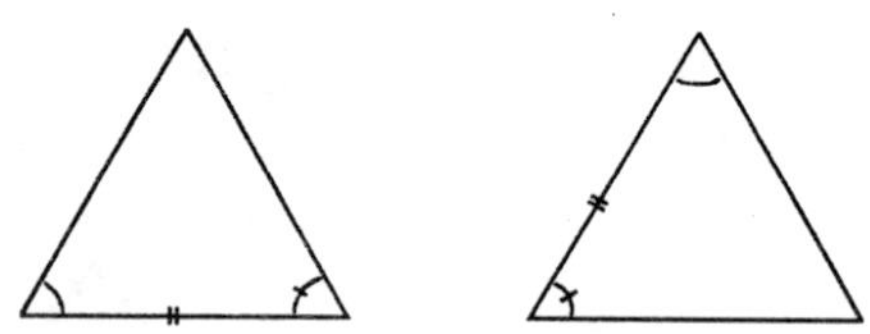

8. Which of the following triangles are congruent?

a.

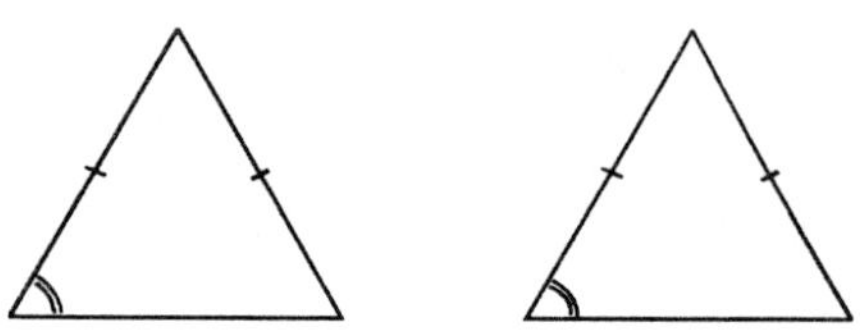

b.

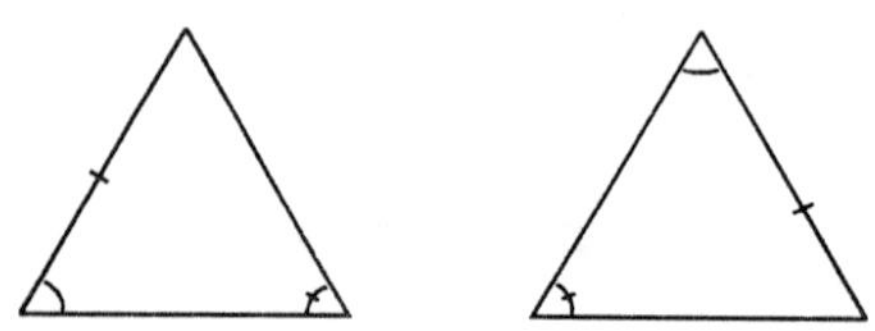

c.

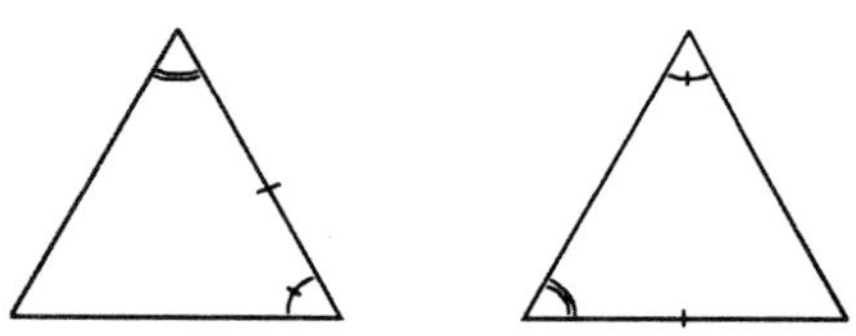

d.

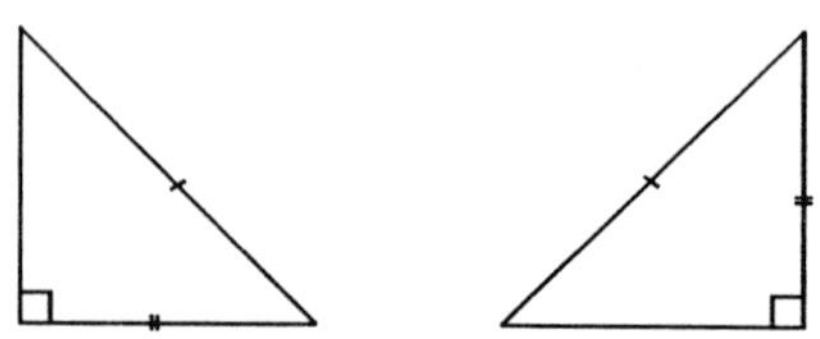

9. Which of the following triangles are congruent?

a.

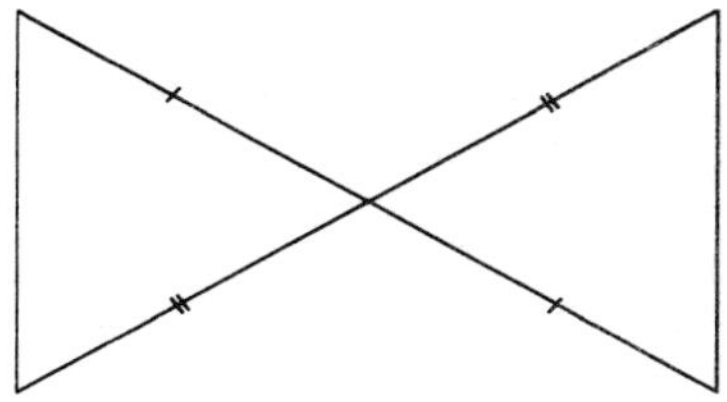

b.

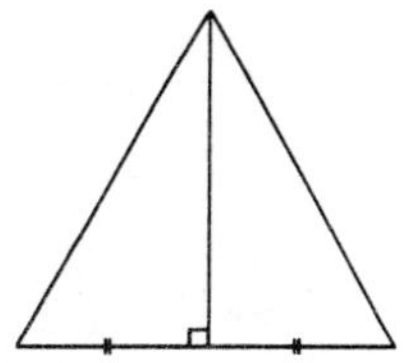

c.

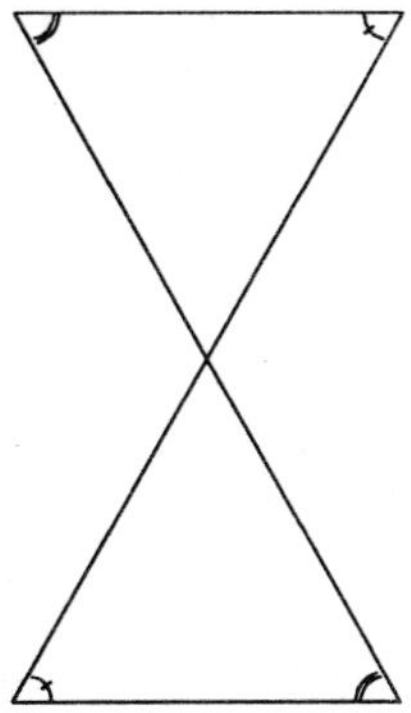

d.

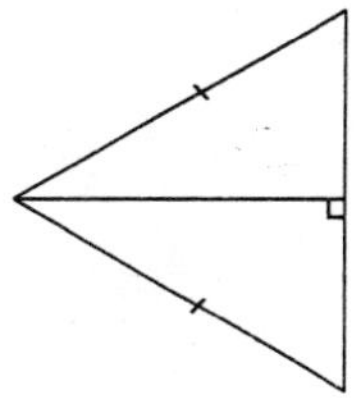

10. Which of the triangles shown in the figure are congruent?

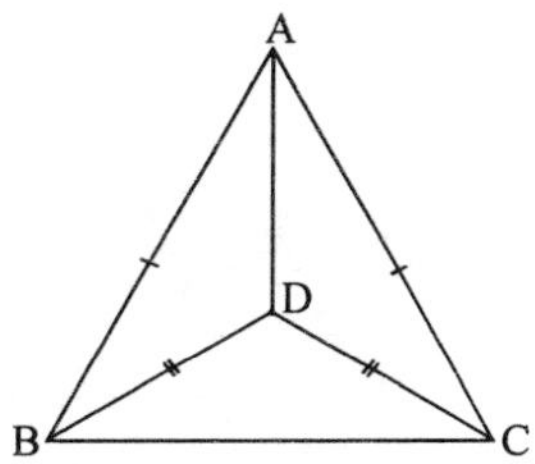

11. Find the length of the hypotenuse of △ ACB.

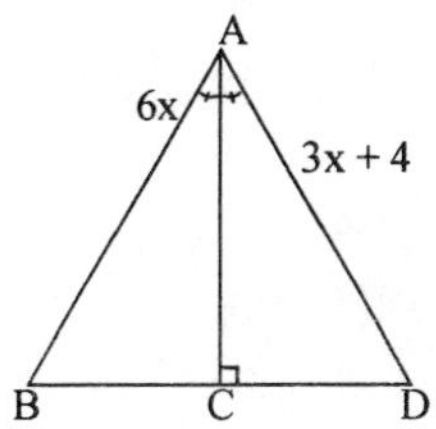

12. What is the length of line segment AB.

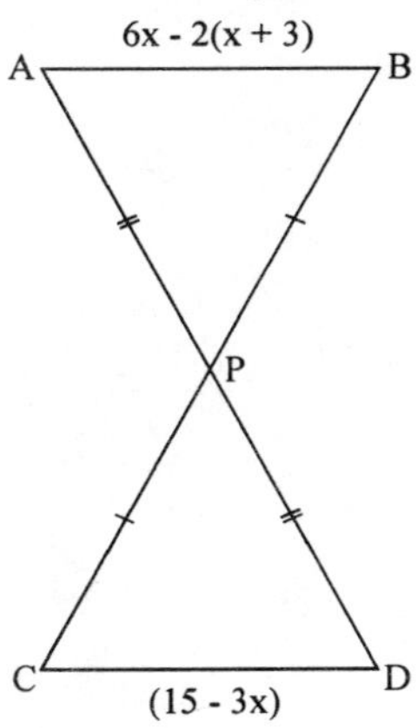

$AB = 6x - 2(x + 3)$

$CD = 15 - 3x$

13. Find the length of the hypotenuse of $\triangle$ ACB.

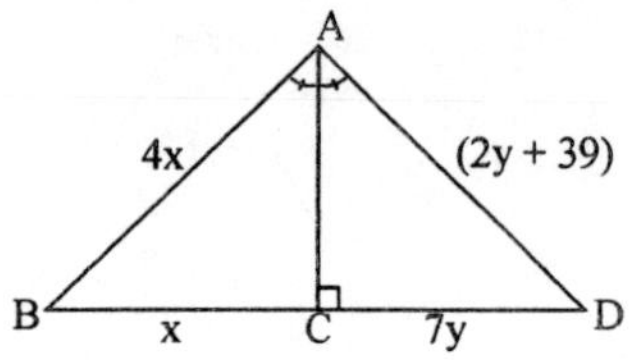

14. Find the length of line segment AC in $\triangle$ ACB.

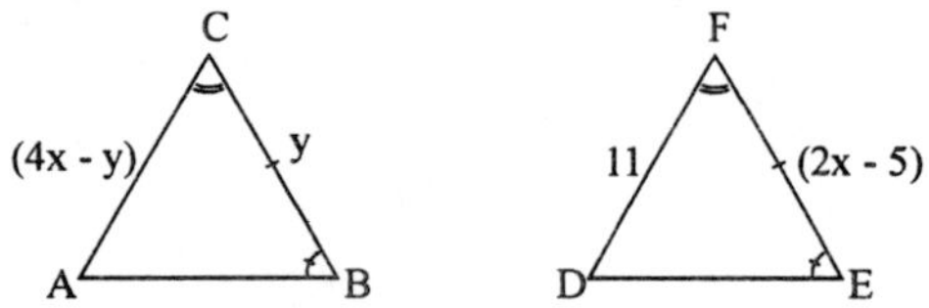

15. Find the values of x and y.

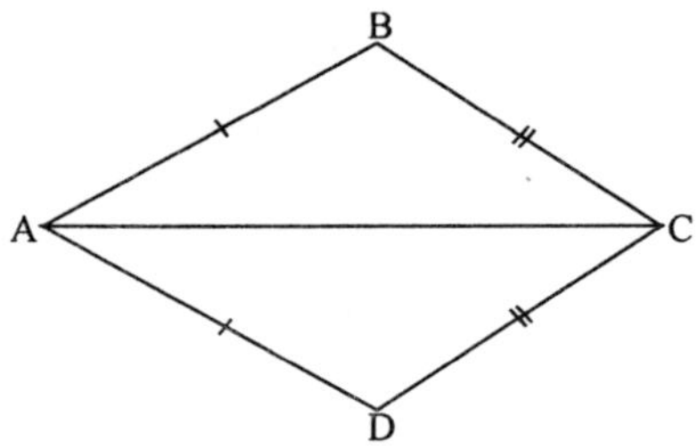

BC = 5x – 2y
DC = 32
AD = 4
AB = 2x – 3y

16. What is the length of line segment NM?

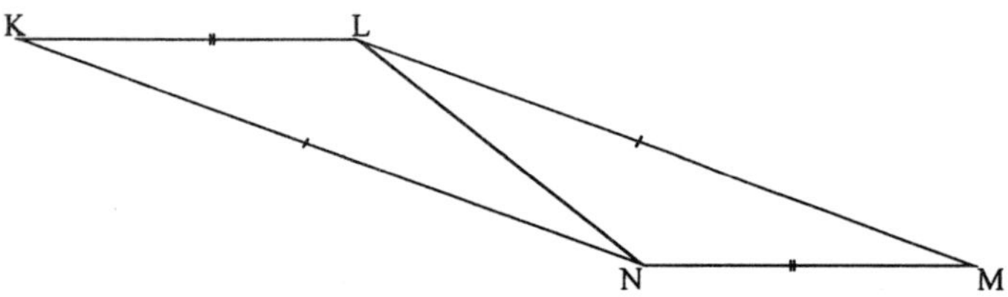

KL = 3x – 11
LM = 13
NM = y
KN = 2x + 5y

17. Explain why $\triangle$ ADC is congruent to $\triangle$ BCD.

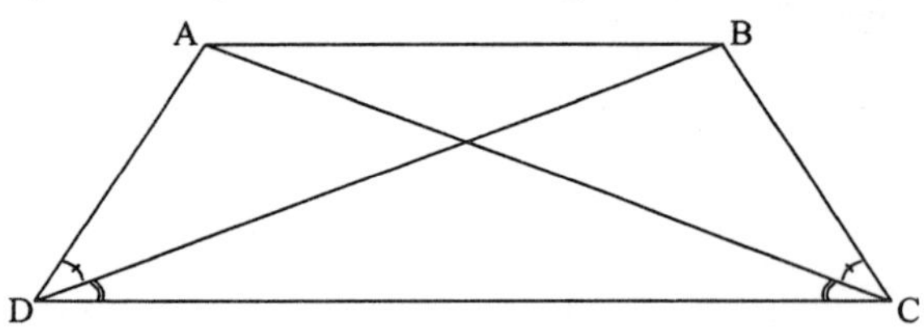

18. Use exercise 17 to fill in the blanks: $\triangle$ CAD $\cong$ $\triangle$ _____ _____ _____

19. a. Which of the following triangles are congruent?

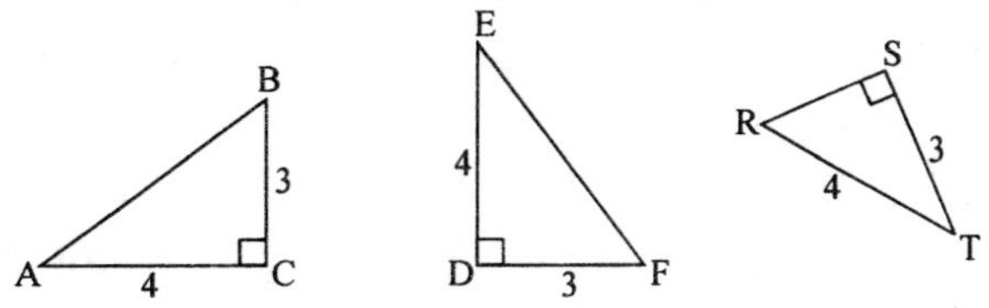

b. Which of the following triangles are congruent?

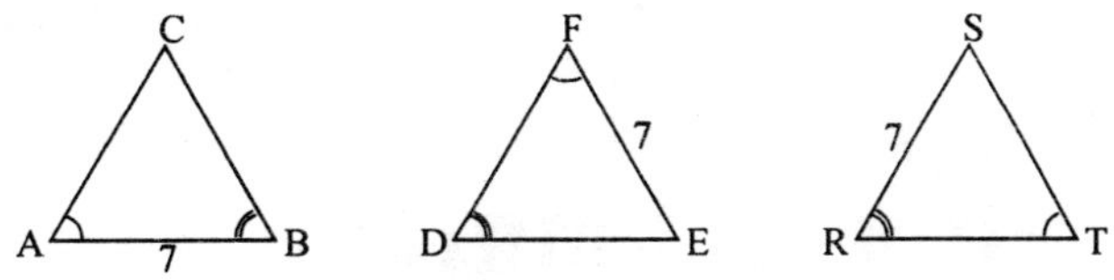

20. a. Graph the following points on a rectangular coordinate system:

A(4, 8)　　　　B(1, 4)　　　　C(1, 8)

b. Draw line segments AB, BC and AC.

c. Graph the following points on the same rectangular coordinate system. Notice that the coordinates of these points were obtained by interchanging the coordinates of points A, B and C.

A'(8, 4)　　　　B'(4, 1)　　　　C'(8, 1)

The little symbol to the right of A, B and C is read "prime."
We read these points *A prime, B prime* and *C prime,* to indicate that these points are related to points A, B and C.

d. Draw line segments A'B', B'C' and A'C'.

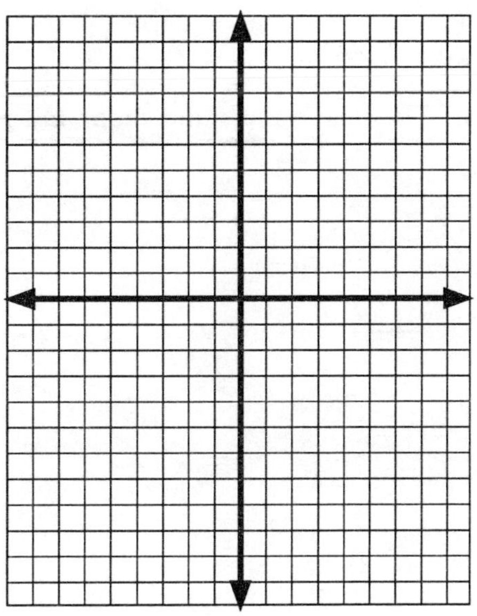

e. Did you get two triangles? Are the triangles congruent? Why?

f. Now graph and connect the following points. Notice that they were obtained by changing the signs of the x coordinates of A, B and C.

A"(–4, 8)   B"(–1, 4)   C"(–1, 8)

g. Did you get a triangle? Is it congruent to triangle ABC?

h. Do you think that you will a get a triangle that is congruent to triangle ABC if you change the signs of the y coordinates of A, B and C?

i. Do you think that you will a get a triangle that is congruent to triangle ABC if you change the signs of both coordinates of A, B and C?

Section 2.3 Similar Triangles

(*You May Need to Study Appendix A Before Beginning This Section*)

Figures that have the same shape but not necessarily the same size are said to be **similar**. Miniature objects, such as doll house furniture, are designed to be smaller models of what they represent. We say that these miniature objects are similar to the objects that they represent. We will agree that two geometric figures are similar if one can be obtained from the other by mentally enlarging or reducing its size without changing its shape.

EXAMPLES

These two pentagons are similar. They have the same shape and the same size. We observe that if two figures are congruent, they must also be similar.

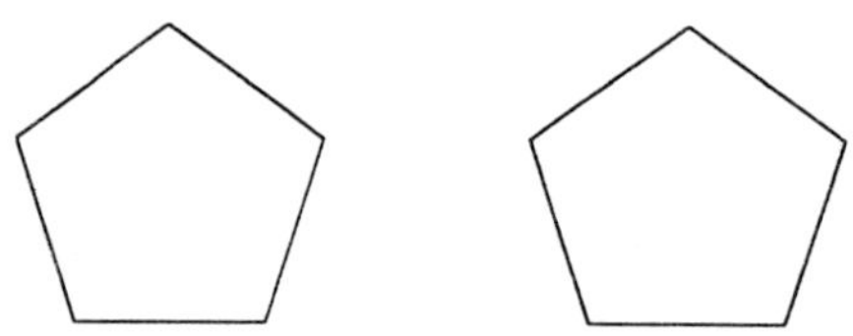

These two hexagons are similar. They have the same shape but not the same size.

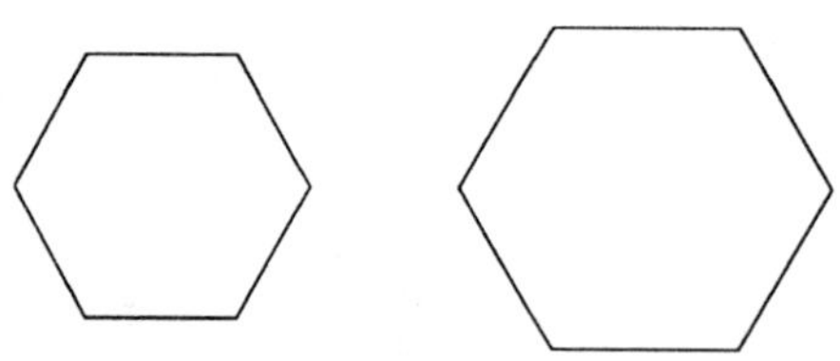

All squares are similar.

All circles are similar.

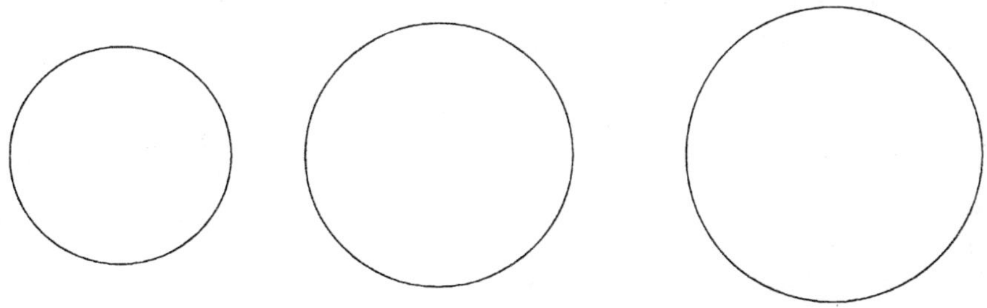

These two figures are not similar. They do not have the same shape.

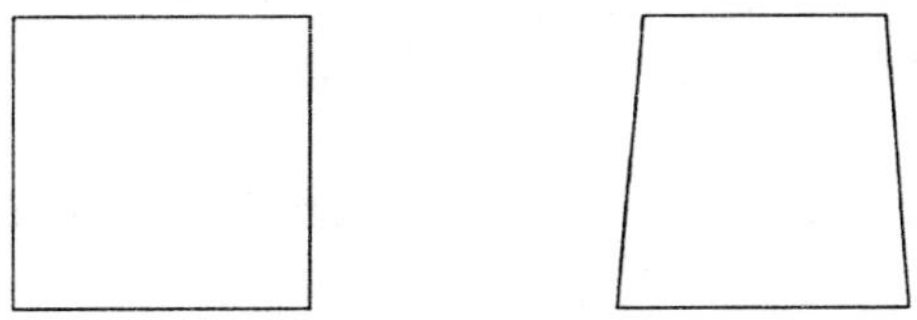

We may conclude that polygons are similar if they appear to have the same shape. It is not always practical, however, to make such a conclusion from observation. Fortunately we can accomplish our objective through indirect procedures, similar to those used to determine congruency of triangles.

We will **define two triangles to be similar** if a correspondence can be set up between their vertices so that all pairs of corresponding sides are proportional, and all pairs of corresponding angles have equal measures, respectively. In the illustration, $\triangle$ ABC is similar to $\triangle$ DEF.

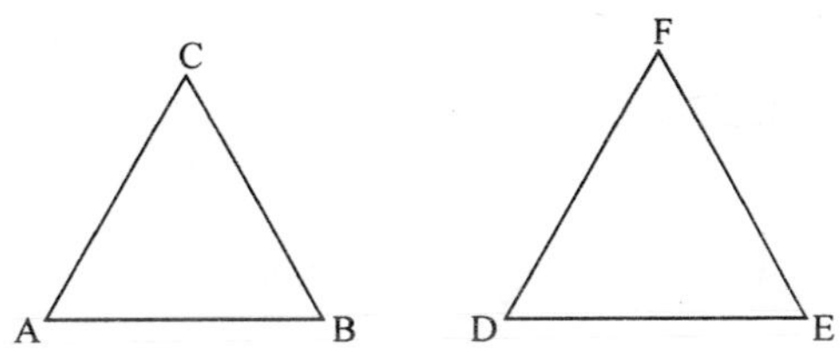

We conclude the following:

correspondence between vertices	**ABC $\longleftrightarrow$ DEF**
corresponding angles	**$\angle A = \angle D$, $\angle B = \angle E$, $\angle C = \angle F$**
Corresponding sides are proportional	$\frac{AB}{DE} = \frac{BC}{EF} = \frac{AC}{DF}$

We write $\triangle$ **ABC ~ $\triangle$ DEF.**

The symbol "~" represents the phrase "is similar to." Note that when the symbol is used, corresponding vertices should be in the same relative positions.

In the special case of triangles, either of the two conditions required by the definition is sufficient to ensure similar triangles, that is, we may conclude that two triangles are similar

If they have *all pairs of corresponding sides proportional*

OR

If they have *three pairs of corresponding angles with equal measures.*

The word OR means that only one of the required conditions has to be true.

In the following example, $\triangle$ PQR is similar to $\triangle$ MNO because they have three pairs of corresponding angles with equal measures.

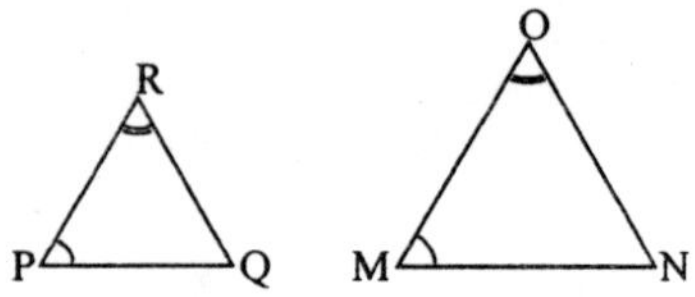

$\angle R = \angle O \qquad \angle P = \angle M \qquad \angle Q = \angle N$

We observe from the figures that the two triangles have two pairs of angles with equal measures. We know from our earlier studies that the remaining pair of angles must have the same measure. We can therefore conclude that two triangles are similar if they have two pairs of angles with equal measures.

In the next figure, we conclude that triangle ABC is similar to triangle ADE.

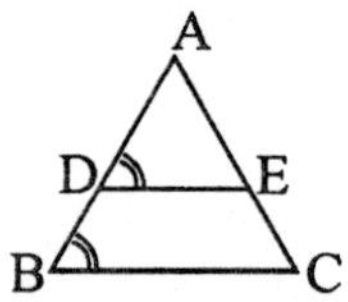

$\angle ADE = \angle B$ and $\angle A = \angle A$ $\qquad$ **$\triangle ABC \sim \triangle ADE$**

In this figure we conclude that triangle ABC is similar to triangle EDC.

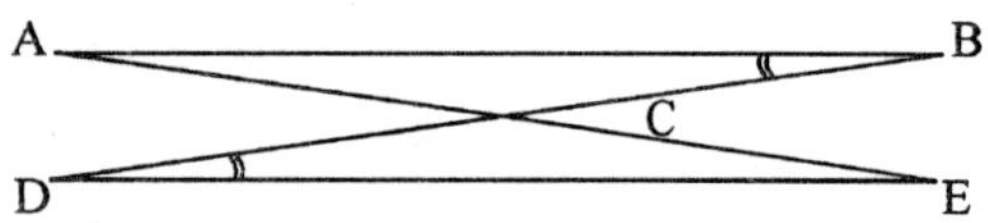

$\angle B = \angle D$ and $\angle ACB = \angle DCE$ (vertical angles)

These triangles are similar because their corresponding sides are proportional.

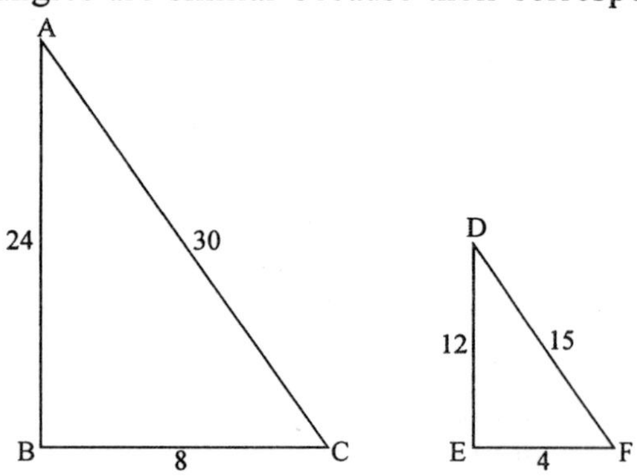

$\triangle$ ABC ~ $\triangle$ DEF BECAUSE

$$\frac{AB}{DE} = \frac{BC}{EF} = \frac{AC}{DF}$$

$$\frac{24}{12} = \frac{8}{4} = \frac{30}{15}$$

In some cases when we know that two triangles are similar, we can use this information to determine the lengths of unknown sides of the triangles. In the following example, triangle ABC is similar to triangle DEF.

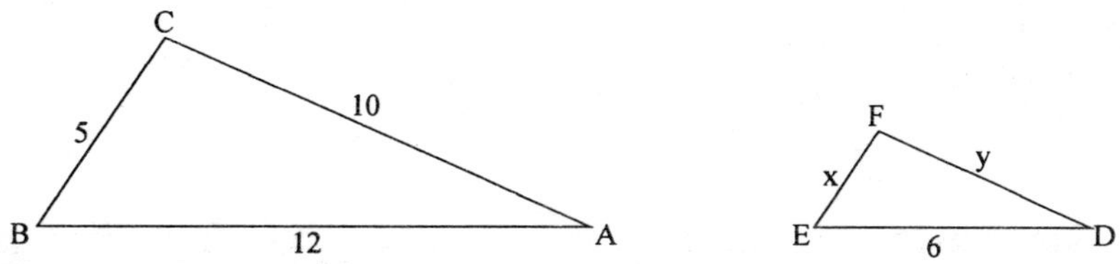

△ ABC ~ △ DEF

Since corresponding sides are proportional:

$$\frac{AB}{DE} = \frac{AC}{DF} \qquad \frac{AB}{DE} = \frac{BC}{EF}$$

$$\frac{12}{6} = \frac{10}{y} \qquad \frac{12}{6} = \frac{5}{x}$$

$$y = 5 \qquad x = \frac{5}{2}$$

Exercise 2.3

The figures are not drawn to scale. Do not use direct measurement.

1. What are similar triangles?

2. How may we show that two triangles are similar?

3. Why are the two triangles similar?

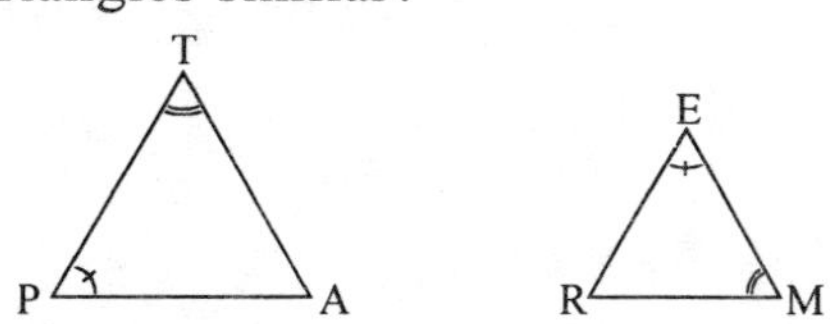

4. If two triangles are congruent, must they be similar?

5. If two triangles are similar, must they be congruent?

6. Are all equilateral triangles similar?

7. Are all squares similar?

8. Are all rectangles similar?

9. Why are the two triangles shown in the figure similar to each other? Which pairs of angles have equal measures? Write the proportions for the three pairs of corresponding sides.

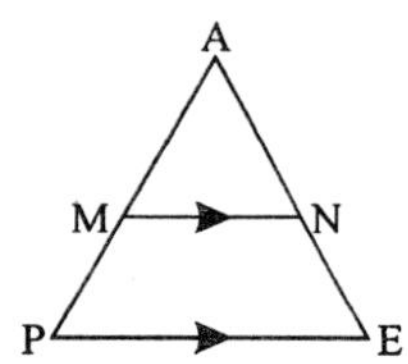

10. Why is △ CAB similar to △ CEA? Write the proportions for the pairs of corresponding sides.

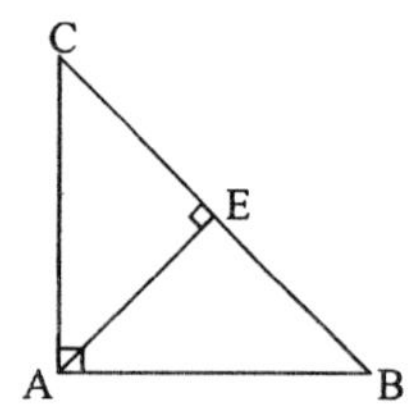

11. Why is △ CAB similar to △ AEB? Write the proportions for the pairs of corresponding sides.

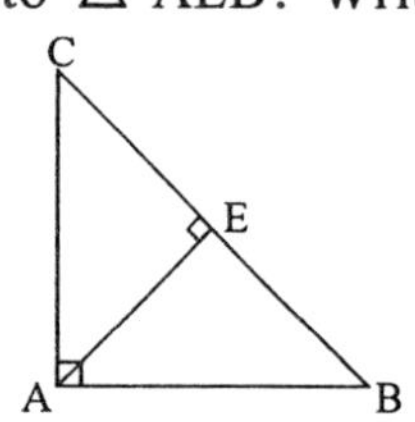

12. Why is △ CEA similar to △ AEB? Write the proportions for the pairs of corresponding sides.

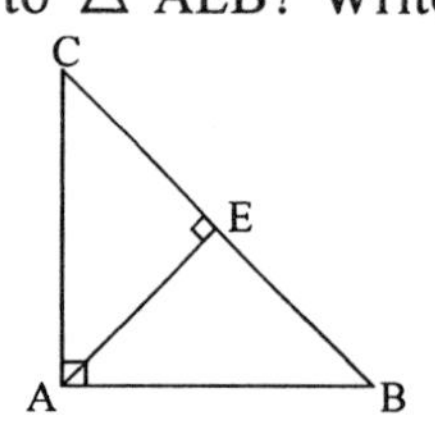

13. Can these triangles be similar? Support your response.

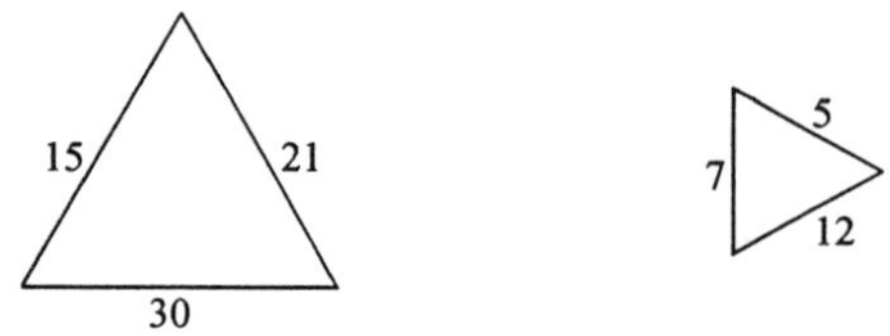

14* Are these triangles similar? Support your response.

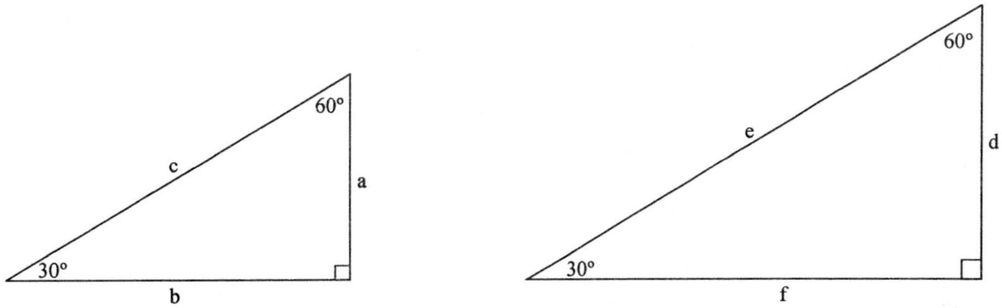

15* Write proportions for the corresponding sides of the triangles in exercise 14.

16* Write the following ratios for both triangles in exercise 14.

> Example: $\dfrac{\text{Hypotenuse}}{\text{side opposite the } 30° \text{ angle}}$
>
> This ratio is $\dfrac{c}{a}$ for the smaller triangle.
>
> This ratio is $\dfrac{e}{d}$ for the larger triangle.
>
> Are these ratios equal? Why?

a. $\dfrac{\text{side opposite to the } 30° \text{ angle}}{\text{hypotenuse}}$

The ratio is ________________ for the smaller triangle.
The ratio is ________________ for the larger triangle.
Explain why the two ratios that you wrote must be equal.

b. $\dfrac{\text{side adjacent to the } 30° \text{ angle}}{\text{hypotenuse}}$

The ratio is ______________ for the smaller triangle.
The ratio is ______________ for the larger triangle.
Explain why the two ratios that you wrote must be equal.

c. $\dfrac{\text{side opposite to the } 30° \text{ angle}}{\text{side adjacent to the } 30° \text{ angle}}$

The ratio is ______________ for the smaller triangle.
The ratio is ______________ for the larger triangle.

Explain why the two ratios that you wrote must be equal.

17* Write the following ratios for both triangles in exercise 14.

a. $\dfrac{\text{side opposite to the } 60° \text{ angle}}{\text{hypotenuse}}$

The ratio is ________________ for the smaller triangle.
The ratio is ________________ for the larger triangle.
Explain why the two ratios that you wrote must be equal.

b. $\dfrac{\text{side adjacent to the } 60° \text{ angle}}{\text{hypotenuse}}$

The ratio is ________________ for the smaller triangle.
The ratio is ________________ for the larger triangle.
Explain why the two ratios that you wrote must be equal.

c. $\dfrac{\text{side opposite to the } 60° \text{ angle}}{\text{side adjacent to the } 60° \text{ angle}}$

The ratio is ________________ for the smaller triangle.
The ratio is ________________ for the larger triangle.

Explain why the two ratios that you wrote must be equal.

The ratios that you wrote in exercises 16 and 17 are important in mathematics. They are studied in the branch of mathematics called trigonometry. These ratios have special names that are given below. Abbreviations are used for the words *opposite, adjacent* and *hypotenuse.*

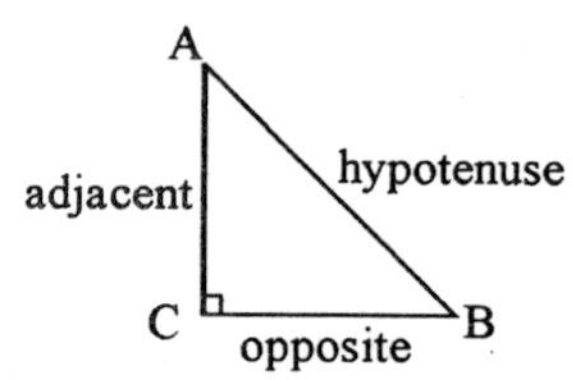

With respect to angle A of triangle ABC

$$\text{sine } A = \frac{\text{opp}}{\text{hyp}} \qquad \text{cosine } A = \frac{\text{adj}}{\text{hyp}} \qquad \text{tangent } A = \frac{\text{opp}}{\text{adj}}$$

18*Find the sine, cosine and tangent of angle A in triangle ABC.

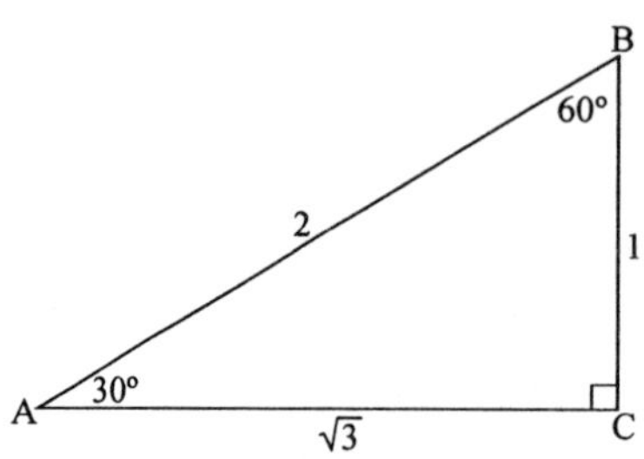

19.*Find the sine, cosine and tangent of angle B in triangle ABC.

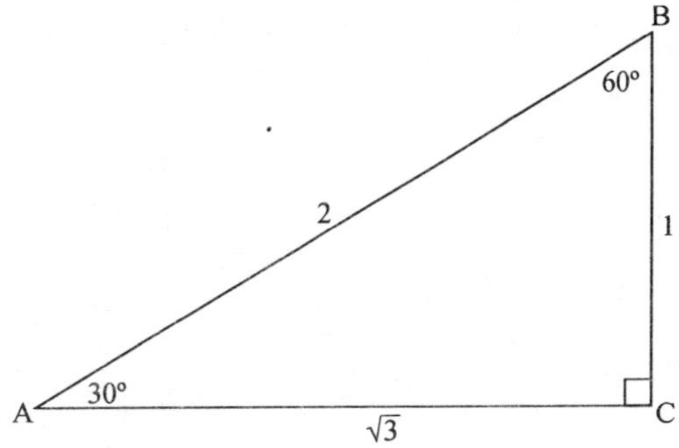

20.*Find the sine, cosine and tangent of angle B in triangle ABC.

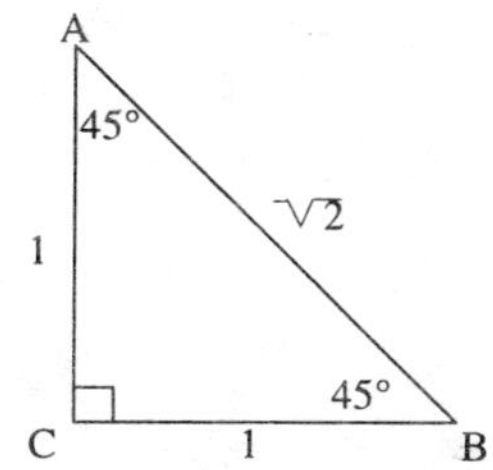

21. Find the value of each variable in the following exercises:

a.

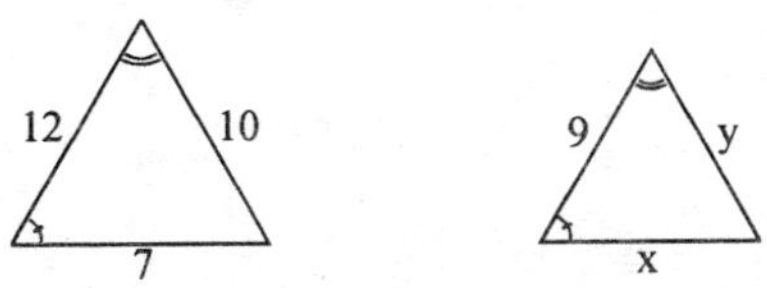

b.

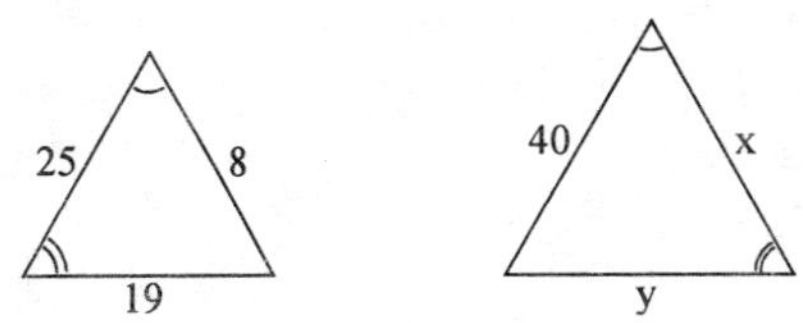

c.

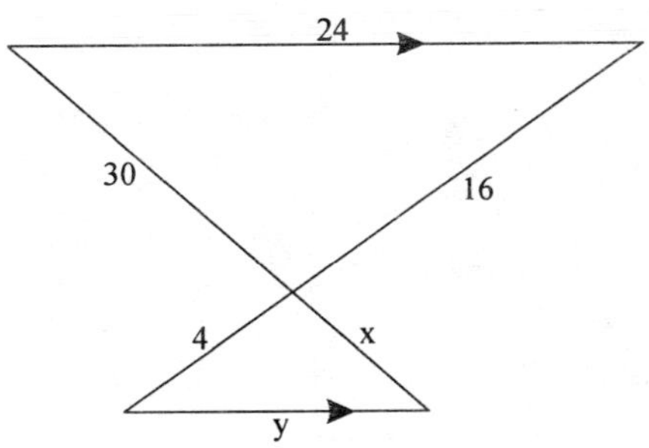

d.

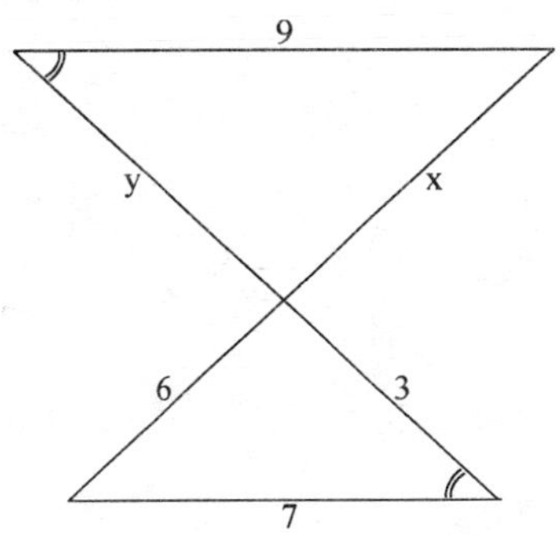

e.

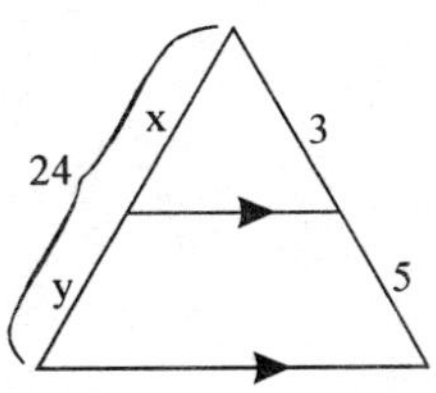

f.

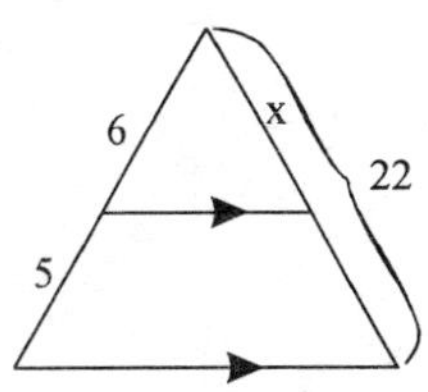

g.

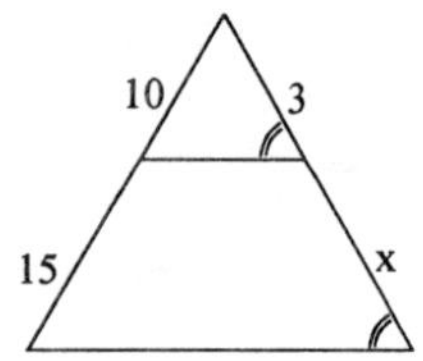
10
3
15
x

h.

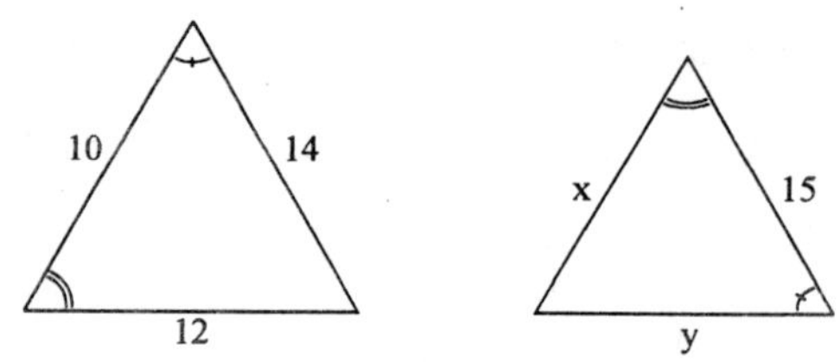
10
14
12
x
15
y

i.

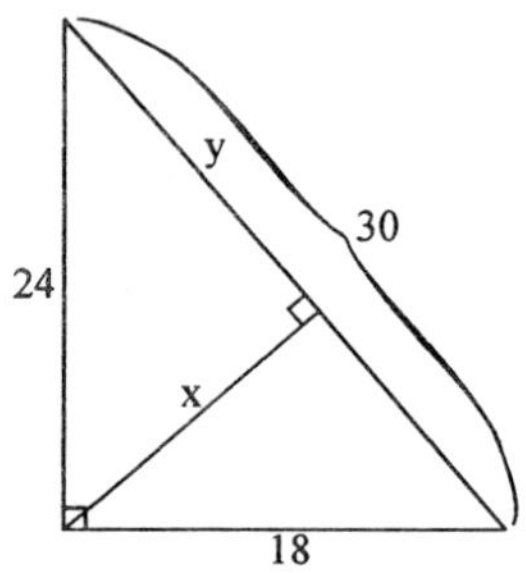
y
30
24
x
18

j.

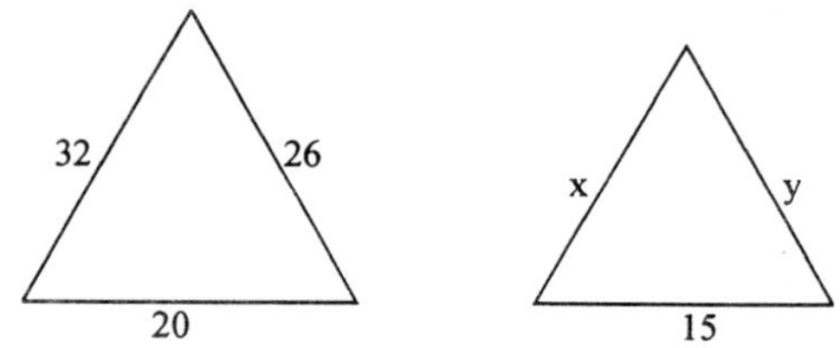
32
26
20
x
y
15

k.

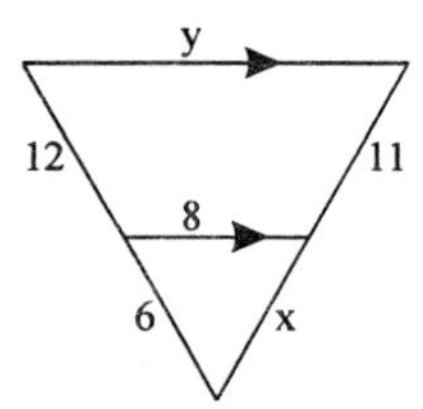
y
12
11
8
6
x

l.

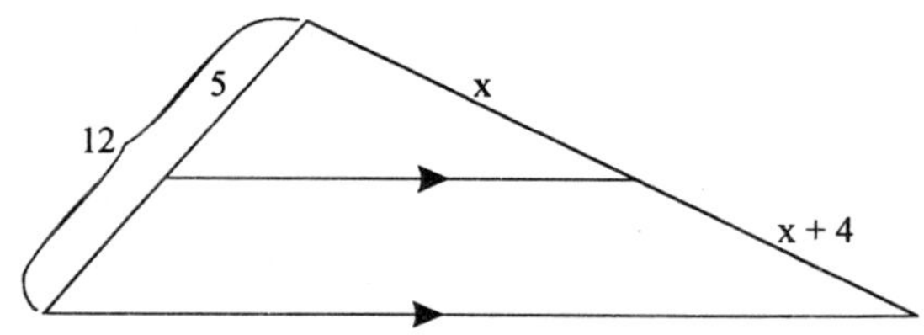
5
x
12
x + 4

m.

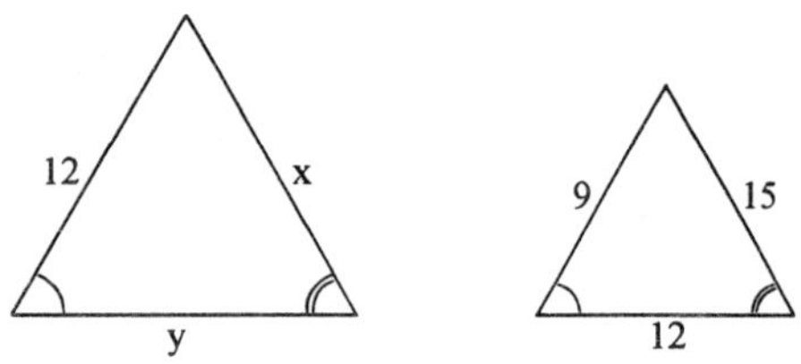
12
x
y
9
15
12

n.

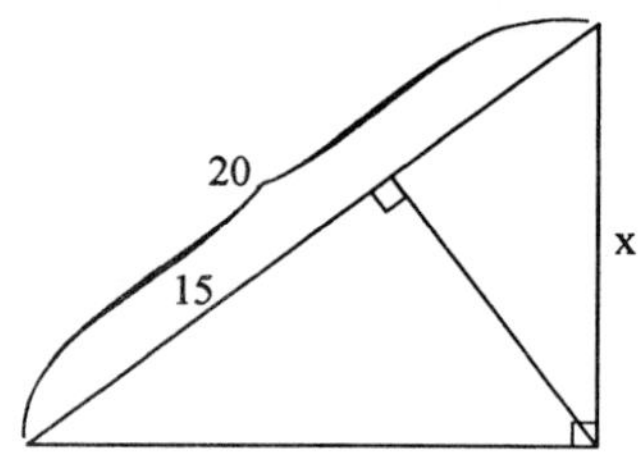
20
15
x

o.

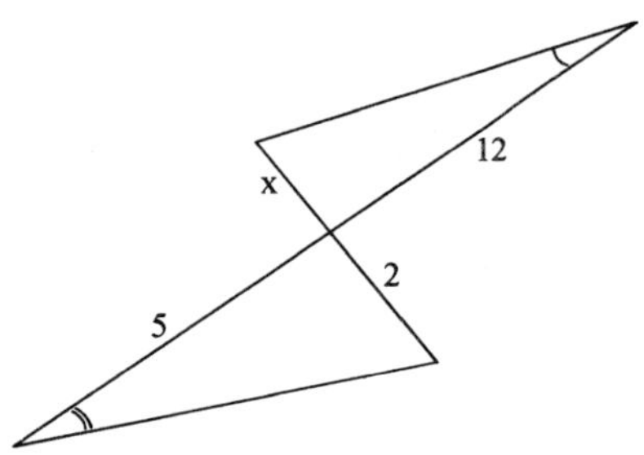
12
x
2
5

p.

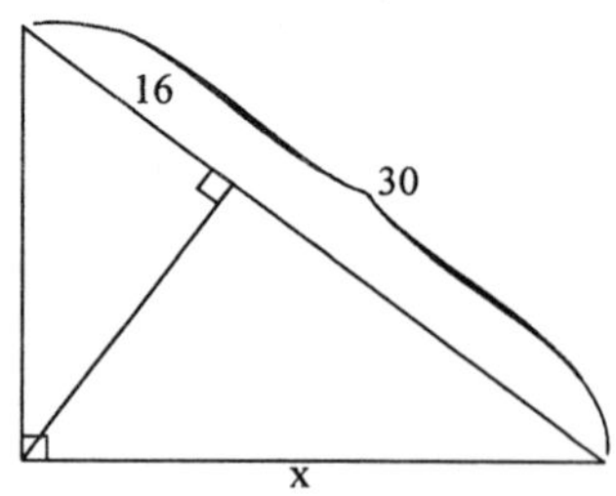
16
30
x

22. Thales of Miletus is said to have found the height of a pyramid in Egypt by measuring the length of its shadow. He may have done this by comparing the length of the shadow of the pyramid to the length of a shadow of an object of known height at the same time of day.

For example:

$$\frac{\text{length of pyramid's shadow}}{\text{length of the objects shadow}} = \frac{\text{height of the pyramid}}{\text{height of the object}}$$

a. A boy casts a 12-foot shadow at the same time that a flagpole casts a shadow 224 feet long. If the boy is 5 feet tall, what is the height of the flagpole?

b. If a flagpole casts a shadow that is 8 feet long at the same time that a child 3 feet tall casts a 2-foot shadow, how high is the flagpole?

Chapter Two Highlights

Vocabulary

polygon
vertices of a polygon
sides of a polygon
diagonals
interior angles of a polygon
exterior angles of a polygon
triangle
quadrilateral
pentagon
hexagon
heptagon
octagon
nonagon
decagon
dodecagon
convex polygon
regular polygon
equiangular
equilateral
square
acute triangle
right triangle
obtuse triangle
hypotenuse
legs of a right triangle
equiangular triangle
equilateral triangle
scalene triangle
isosceles triangle
legs of an isosceles triangle
base of an isosceles triangle
base angles (isosceles triangle)
vertex angle (isosceles triangle)
trapezium
trapezoid
bases of a trapezoid
legs (trapezoid)
base angles (trapezoid)
isosceles trapezoid
parallelogram
rectangle
rhombus
congruent triangles
similar triangles

Required skills

Recognize notation for similar triangles
Recognize notation for congruent triangles

Conclusions

1. The sum of the measures of the angles of a triangle is 180 degrees.
2. The sum of the measures of the angles of an N-gon is 180(N – 2) degrees
3. The measure of each angle of a regular N-gon is $\frac{180(N-2)}{N}$ degrees.
4. The sum of the lengths of any two sides of a triangle is always greater than the length of the remaining side of the triangle. (The Triangular Inequality Observation)
5. If two sides of a triangle have the same length, then the angles opposite those sides have the same measure.
6. If a triangle is equilateral, then it is equiangular.
7. If two angles of a triangle have the same measure, then the sides opposite those angles have the same length.
8. If a triangle is equiangular, then it is equilateral.
9. Each angle of an equiangular triangle has a measure of 60 degrees.
10. Each angle of a square has a measure of 90 degrees.
11. Each angle of a rectangle has a measure of 90 degrees.
12. If two angles of a triangle have the same measures, respectively, as two angles of another triangle, then the remaining two angles of the triangles have the same measure.
13. If the lengths of three sides of one triangle equal, respectively, the lengths of three sides of another triangle, then the triangles are congruent. SSS

14. If two sides and the included angle of one triangle have lengths and measure equal, respectively, to the lengths and measure of two sides and the included angle of another triangle, then the triangles are congruent. SAS

15. If two angles and the included side of one triangle have measures and length equal, respectively, to the measures and length of two angles and the included side of another triangle, then the triangles are congruent. ASA

16. If two angles and the side opposite one of them have measures and length equal, respectively, to the measures and length of two angles and the corresponding side of another triangle, then the triangles are congruent. AAS

17. If the hypotenuse and a leg of one right triangle have lengths equal, respectively, to the lengths of the hypotenuse and a leg of another right triangle, then the triangles are congruent. HL

18. If the hypotenuse and an acute angle of one right triangle have length and measure equal, respectively, to the length and measure of the hypotenuse and an acute angle of another right triangle, then the triangles are congruent. HA

    Note: We can not conclude that triangles are congruent by AAA.
    Note: We can not conclude that triangles are congruent by SSA.

19. Two triangles are similar if the measures of two angles of one triangle equal, respectively, the measures of two angles of the other triangle.

20. Two triangles are similar if their corresponding sides are proportional.

Chapter Three

Circles and Constructions

Objectives

Recognize and use basic terminology and notation for circles and arcs
Use a compass and straightedge to do basic constructions
Determine the measure of central and inscribed angles, major and minor arcs and semicircles
Know the relationship for medians, angle bisectors, perpendicular bisectors and altitudes of a triangle

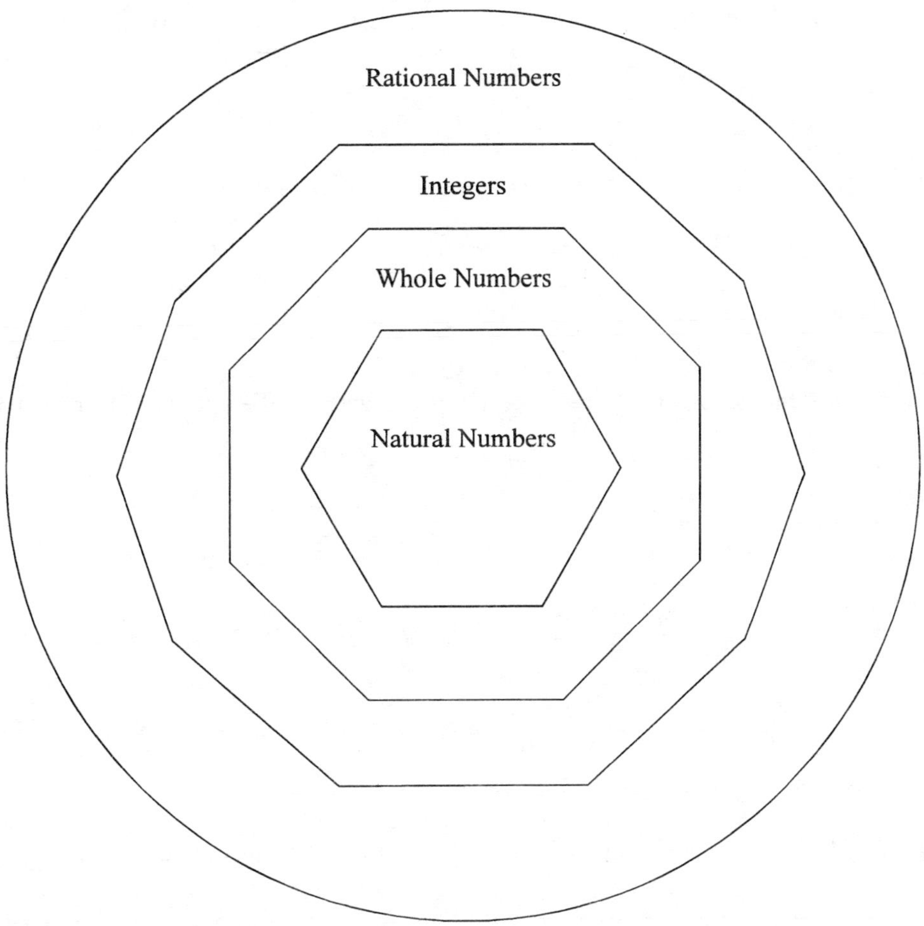

Section 3.1 Introduction to Circles

A **circle** is a set of points in a plane that are the same distance from a fixed point called the **center** of the circle. The fixed distance is called the **radius** of the circle. Any line segment that joins the center of a circle to a point on the circle is also called a **radius** of the circle. The term "radius" is thus used to represent both a line segment and its length. We should note that all radii (plural of radius) of a given circle must have the same length. We may *name a circle by using the name of its center.* In the illustration, we have:

circle O
center O
radius $\overline{OA}$
radii $\overline{OA}$, $\overline{OB}$, $\overline{OC}$
OA = OB = OC

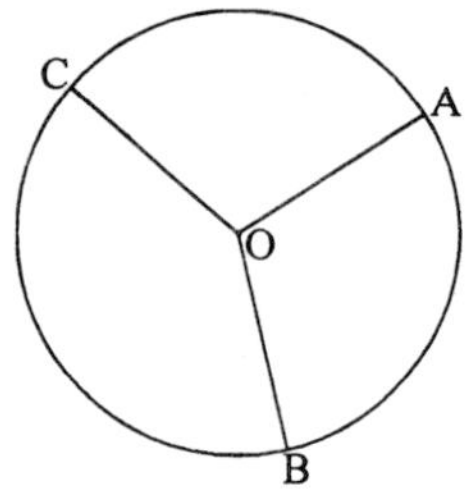

If the distance from a point in the plane to the center of a circle is less than the radius then we say that the **point is in the interior of the circle**. If the distance from a point in the plane to the center of a circle is greater than the radius, then we say that the **point is in the exterior of the circle.** In the following figure, P is in the interior of circle O while Q is in the exterior of the circle.

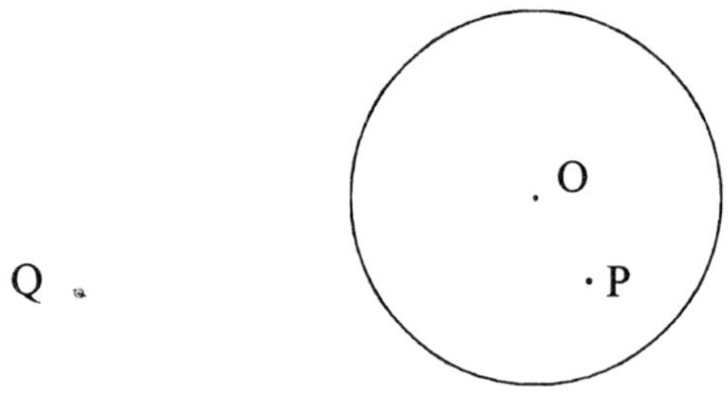

Two or more circles that have the same center but different radii are called **concentric circles.**

concentric circles

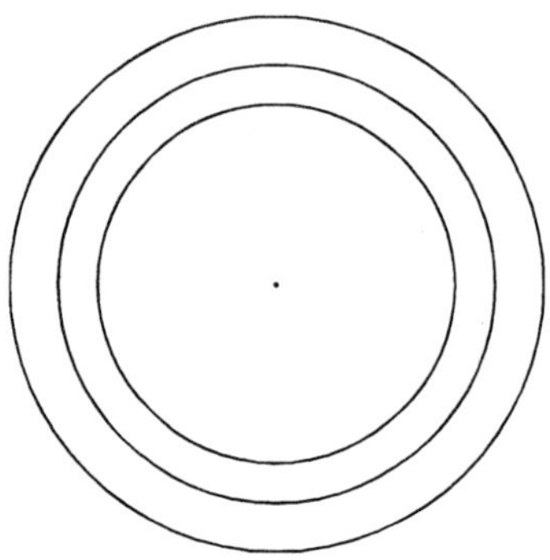

A **chord** of a circle is a line segment that joins two points of the circle. If the chord contains the center of the circle then it is called a **diameter** of the circle. The term "diameter" can also be used to represent the length of such a segment. Notice that the length of a diameter is always twice the length of a radius of the same circle. Since all radii of a circle are equal in length we may conclude that all diameters of a circle are also equal in length.

circle O with chord $\overline{MN}$
circle O with diameters $\overline{AB}$ and $\overline{CD}$
AB = CD

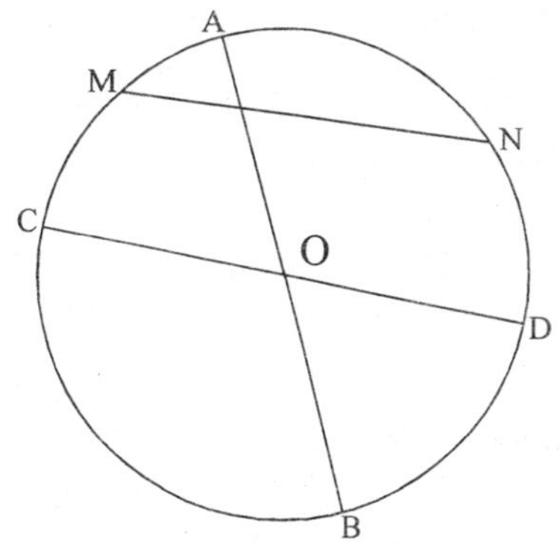

Secants and tangents are special lines that are associated with a circle. A **secant** is a line that intersects the circle in two points; while a **tangent** is a line that intersects the circle in exactly one point. The point where a tangent intersects a circle is called the **point of tangency** or the **point of contact.**

circle O
secant m
tangent n
point of tangency P

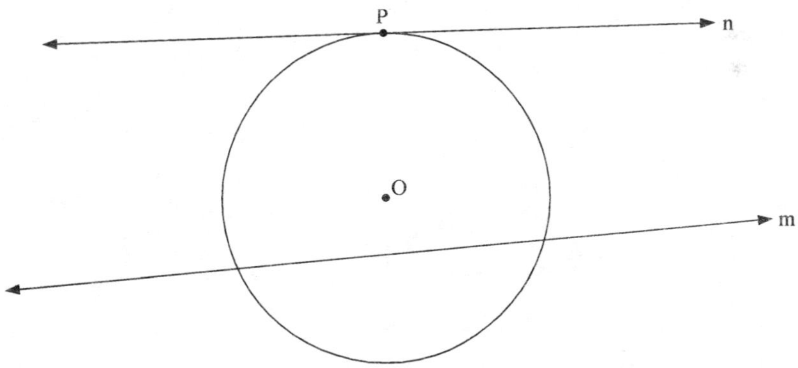

A segment of the circle itself is called an **arc** of the circle. An **arc** then consists of two points of the circle (called endpoints of the arc) together with all the points of the circle that lie between those two points. If the endpoints of an arc are also the endpoints of a diameter of the circle, then the arc is called a **semicircle.** Notice that any two points of a circle will always separate the circle into two arcs that have those two points as endpoints.

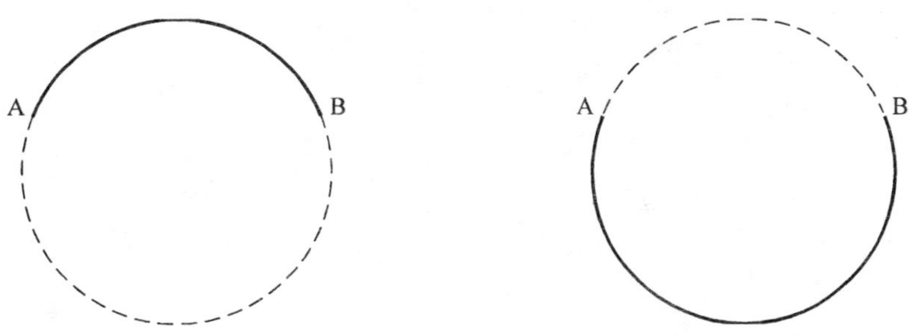

We may *name an arc by using the names of its two endpoints. A miniature arc may then be placed above the two letters naming the arc.* Naming an arc in this way, however, is always ambiguous since every two points of a circle are endpoints for two different arcs of the circle.

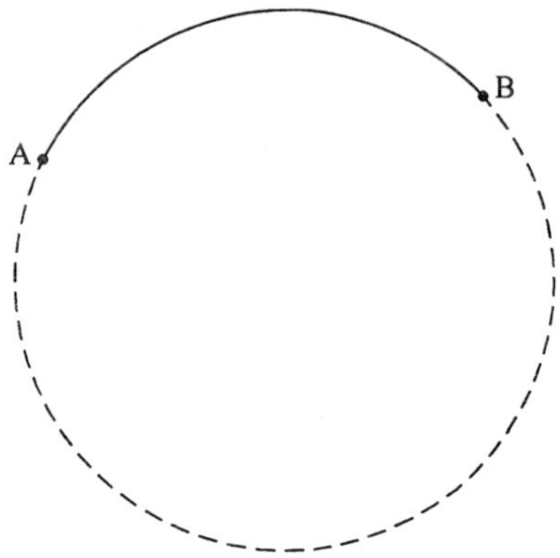

solid line $\overset{\frown}{AB}$

broken line $\overset{\frown}{AB}$

Whenever we are not sure from context which of two arcs is indicated, we agree to name another point on the arc. Otherwise, we simply use the names of the endpoints.

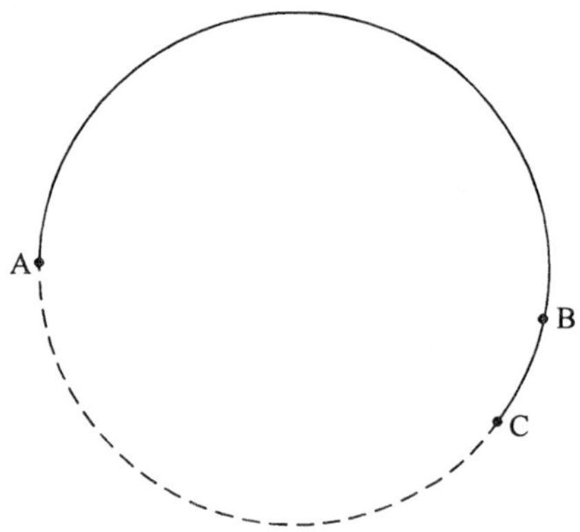

The endpoints of the arc must be the first and last points named $\overset{\frown}{ABC}$

We also have special angles that are related to circles. One such angle is called a central angle. A **central angle** is an angle that has the center of the circle as its vertex. The sides of a central angle intercept or "cut off" an arc of the circle. Except for its endpoints, all the points of the intercepted arc lie in the interior of the angle. In the figure ∠ AOB is a central angle of circle O. Notice that its sides contain the endpoints of arc AB. We say that central angle AOB intercepts arc AB. We also say that arc AB is intercepted by the central angle AOB.

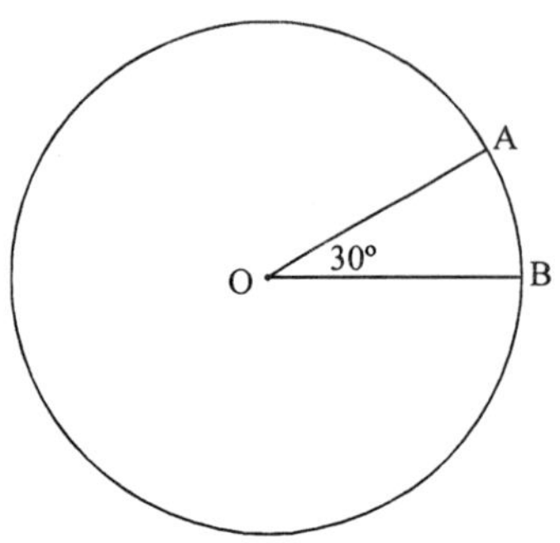

Since every central angle intercepts a unique arc, it is reasonable to define the **measure of the arc** to be equal to the measure of its central angle. The **measure of the other arc that shares the same**

endpoints as the arc intercepted by a central angle can then be defined as the positive difference between 360 and the measure of the arc intercepted by the central angle. In the first circle with center O, $\angle$ AOB = 30, so we conclude that the measure of arc AB is also 30. Furthermore, we conclude that the measure of arc APB is 330 (360 – 30). In the second circle with Center O, we conclude that $\angle$ COD = 60 since the measure of arc CD is 60. *Note the use of the letter "m" to represent the measure of an arc.* We may use this notation to distinguish an arc from its measure.

$m\,\widehat{AB} = 30$
$m\,APB = 330$
$\angle\,\widehat{AOB} = 30$

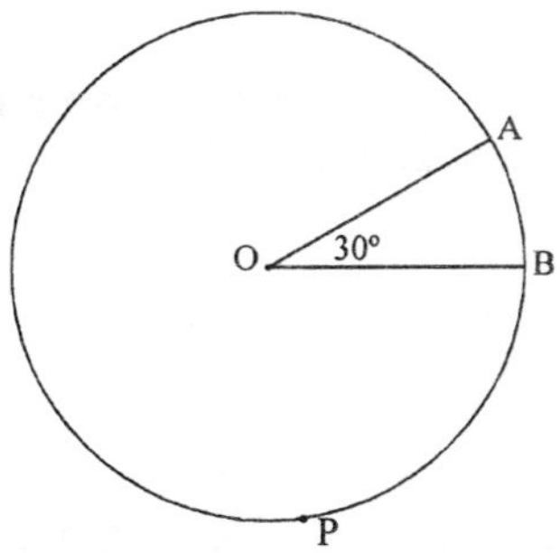

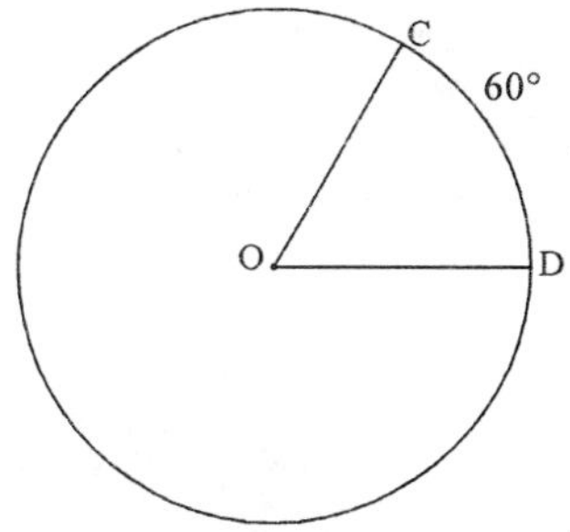

$m\,\widehat{CD} = 60$
$\angle\,COD = 60$

We also notice that the degree measure of an arc is not affected by the size of the circle containing the arc. For the concentric circles in the figure, corresponding arcs all have the same measure.

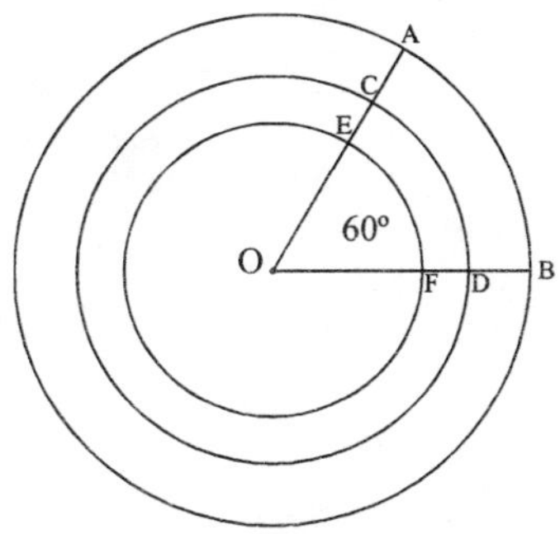

Circle O

$\mathbf{m\,\widehat{AB} = m\,\widehat{CD} = m\,\widehat{EF}}$

Now we will separate arcs into three categories: **semicircles, major arcs** and **minor arcs.** If we assign to a semicircle a measure of 180 degrees, then we can define a major arc as one that has a measure that is greater then 180 degrees; and we can define a minor arc as one that has a measure that is less than 180 degrees.

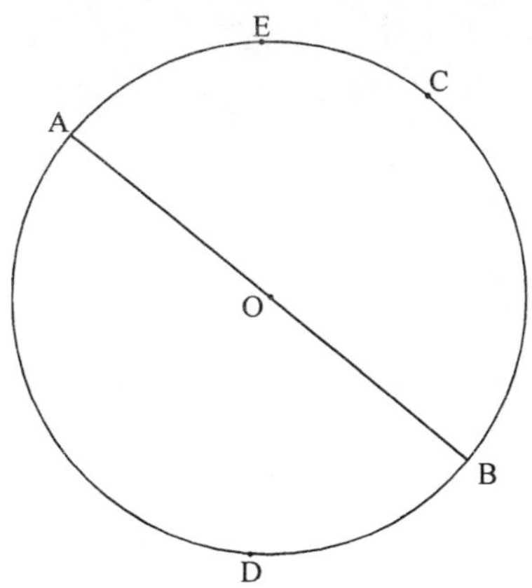

circle O
semicircle $\widehat{AB}$
minor arc $\widehat{AEC}$
major arc $\widehat{CBA}$

It is reasonable to conclude for arcs that "the whole is equal to the sum of its parts" just as we made the same conclusion for line segments and angles. In the figure $m\,\widehat{ABC} = m\,\widehat{AB} + m\,\widehat{BC}$.

Now let us discuss inscribed and circumscribed polygons.

A **polygon is circumscribed about a circle** if each of its sides is tangent to the circle ("tangent to" means each side touches the circle in one point"). In the illustration, polygon ABCD is circumscribed about circle O. Also, we may say that circle O is inscribed in polygon ABCD.

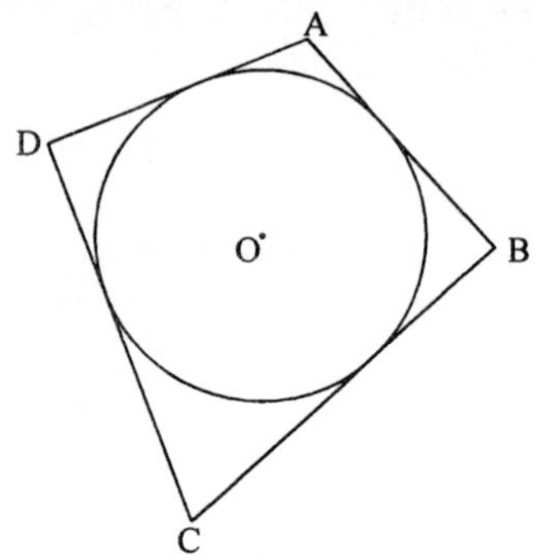

polygon ABCD is circumscribed about circle O

A **polygon is inscribed in a circle** if each of its vertices lies on the circle. Triangle ABC is inscribed in circle O. Also, we may say that circle O is circumscribed about triangle ABC.

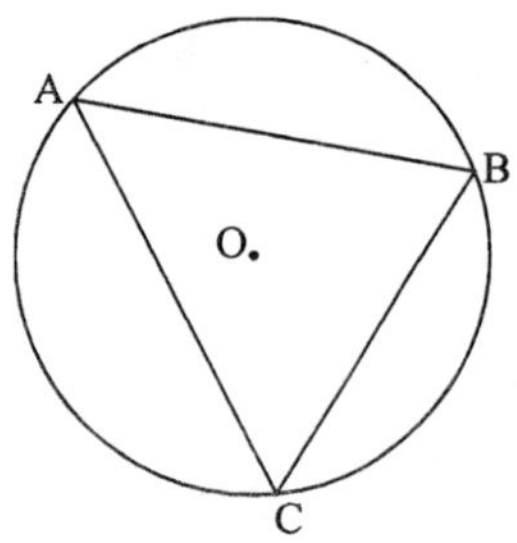

Triangle ABC is inscribed in circle O

Exercises 3.1

Answers to selected exercises are given in the back of the book.

1. How many radii does a circle have?
2. How many diameters does a circle have?
3. Give an example of a polygon that is inscribed in a circle.

4. Give an example of a polygon that is circumscribed about a circle.

5. What do we call each point or set of points for circle O?

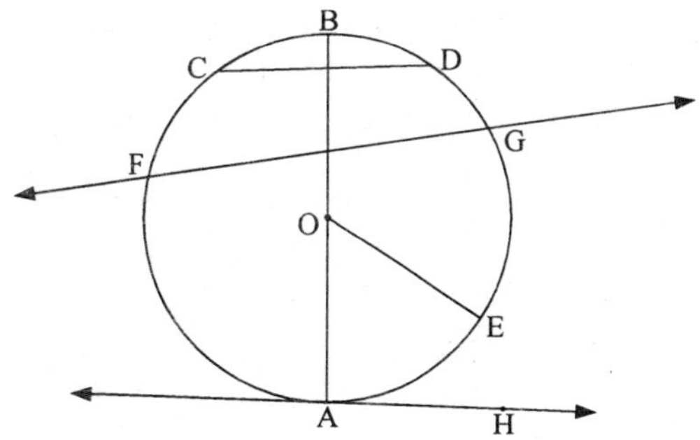

a. $\overline{CD}$
b. O
c. $\overline{OE}$
d. $\overleftrightarrow{FG}$
e. $\overleftrightarrow{AH}$
f. $\overline{AB}$
g. A

6. Identify each of the following for circle P.

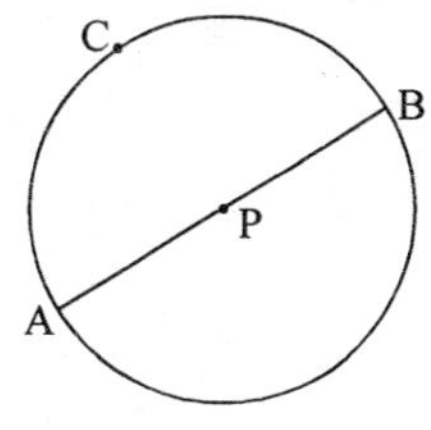

a. a minor arc
b. a major arc
c. a semicircle

7. Determine each of the following for circle O.

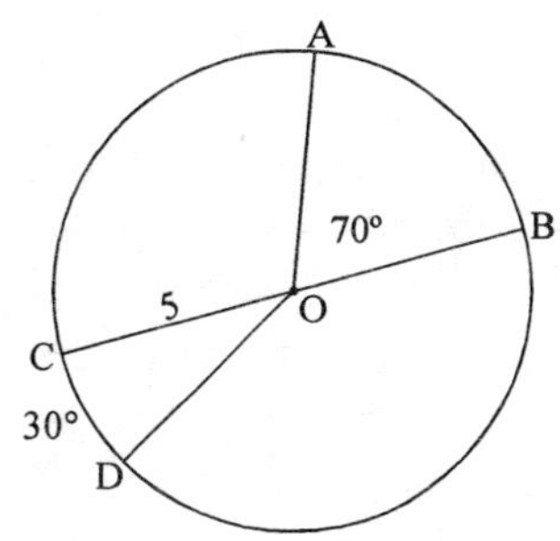

a. measure of $\overset{\frown}{AB}$ b. measure of $\overset{\frown}{ADB}$
c. measure of $\overset{\frown}{CAB}$ d. CB
e. ∠ COD f. $\overset{\frown}{DB}$
g. $\overset{\frown}{ACD}$ h. OB

8. Respond true or false:

a. All radii of a circle have equal lengths. ________
b. All diameters of a circle have equal lengths. ________
c. The radius of a circle is ½ the diameter of the circle. ________
d. The measure of a semicircle is 180. ________
e. The measure of a central angle is equal to ½ the measure of the arc that it intercepts. ________

9. Explain why ∠ 1 and ∠ 2 have the same measure in circle P.

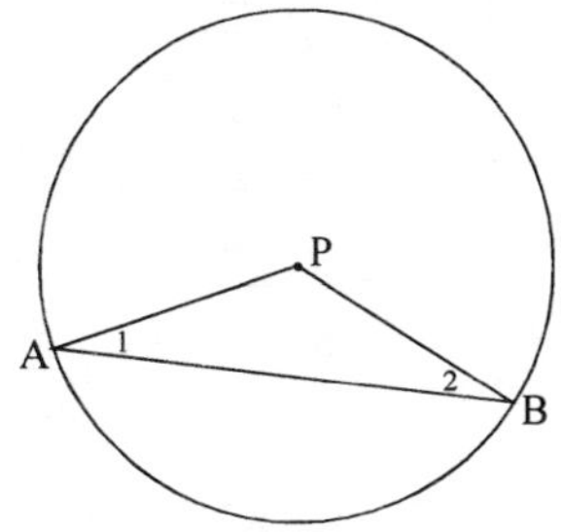

10. If OA = 5, EB = 8 and CD = 6, find the length of $\overline{OD}$ for circle O.

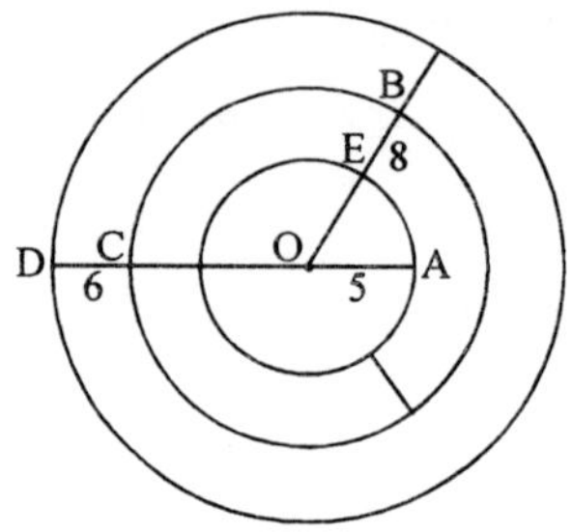

11. In the figure, B is in the interior of circle O and C is in the exterior of circle O. The length of the radius of circle O is represented by r. Use the symbols <, > or = to compare each of the following:

a. OB to r
b. OC to r
c. OA to r

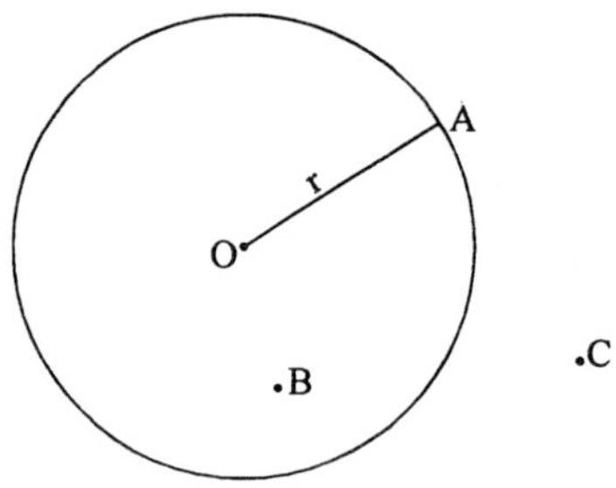

12.* An angle is called an **inscribed angle of a circle** if its vertex is a point of the circle and if each of its sides contains a chord of the circle.
In the figure ∡ APB is an inscribed angle of circle O, and ∡ AOB is a central angle of circle O. Both angles intercept the same arc, $\overset{\frown}{AB}$.
Can you use the measures given for ∡ AOB in the chart to fill in the corresponding measures of $\overset{\frown}{AB}$ and ∡ APB? If you need a hint, refer to exercise 9.

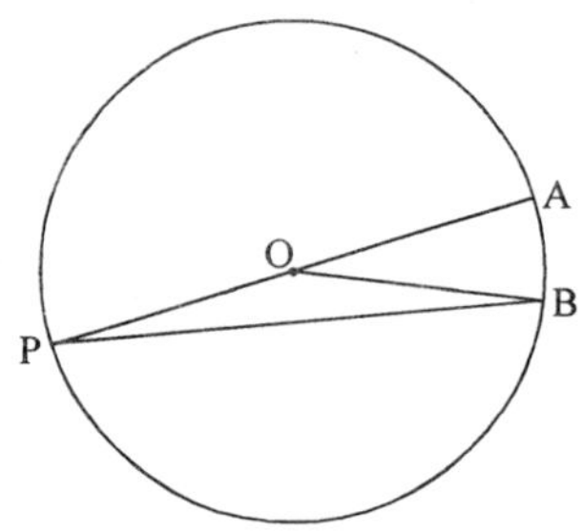

measure of ∡ AOB	30	40	56	19	12	x
measure of $\overset{\frown}{AB}$						
measure of ∡ APB						

13.* Can you use the results from exercise 12 to make a conclusion about an inscribed angle of a circle? Fill in the blank in the statement below.

The measure of an inscribed angle of a circle equals ____________ *the measure of the arc that it intercepts.*

In Chapter 10, we will prove that this conclusion is true even when a side of the inscribed angle does not contain the center of the circle.

14.* Angle P, angle Q and angle R are inscribed angles of circle O. Determine the measure of each of these angles.

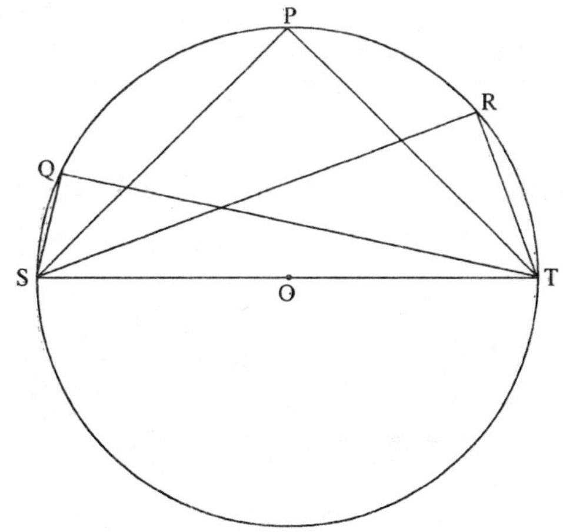

15.* Use the result of exercise 14 to complete the following statement:

If an angle is inscribed in a __________________ *, then the angle must be a* __________________ *angle.*

16. In the figure, ∠ APB has a measure of 35. What is the measure of ∠ ARB? Why?

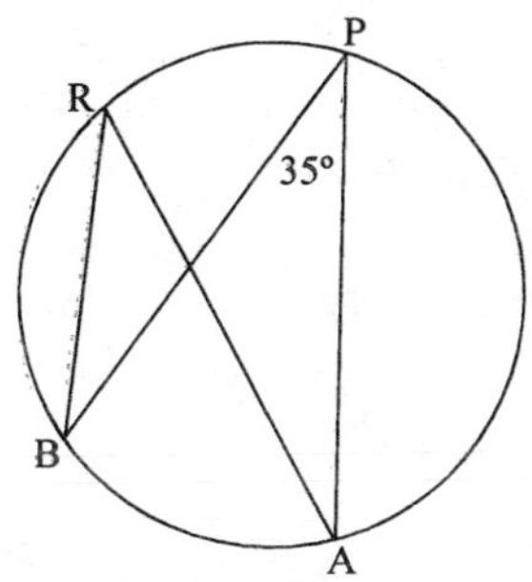

17. In circle O, $\overset{\frown}{PQ}$ has a measure of 72 , and $\overset{\frown}{PQR}$ has a measure of 160. Find the measure of each of the following:

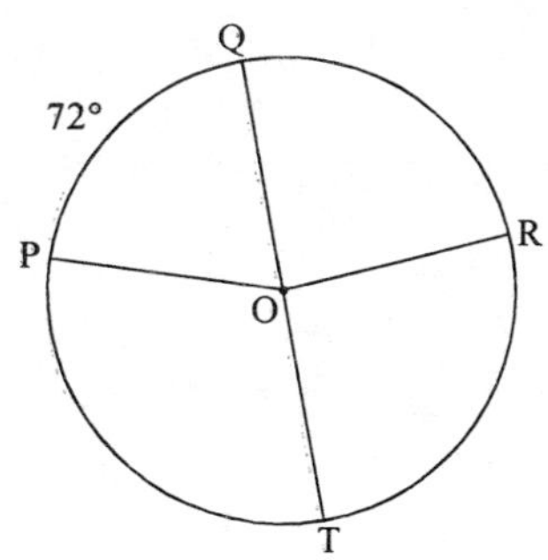

a. $\overset{\frown}{QR}$

b. ∠ QOR

c. ∠ QOP

d. ∠ POR

e. ∠ PQR

f. ∠ PQT

g. $\overset{\frown}{PT}$

h. ∠ QPT

18. What is the measure of a central angle if it intercepts an arc with the following measure?

a. 52°   b. 85°   c. x°   d. (3x + 2y)°

19. What is the measure of an inscribed angle if it intercepts an arc with the following measure?

a. 52°   b. 85°   c. x°   d. (3x + 2y)°

20. In circle O, $\overline{OP}$ has length (5x – 3y), $\overline{OR}$ has length (7x + 2y) and $\overline{OS}$ has length 8. Find the values of x and y.

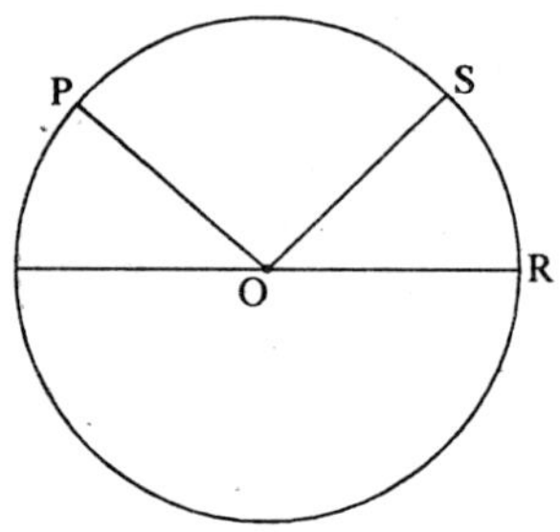

Section 3.2 Constructions

Geometric **constructions** are drawings done only with two tools: a straightedge and a compass. A **straightedge** is a tool used for drawing straight lines. Unlike a ruler, it can not be used for measuring. A **compass** is a tool for drawing circles and arcs.

In this section, we will learn how to do some basic constructions. As you work through the constructions, try to explain why each construction is valid. Be sure to do your own constructions as you work through the text.

A. CONSTRUCT A SEGMENT EQUAL IN LENGTH TO A GIVEN SEGMENT

Construct $\overline{PQ}$ so that PQ = AB

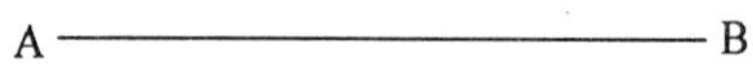

1. Use your straightedge to draw ray PR.

2. Using your compass with AB as a radius and P as a center, draw an arc that intersects $\overrightarrow{PR}$. Label the point of intersection Q. Note that PQ = AB.

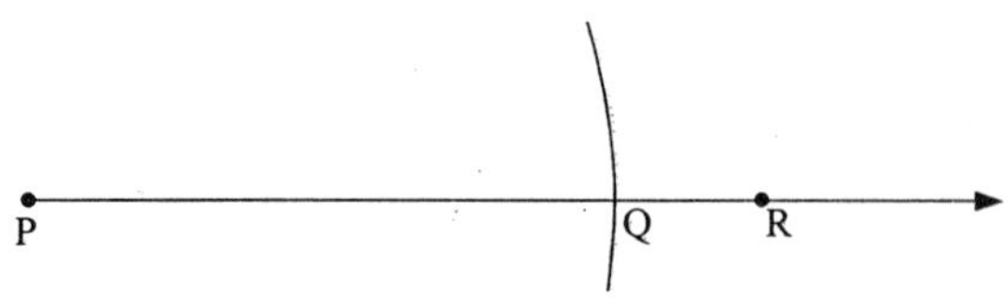

B. CONSTRUCT AN ANGLE EQUAL IN MEASURE TO A GIVEN ANGLE

Construct ∡ QPR so that ∠ ABC = ∠ QPR

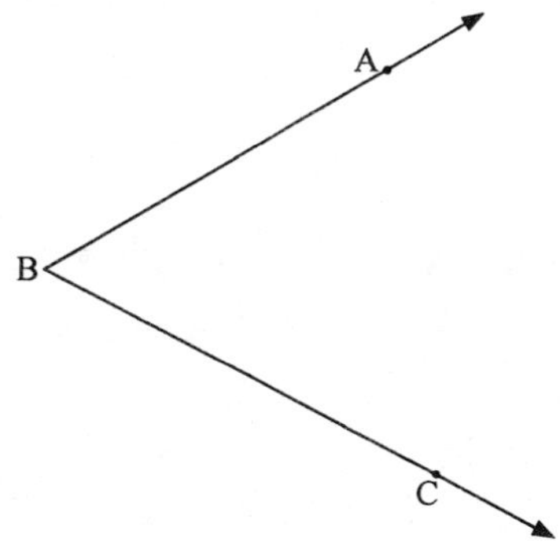

1. Use your straightedge to draw $\overrightarrow{PS}$.

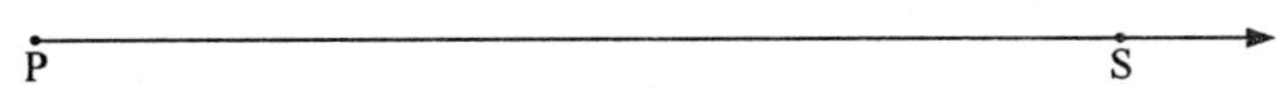

2. With B as center and any radius, use your compass to draw an arc that intersects the two sides of ∡ ABC. Label the points of intersection D and E.

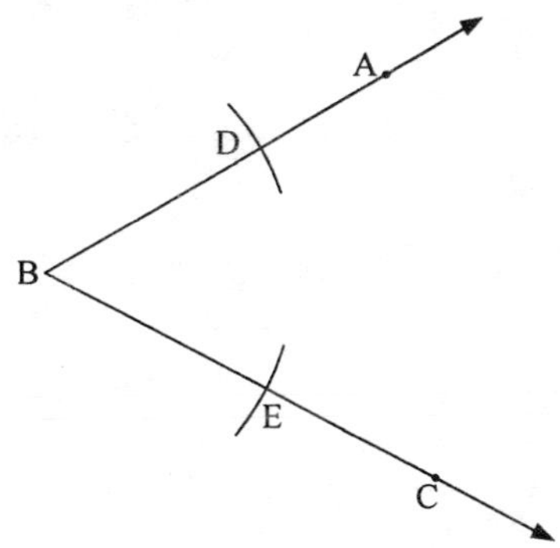

3. With your compass set at the same radius and with P as a center, draw an arc intersecting $\overrightarrow{PS}$. Label the point of intersection Q.

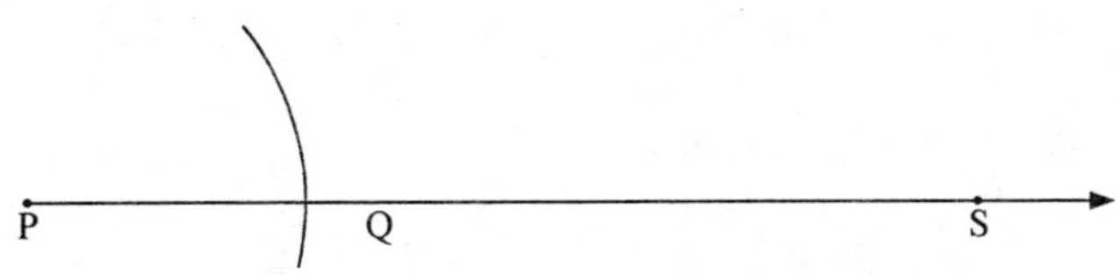

4. Now using Q as a center and DE as a radius, draw an arc that intersects the last arc (that is, the arc drawn in step 3). Label the point of intersection R.

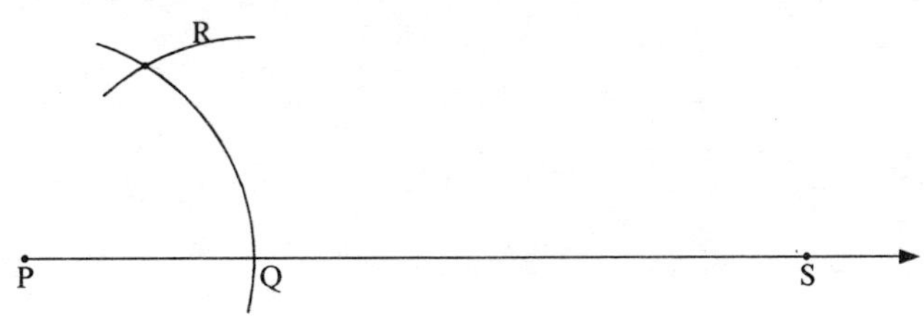

5. Draw $\overrightarrow{PR}$. ∠ QPR = ∠ ABC

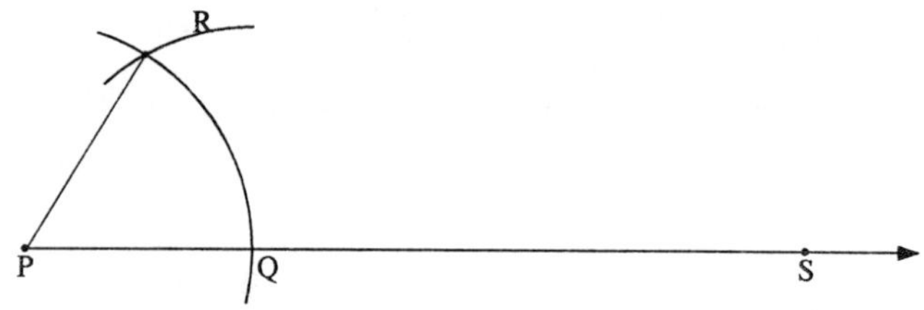

C. **CONSTRUCT A TRIANGLE THAT IS CONGRUENT TO ANOTHER TRIANGLE**

Construct △ PQR so that △ PQR ≅ △ ABC

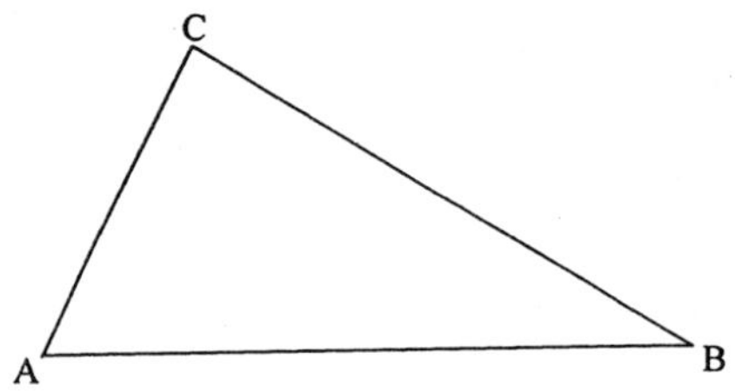

1. Use your straightedge to draw line *ℓ*. Locate any point on the line and label the point P.

2. With your compass set at radius AB and center at P, draw an arc intersecting line *ℓ*. Label the point of intersection Q.

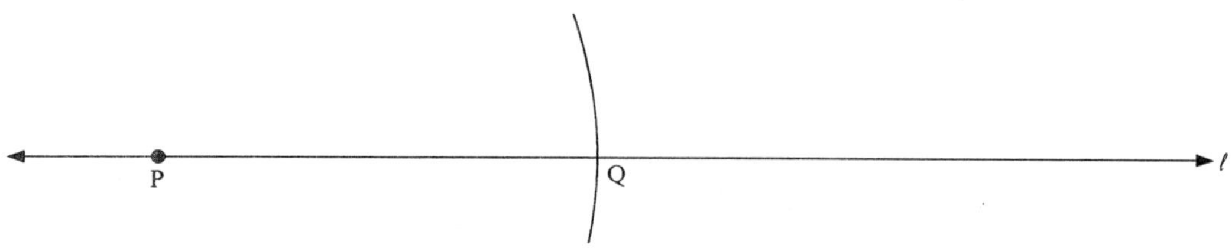

3. With your compass set at radius BC and with Q as a center, draw an arc.

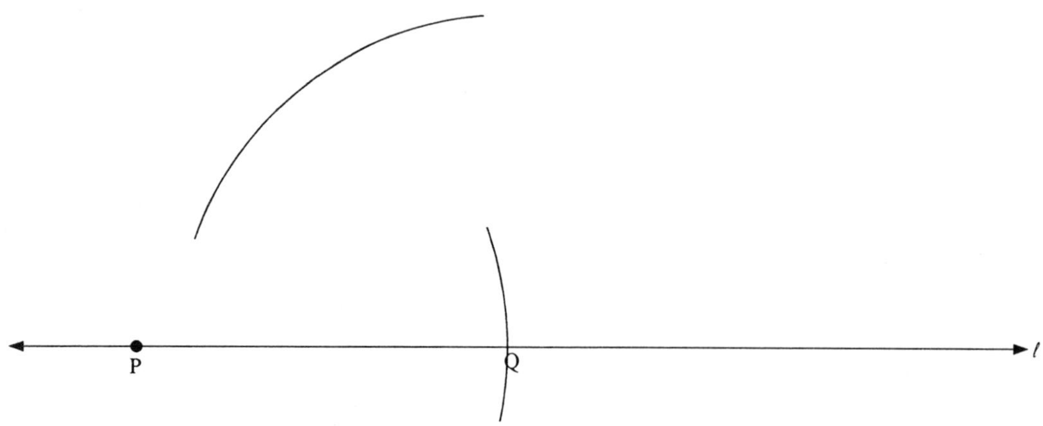

4. Now with your compass set at radius AC with P as a center, draw another arc. This arc should intersect the arc drawn in step 3. Label the point of intersection R.

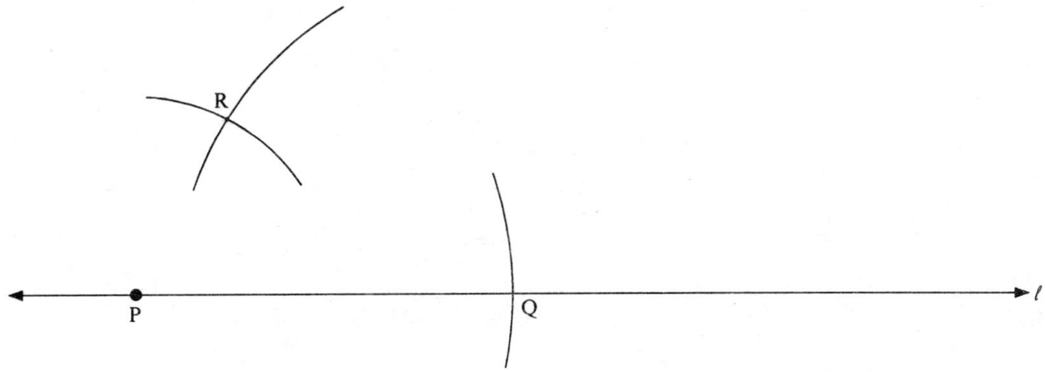

5. Use your straightedge to draw $\overline{PR}$ and $\overline{RQ}$. △ PQR ≅ △ ABC

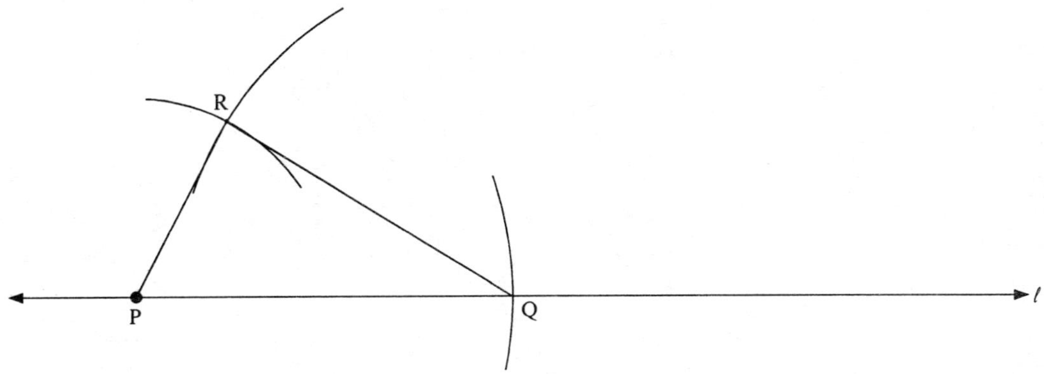

D. **CONSTRUCT THE BISECTOR OF AN ANGLE**

Construct the **bisector** of ∠ ABC.

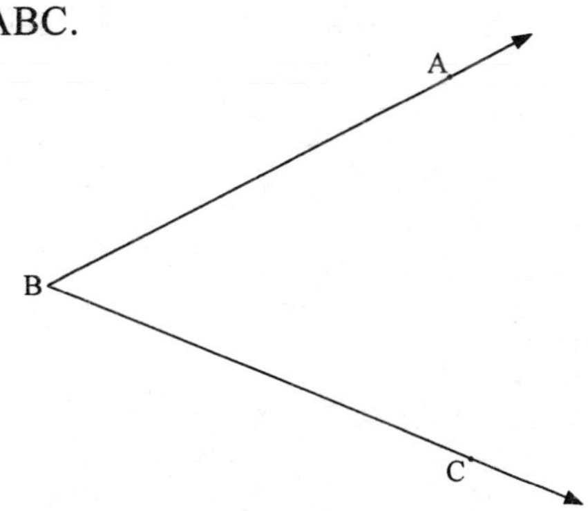

1. Using B as a center with any radius, use your compass to draw an arc that intersects both $\overrightarrow{BA}$ and $\overrightarrow{BC}$. Label the points of intersection P and Q.

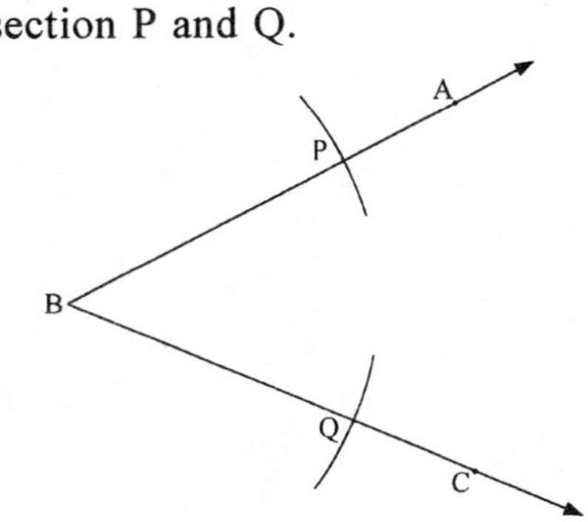

2. Using Q as a center with a radius greater than ½ PQ, draw an arc.

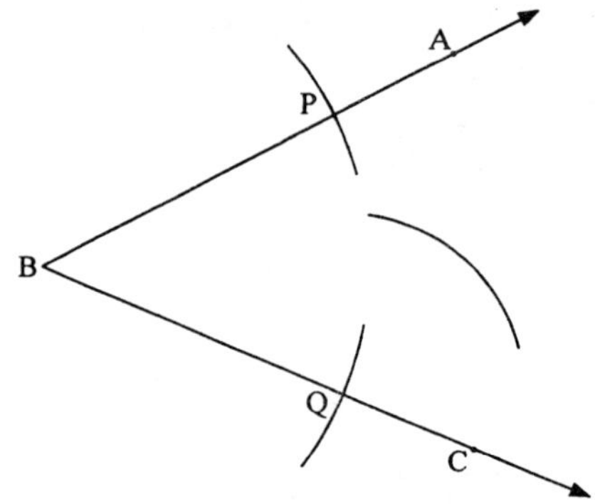

3. Using P as a center with the same radius, draw another arc intersecting the arc drawn in step 2. Label the point of intersection R.

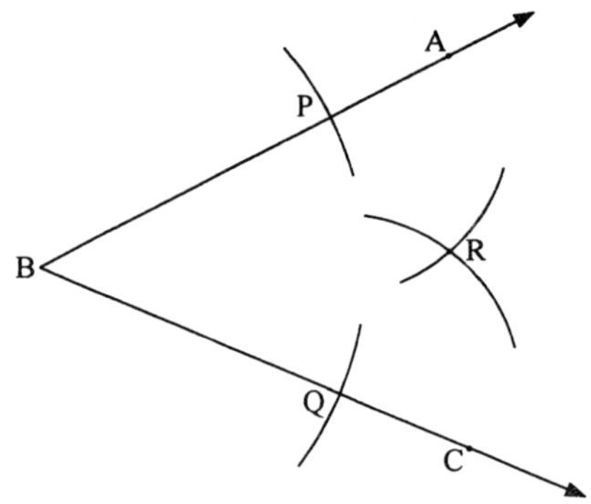

4. Use your straightedge to draw $\overrightarrow{BR}$. $\overrightarrow{BR}$ is the bisector of ∠ ABC.

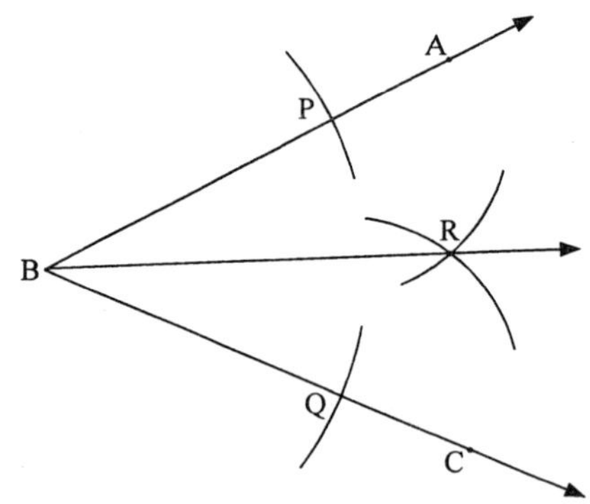

E. **BISECT A GIVEN SEGMENT**

Bisect $\overline{AB}$.

1. Set your compass to a radius that is more than ½ AB, then with A as a center,draw an arc intersecting $\overline{AB}$.

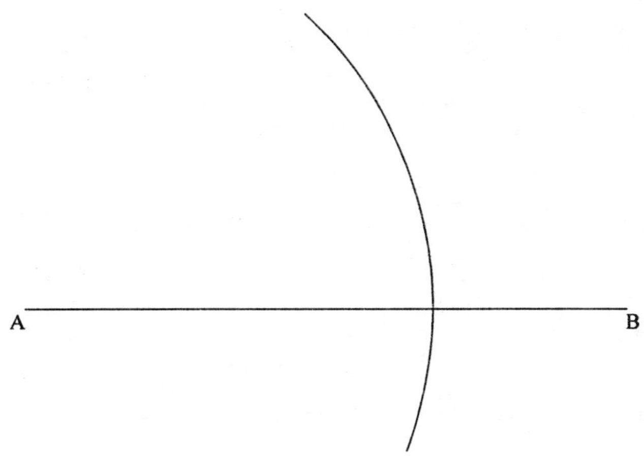

2. With B as a center and the same radius, draw another arc. This arc should intersect the first arc in two points. Label the points of intersection P and Q.

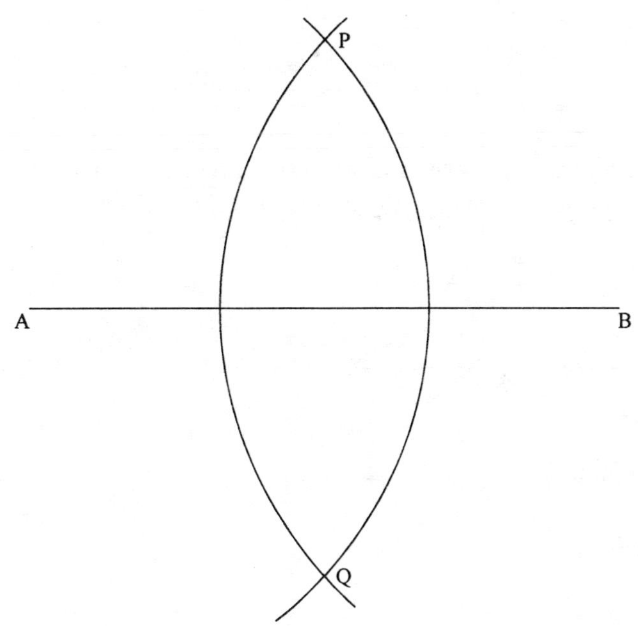

3. Use your straightedge to draw $\overline{PQ}$. PQ will intersect $\overline{AB}$. Label the point of intersection R. R is the midpoint of $\overline{AB}$ and $\overleftrightarrow{PR}$ is perpendicular to $\overline{AB}$.

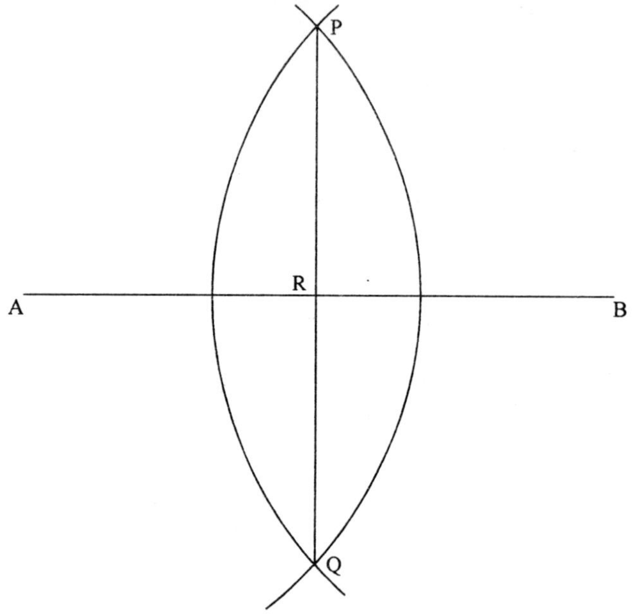

F. CONSTRUCT A PERPENDICULAR FROM A POINT NOT ON A LINE TO THE LINE

Construct the perpendicular from P to ℓ.

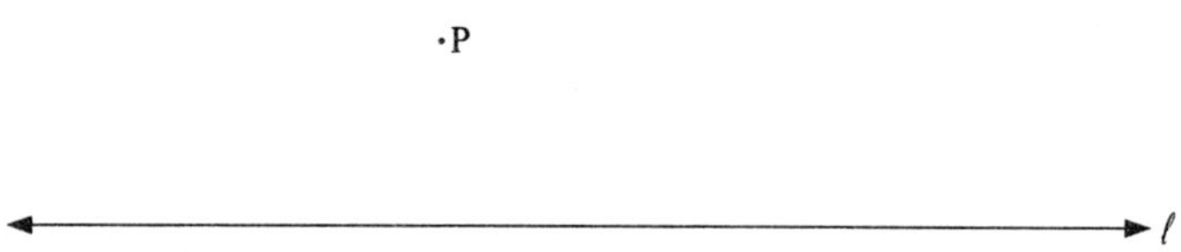

1. With P as a center and a radius that is greater than the distance from P to ℓ, draw an arc that intersects ℓ in two points. Label the two points of intersection R and S.

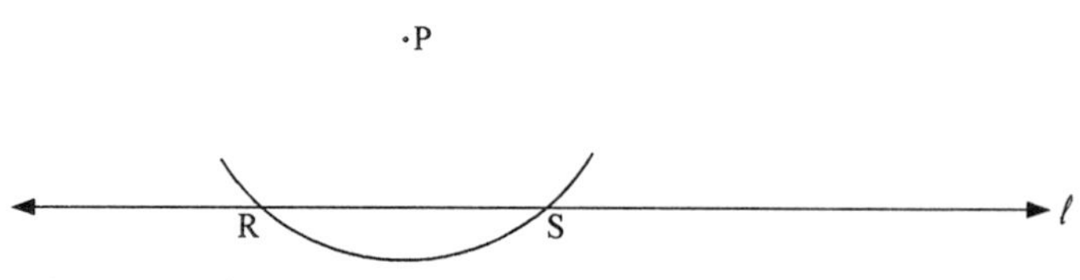

2. With the same radius and with R as center draw an arc.

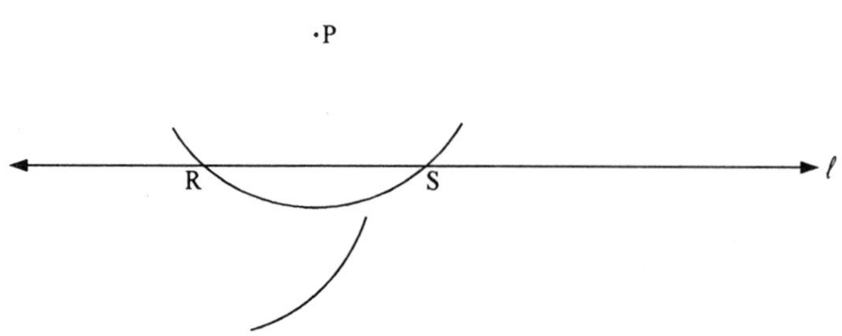

3. With S as center and the same radius, draw another arc intersecting the arc drawn in step 2. Label the point of intersection Q.

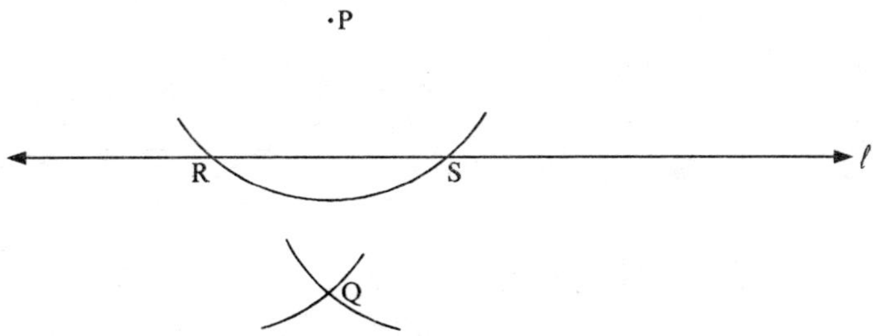

4. Use your straightedge to draw $\overleftrightarrow{PQ}$. $\overleftrightarrow{PQ}$ is the perpendicular from P to ℓ.

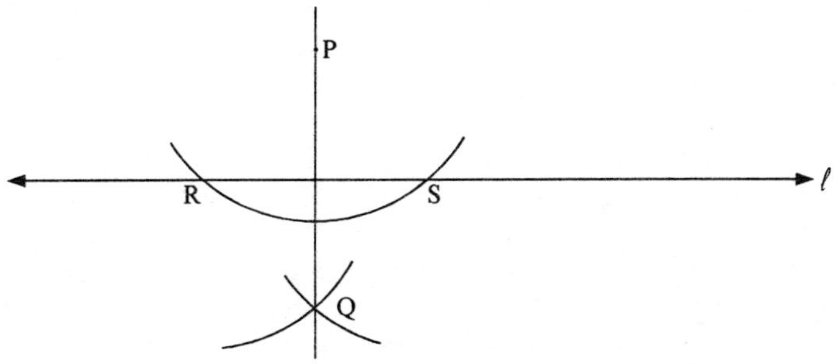

G. **GIVEN A POINT ON A LINE, CONSTRUCT THE PERPENDICULAR TO THE LINE THROUGH THE POINT.**

Construct the perpendicular to ℓ through P.

1. With P as a center and any radius, draw an arc that intersects ℓ in two points. Label the points of intersection Q and R.

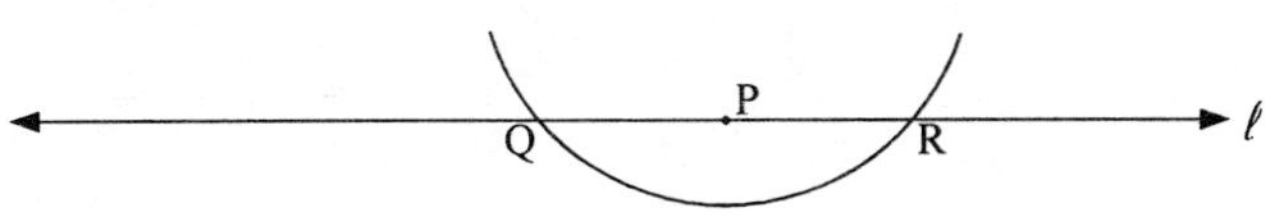

2. With Q as a center and with a radius greater than QP, draw an arc. With R as a center and the same radius, draw an arc intersecting the last drawn arc. Label the point of intersection S.

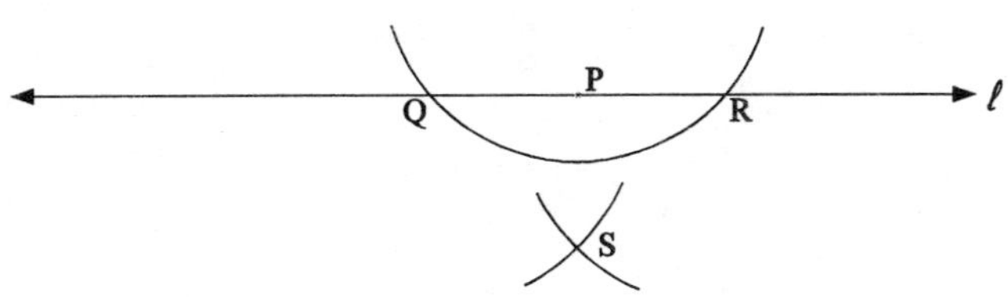

3. Use your straightedge to draw line PS. Line PS is the perpendicular through P to ℓ.

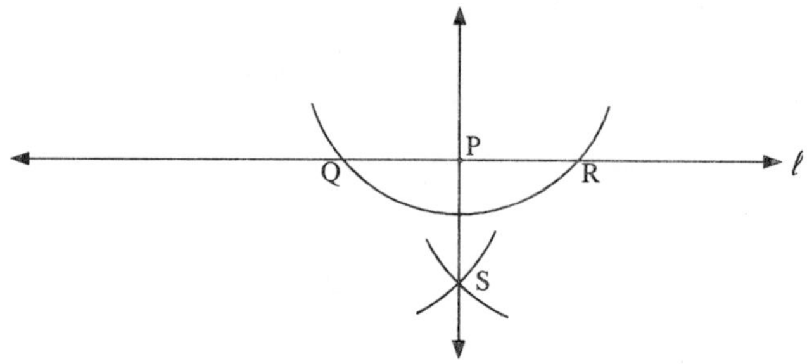

Exercises 3.2

1. Use your compass and straightedge to bisect line segment AB.

2. Use your compass and straightedge to bisect angle ABC.

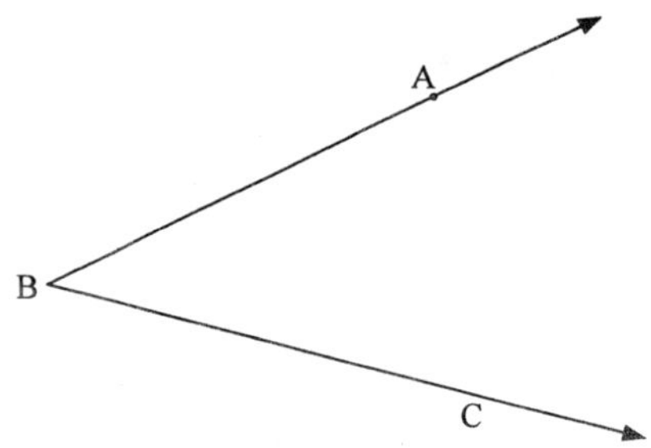

3. Construct a perpendicular from P to $\overline{AB}$.

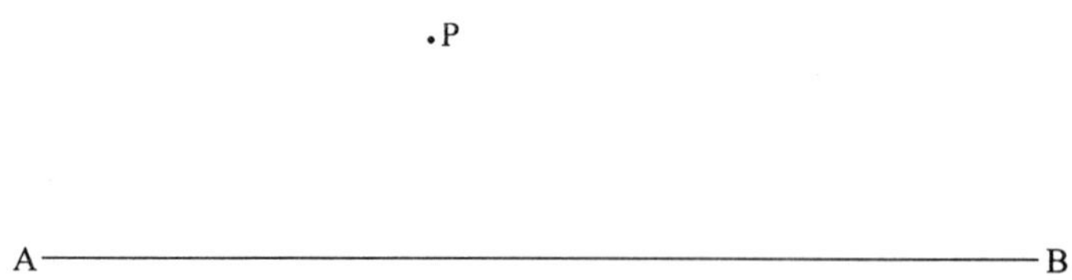

4. Construct one triangle with sides equal in length to each of the following segments:

5. Construct a triangle that is congruent to the following triangle:

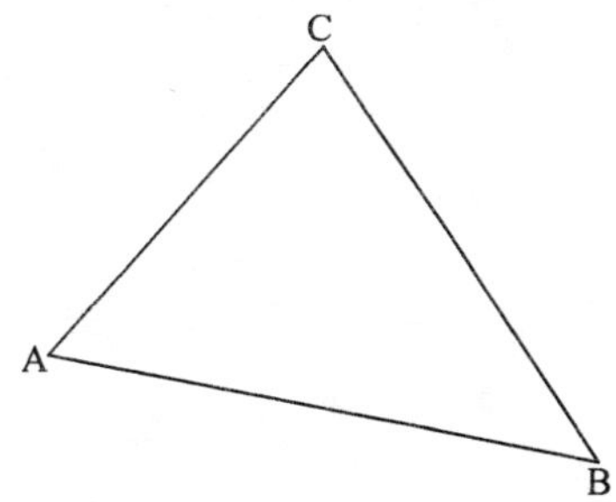

6. Construct an angle that is congruent to the following angle:

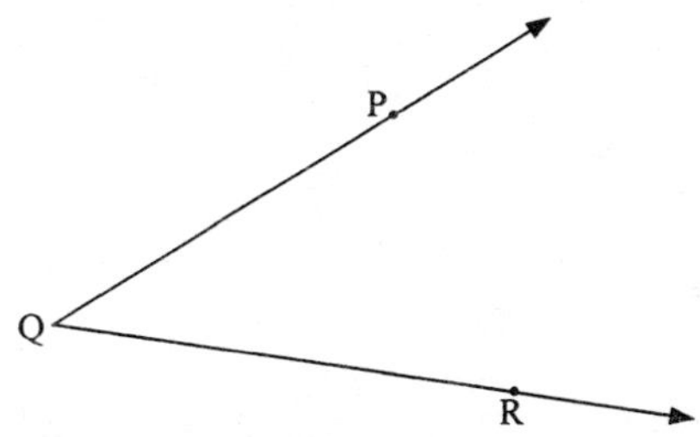

7. Use your compass and straightedge to construct an equilateral triangle.

8. Construct a hexagon by following the directions below:

a. Construct a circle and label its center O
b. Choose a point on the circle and name it A
c. Leaving the compass adjusted to the radius of the circle, mark off equal arcs along the circle, starting at the point called A.
d. Label the points at which the arcs intersect the circle B, C, D, E, and F.
e. Connect the points in order to form a regular hexagon.

9. Construct another circle and repeat the procedure given in exercises 8a, 8b, 8c, and 8d. Rather than connecting all six points in order,connect only points A, C and E to form a regular polygon. What regular polygon is formed?

10.* A line segment drawn from a vertex of a triangle to the midpoint of the opposite side of the triangle is called a **median.** Construct the three medians of a scalene triangle. What do you notice?

The medians of a triangle will always intersect in a common point.

This point is called the **centroid** of the triangle. It represents the point of gravity for the triangle. Use your ruler to verify that the centroid is two-thirds of the distance from any vertex of the triangle to the midpoint of the side of the triangle opposite that vertex.

11.*Construct the **perpendicular bisectors** of the sides of a scalene triangle. What do you notice? The perpendicular bisectors of the sides of a triangle meet in a point called the **circumcenter** of the triangle. Use your ruler to verify that this point is equidistant from the vertices of the triangle.

12.*Construct the **angle bisectors** of a scalene triangle. What do you notice? The bisectors of the angles of a triangle meet in a point which is equidistant from the sides of the triangle. It is called the **incenter** of the triangle.

13.* An **altitude** of a triangle is a perpendicular line or line segment drawn from a vertex of the triangle to the line containing the side of the triangle opposite that vertex. Construct the altitudes from each vertex of a scalene triangle. What do you notice? The altitudes of a triangle meet in a point that is called the **orthocenter** of the triangle.

When three or more lines or line segments intersect in a common point, they are said to be **concurrent.**

We may conclude the following:

The medians of a triangle are concurrent.
The perpendicular bisectors of a triangle are concurrent.
The angle bisectors of a triangle are concurrent.
The altitudes of a triangle are concurrent.

14. Construct triangles with sides having the following lengths. You may want to use your ruler to mark off the lengths before doing the constructions. You may wish to use centimeter units.

a. 3 – 4 – 5 b. 5 – 5 – 5 c. 2 – 3 – 6

Were you able to construct triangles in each case? Did you notice that the first triangle appears to be a right triangle?

15. Suppose that you have two compasses, a straightedge and a pencil. One compass is set with a radius of 3 units, while the other compass is set to a radius of 8 units.

   a. Without changing the radius of either compass, can you mark off a segment that is 50 units long?

   b. Without changing the radius of either compass, can you mark off a segment that is 1 unit long?

   c. Without changing the radius of either compass, can you mark off a segment that is N units in length if N represents any counting number?

16. Draw (you may use any tools that you wish) a regular pentagon on your paper. Extend all five of its sides (Extend each until it meets another extended side). What figure is formed? Use indirect measurement to find the measure of the angle at each point of the star.

17. Draw a regular hexagon. Extend all 6 of its sides. What figure is formed? Use indirect measurement to find the measure of the angle at each point of the star.

18. Use the following directions to circumscribe a circle about triangle PQR.

a. Construct the perpendicular bisectors of any two sides of the triangle. Label the point of intersection O.

b. Using point O as a center and OP as a radius, draw a circle.

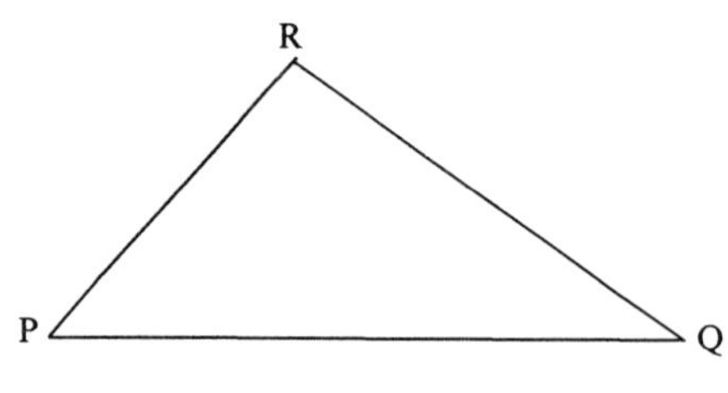

19. Use the following directions to inscribe a circle in the triangle PQR.

a. Construct the bisectors of ∠ P and ∠ Q. Label the point of intersection, point O.

b. Construct a perpendicular from point O to $\overleftrightarrow{PQ}$. Label the point where the perpendicular intersects $\overleftrightarrow{PQ}$, point S.

c. Using point O as a center and OS as a radius, draw a circle.

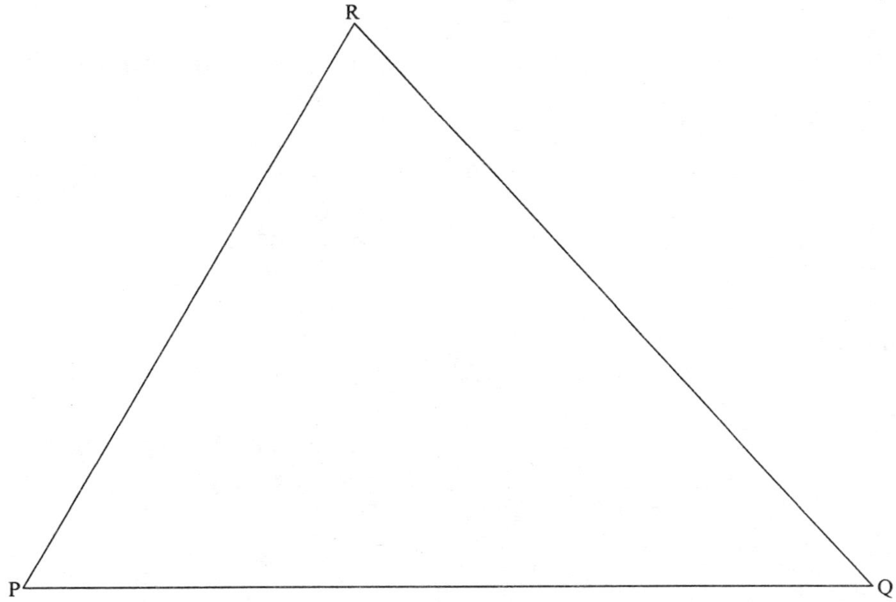

20.*Construct a right triangle so that the length of its hypotenuse is twice the length of one of its legs.

Use your protractor to measure the two acute angles of the triangle. Measure as accurately as you can.

What is the measure of the smaller acute angle? ________

What is the measure of the larger acute angle? ________

Do you believe that the following statement is true?

In a right triangle, if the hypotenuse has a length that is twice the length of one of its legs, then the two acute angles of the triangle must have measures of 30 and 60 degrees, respectively.

Chapter Three Highlights

Vocabulary

circle
center of a circle
radius of a circle
concentric circles
chord of a circle
interior of a circle
exterior of a circle
diameter
secant perpendicular
tangent
point of tangency
arc
semicircle
central angle
major arc
minor arc
inscribed polygon
circumscribed polygon
construction
straightedge
compass
median*
centroid*
angle bisector*
incenter*
bisector*
circumcenter*
altitude*
orthocenter*
inscribed angle*

*Check exercises

Required Skills

Recognize and use notation for arcs and measures of arcs
Construct a segment equal in length to a given segment
Construct an angle equal in measure to a given angle
Construct a triangle that is congruent to another triangle
Construct the bisector of an angle
Bisect a given segment
Construct a perpendicular from a point not on a line to the line
Given a point on a line, construct the perpendicular to the line through the point
Construct an equilateral triangle with sides of any given length*
Construct the medians of a triangle*
Construct the angle bisectors of a triangle*
Construct the perpendicular bisectors of a triangle*
Construct the altitudes of a triangle*
Circumscribe a circle about a triangle*
Inscribe a circle in a triangle*
*Check exercises

Conclusions

1. All radii of a circle are equal.
2. The diameter of a circle is twice the radius of the circle.
3. The measure of a central angle is equal to the measure of the arc that it intercepts.
4. The measure of a semicircle is 180 degrees.
5. The measure of a minor arc is equal to the measure of the central angle that intercepts the arc.
6. The measure of a major arc is obtained by subtracting the measure of the corresponding minor arc from 360 degrees.
7. The medians of a triangle intersect in a common point called the centroid. This point is $2/3$ of the distance from any vertex of the triangle to the midpoint of the side of the triangle opposite that vertex.
8. The angle bisectors of a triangle meet in a point called the incenter. This point is equidistant from the sides of the triangle.
9. The perpendicular bisectors of a triangle meet in a point called the circumcenter. This point is equidistant from the vertices of the triangle.
10. The altitudes of a triangle meet in a point called the orthocenter of the triangle.
11. The measure of an inscribed angle is equal to $\frac{1}{2}$ the measure of its intercepted arc.
12. In a right triangle, if the hypotenuse has a length that is twice the length of one of its legs, then the two acute angles of the triangle must have measures of 30 and 60 degrees, respectively.

Chapter Four

Perimeter and Area Pythagorean Theorem

Objectives

Use formulas to find the areas of parallelograms, rectangles, squares, triangles, trapezoids and circles
Find the surface area of right rectangular and triangular prisms and right circular cylinders
Know and apply the Pythagorean Theorem and its converse

Section 4.1 Perimeter and Area

The **perimeter** of a polygon is the distance around the polygon. In other words, it is the sum of the lengths of the sides of the polygon. The formulas for finding perimeters of squares and rectangles are given below. Notice that the formula for the perimeter of a square is the same as that of a rectangle. This is so because a square (as you recall) is also a rectangle. Since all the sides of a square have the same length, we generally write the formula for its perimeter as simply $P = 4S$ where S represents the length of a side of the square.

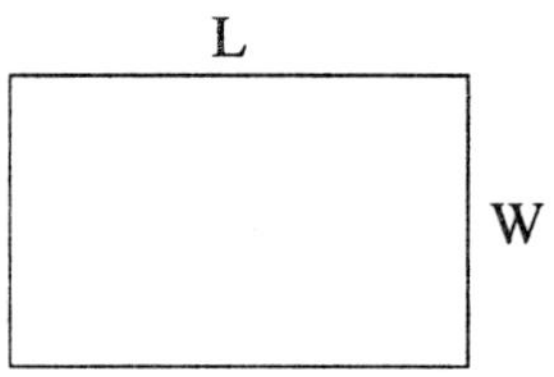

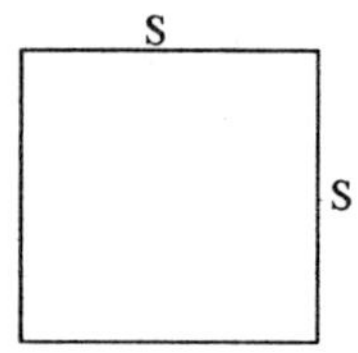

P (perimeter)
L (length)
W (width)
$P = 2L + 2W$

$P = 2L + 2W$
$P = 2S + 2S$
$P = 4S$

$P = 2L + 2W$ Perimeter of a rectangle	$P = 4S$ Perimeter of a square

EXAMPLE1:
Find the perimeter of a rectangle that has a length of 6 cm and a width of 8 cm.

$P = 2L + 2W$
$P = 2(6) + 2(8)$
$P = 12 + 16$
$P = 28$ The perimeter is 28 cm.

EXAMPLE 2:
Find the perimeter of a square with a side of length 25 cm.

$P = 4S$
$P = 4\ (25)$
$P = 100$ The perimeter of the square is 100 cm.

The distance around a circle is called its **circumference.** The ancient Egyptians knew that the ratio of the circumference of a circle to its diameter is always a constant. Today we use the **Greek letter π (pi)** to represent this constant. The approximate value of this constant is 3.14 or 22/7. Keep in mind that these values are not exact replacements for π, so they should only be used when approximations are desired. Otherwise, we should simply keep the symbol, π, to represent the ratio. We now express the relationship between the circumference (C) of a circle and its diameter (d) using π:

$$\frac{C}{d} = \pi$$

Solving for C in this equation, we obtain the formula for finding the circumference of a circle when given its diameter.

$$C = \pi d$$

Circumference of a Circle

Since the diameter of a circle is equal to twice the radius (r) of the circle, we can replace d with 2r thereby obtaining an alternate formula for finding the circumference:

$$C = 2\pi r$$

Circumference of a Circle

This formula may be used to find the circumference of a circle when we know the radius of the circle.

EXAMPLE1:

Find the circumference of a circle with a radius of 6 cm.

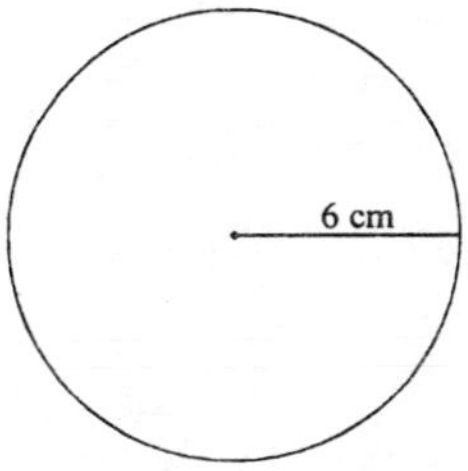

$C = 2\pi r$
$C = 2\pi(6)$
$C = 12\pi$

The circumference is 12π cm.

EXAMPLE 2:

Find the circumference of a circle with a diameter of 10 cm.

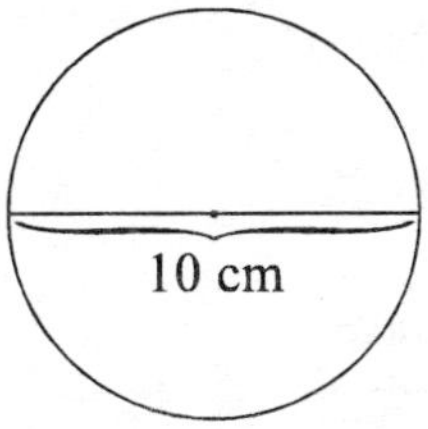

$C = \pi d$
$C = \pi(10)$
$C = 10\pi$

The circumference is 10π cm.

We think of **area** as the number of **square units** necessary to cover a surface. We use square units because such units can cover a surface without leaving gaps and without overlapping. Although other plane figures such as triangles and hexagons may also be used to cover a surface, it is generally agreed that the square is the simplest unit for doing the job. Following are some units that are used to measure area. The square unit actually used depends upon the size of the surface being measured.

square decimeter (sq. dcm or dcm^2)

square inch (sq. in. or in^2)

square centimeter (sq. cm or cm^2)

To find the area of a surface, we must determine the number of square units needed to cover that surface. We can do this by actually counting square units or by using formulas in some cases. To find the area of the following rectangular surface in square centimeters, all we need to do is divide the surface into squares, each of which is one centimeter on a side, and then count the squares. We count 16 squares, so we conclude that the area of the rectangular surface is 16 square centimeters. Notice that counting is not necessary if we observe that the total number of square centimeters is equal to the product of the number of squares in one row and the number of squares in one column.

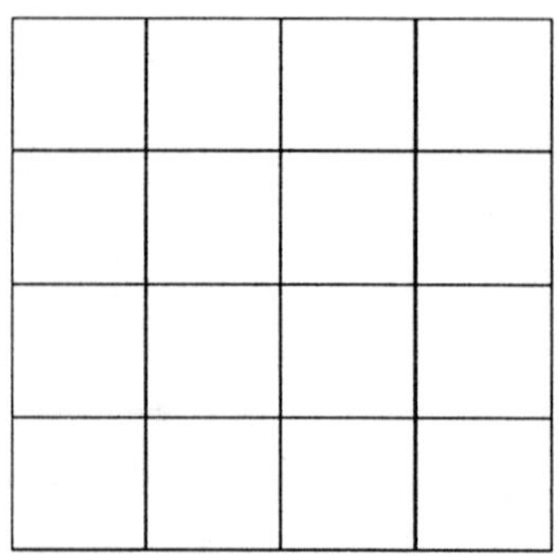

The area of the rectangular surface is 4 x 4 or 16 square centimeters.

We generalize that the area of any rectangular surface is the product of its length and width, Formulas for finding the area of a rectangle and the area of a square are given below. Notice that the area formula for a square is the same as the area formula for a rectangle. Of course you know why!

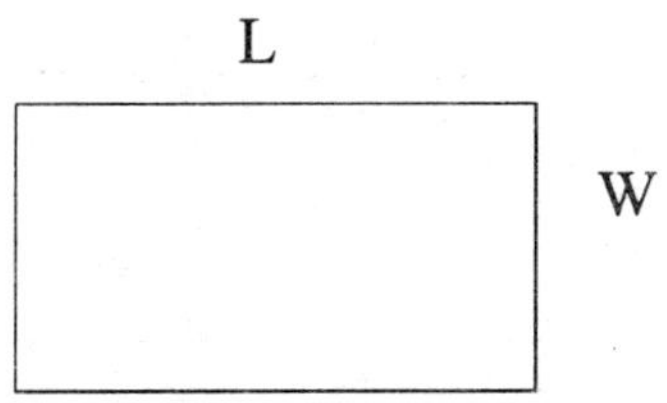

rectangle

A represents area
L represents length
W represents width

A = L x W
Area of a Rectangle

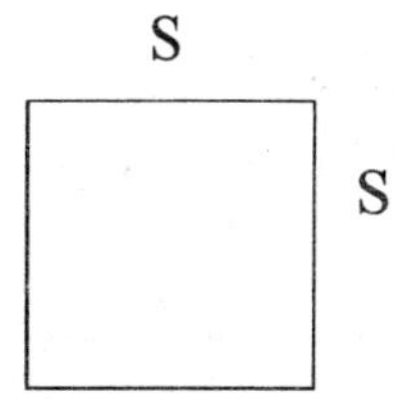

square

S represents a side of the square

A = S^2
Area of a Square

EXAMPLE 1:

Find the area of a rectangle with a length of 5 cm and a width of 3 cm.

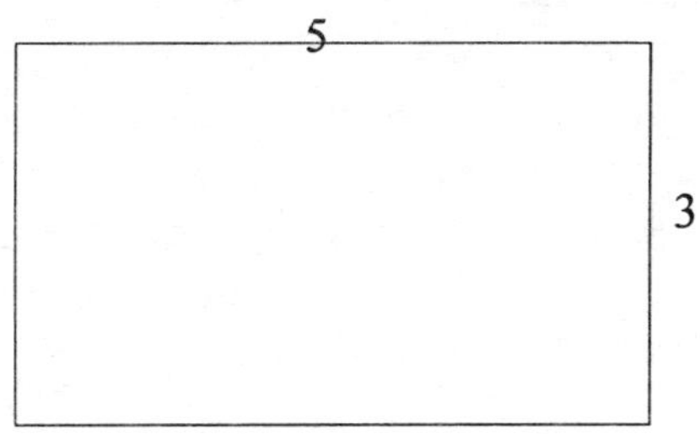

A = L x W
A = 5 x 3
A = 15

The area is 15 square centimeters.

EXAMPLE 2:

Find the area of a square with a side of 12 inches.

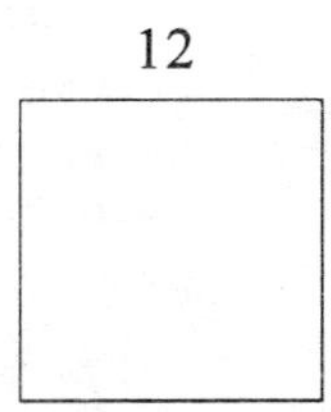

A = S^2
A = 12^2
A = 144

The area is 144 square inches.

In an earlier section, we learned that a **parallelogram** is a quadrilateral with two pairs of parallel sides. Either of two parallel sides of a parallelogram may be referred to as the **base (b) of the parallelogram**. The distance between these two parallel sides is then referred to as the **corresponding height (h) or altitude of the parallelogram.** The distance between two parallel lines is the length of any perpendicular line segment joining the two lines. Recall that rectangles, squares and rhombuses are also parallelograms, and as such, they assume all the properties of a parallelogram.

To find the area of a parallelogram, we can think of cutting off part of it, as indicated in the following figure. We can then place the part as shown to form a rectangle. Notice that the area of the rectangle is the product of b and h. This product is also the area of the parallelogram. We then conclude that the area of a parallelogram is the product of its base (b) and corresponding height (h).

$A = bh$ Area of A Parallelogram

EXAMPLE:

Find the area of a parallelogram with a base of 15 cm and a corresponding height of 6 cm.

$A = bh$
$A = 15(6)$
$A = 90$

The area is 90 square centimeters.

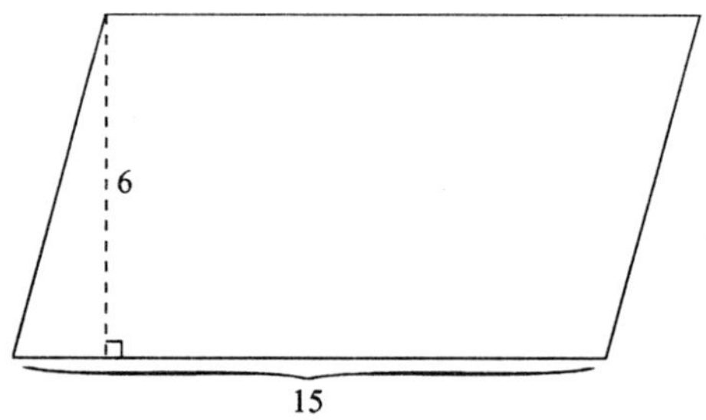

To find the area of a **trapezoid** (a quadrilateral with exactly one pair of parallel sides), we can think of cutting out another trapezoid of the same shape and size and placing the two congruent trapezoids together, as shown in the following figure. The resulting figure is a parallelogram with area $h(b_1 + b_2)$. The area of each trapezoid is exactly one-half the area of the parallelogram. The two parallel sides of a trapezoid are called its **bases,** and the distance between these bases, is called the **height of the trapezoid.** We conclude that the area of a trapezoid is one-half the product of its height and the sum of its bases.

$A = 1/2\ h\ (b_1 + b_2)$ Area of a Trapezoid

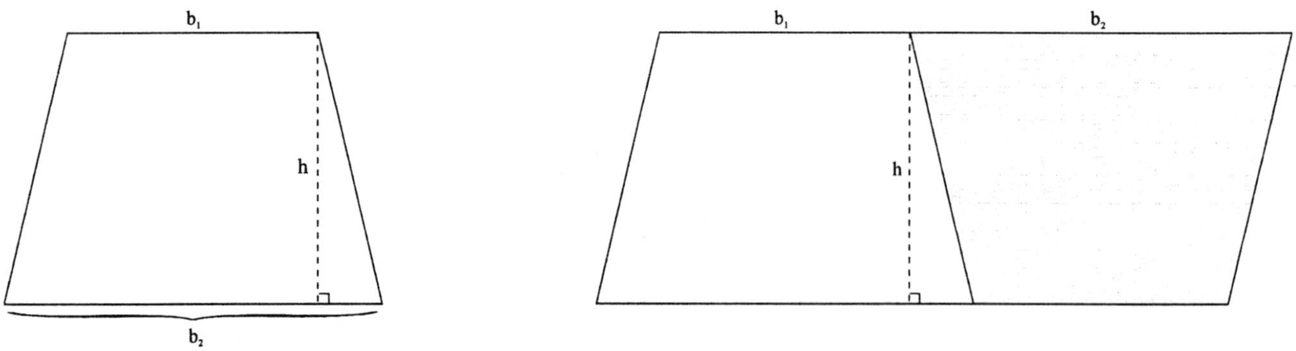

To find the area of a **triangle,** we can think of cutting out another triangle of the same size and shape, and placing them together as shown in the illustration. The resulting figure is a parallelogram with area equal to the product of b and h. Notice that the area of each triangle is 1/2 bh. For a triangle, b or **base** represents any side of the triangle and h represents the height corresponding to that side. This corresponding **height (or altitude)** is the length of the perpendicular line segment joining the vertex opposite the base to the line containing the base of the triangle.

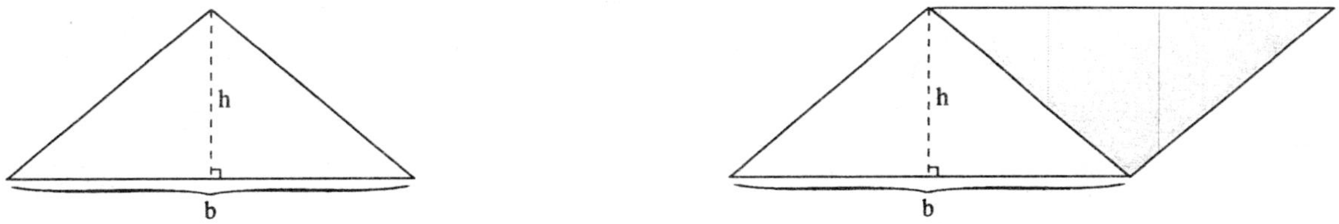

Area of a Triangle $A = 1/2\ bh$

EXAMPLE:
Find the area of a triangle with base 12 cm and height 4 cm.

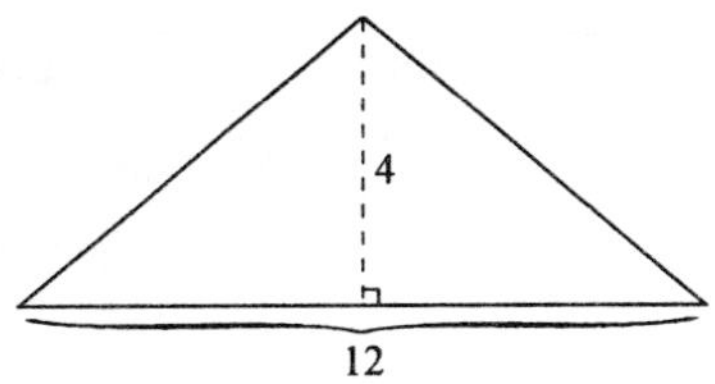

$A = 1/2\ (12)\ (4)$
$A = 1/2\ (48)$
$A = 24$

The area is 24 square centimeters.

To derive a formula for finding the area of a **circle,** we examine a circle that is divided into pie-shaped sectors as shown. We then arrange the sectors as shown in the figure on the right.

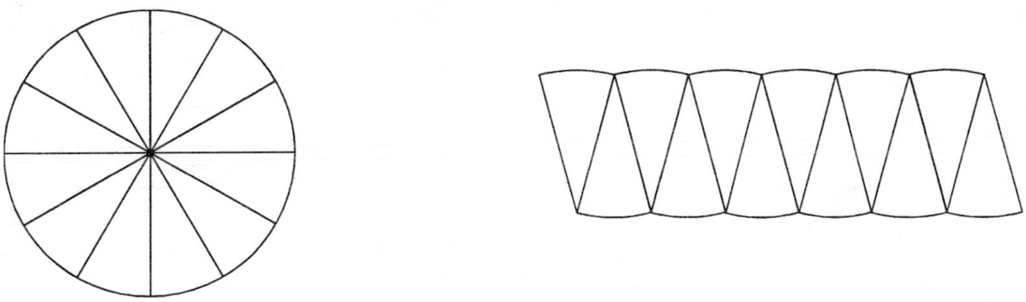

Notice that by cutting the circle into smaller and smaller pie-shaped sectors, the sectors may be arranged to form a figure that is almost rectangular in shape. The height of this rectangle approximates the radius of the circle. The base of the rectangle approximates 1/2 the circumference of the circle, or 1/2 of $2\pi r$ which is πr. We conclude that the area of the rectangle is the product of r and πr or $A = \pi r^2$.

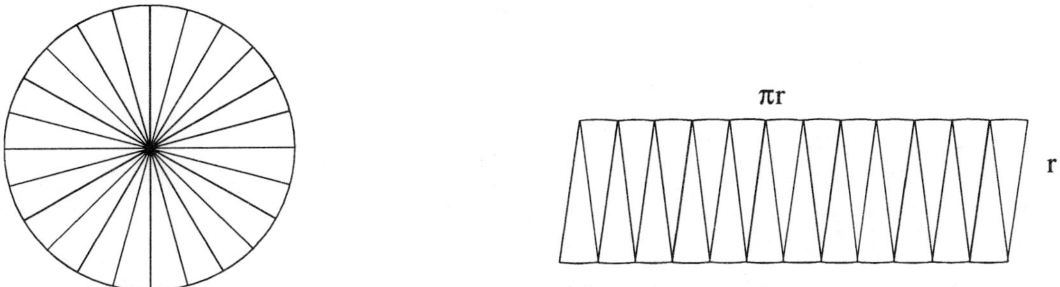

Area of a Circle $A = \pi r^2$

EXAMPLE1:

Find the area of a circle with a radius of 7 yards.

$A = \pi r^2$
$A = \pi (7^2)$
$A = 49 \pi$

The area of the circle is 49 π square yards.

EXAMPLE 2:

The area of a circle is 80 cm^2. What is the diameter of the circle?

$\pi r^2 = A$
$\pi r^2 = 80$
$r^2 = 80/\pi$
$r = \pm \sqrt{80/\pi}$

$r = \sqrt{80/\pi}$ We use the positive value, since length can't be negative.

$r = \dfrac{4\sqrt{5\pi}}{\pi}$ Simplifying the radicand and rationalizing the denominator

$d = \dfrac{8\sqrt{5\pi}}{\pi}$ We are asked to find the diameter which is twice the radius.

Exercises 4.1

1. Find the area of a rectangle with base b and altitude h if:

a. $h = 10$ and $b = 6$

c. $h = 3$ and $b = \sqrt{8}$

b. $h = 3/5$ and $b = 10$

d. $h = 6$ and $b = \sqrt{7}$

2. Find the area of a square with side s if:

a. $s = 8$

c. $s = \sqrt{7}$

b. $s = 3/5$

d. $s = 5\sqrt{3}$

3. Find the area of a parallelogram with base b and altitude h if:

a. $h = 6$ and $b = 3$

d. $h = 5\sqrt{6}$ and $b = \sqrt{15}$

b. $h = 1/2$ and $b = 7/2$

e. $h = x + 3$ and $b = x - 3$

c. $h = 2\sqrt{3}$ and $b = \sqrt{3}$

4. Find the area of a triangle with base b and corresponding altitude h if:

a. $h = 8$ and $b = 3$

b. $h = 1/2$ and $b = 7/2$

c. $h = 3x + 7$ and $b = 5x + 8$

d. $h = 2\sqrt{3}$ and $b = \sqrt{3}$

e. $h = 5\sqrt{6}$ and $b = 15$

5. Find the area of a trapezoid with bases b_1 and b_2 and corresponding altitude h if:

a. $h = 8$, $b_1 = 12$, $b_2 = 9$

b. $h = 18$, $b_1 = 5\sqrt{3}$, $b_2 = 7\sqrt{3}$

6. Find the area of each figure when possible:

a.

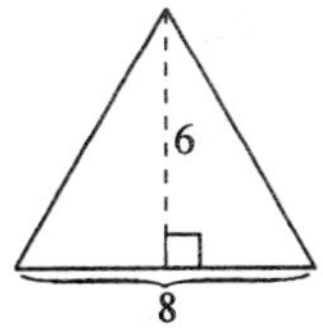

b.

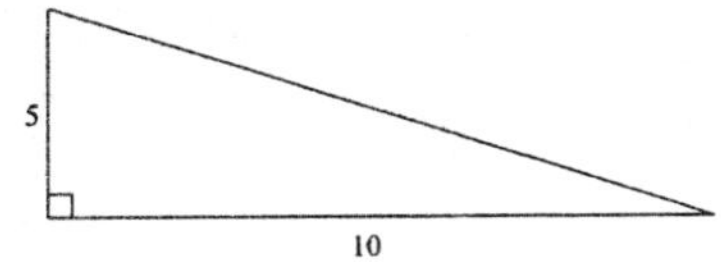

c.

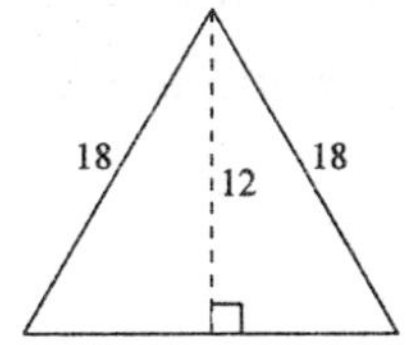

d.

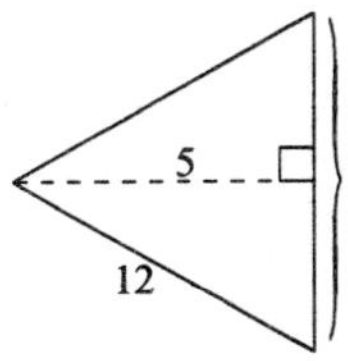

e.

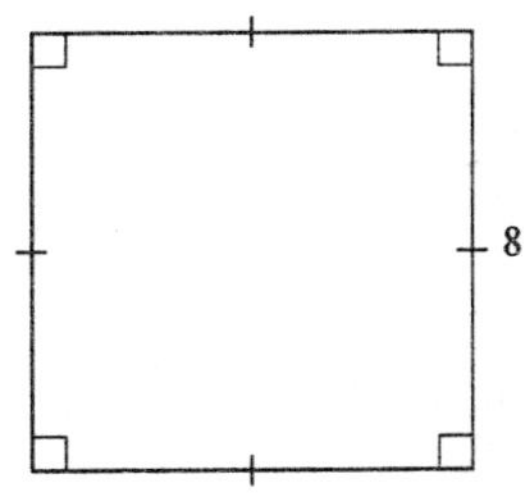

f.

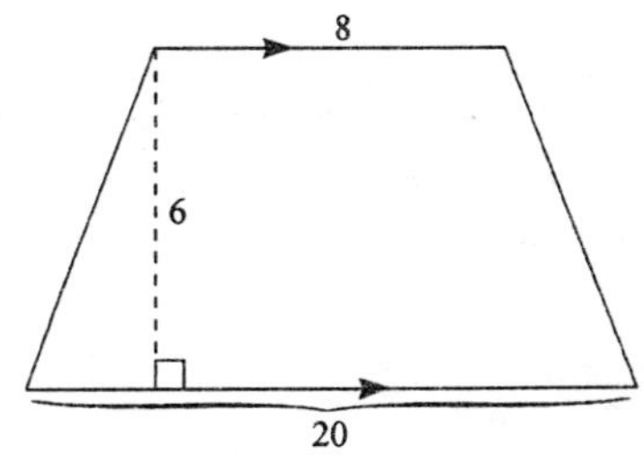

g.

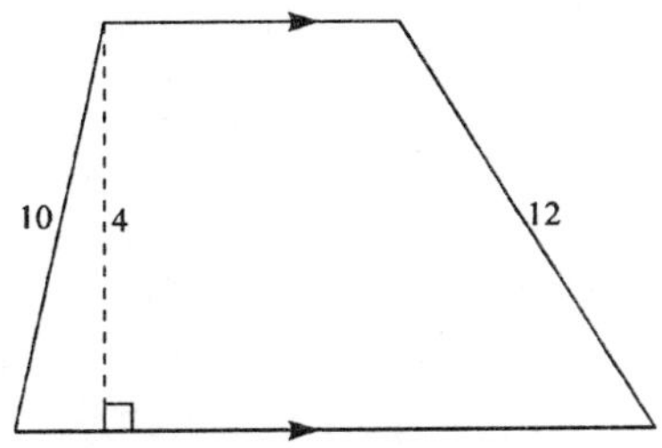

h.

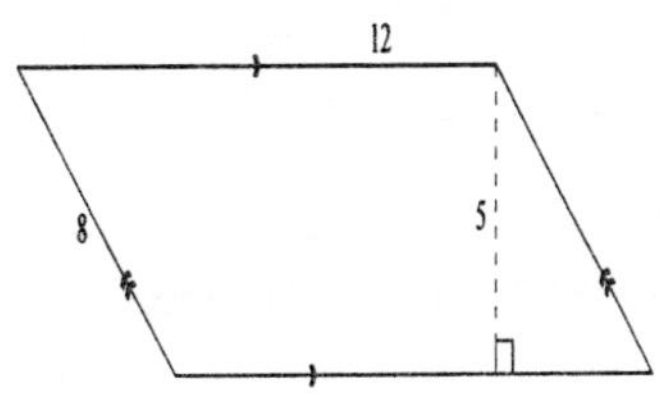

i.

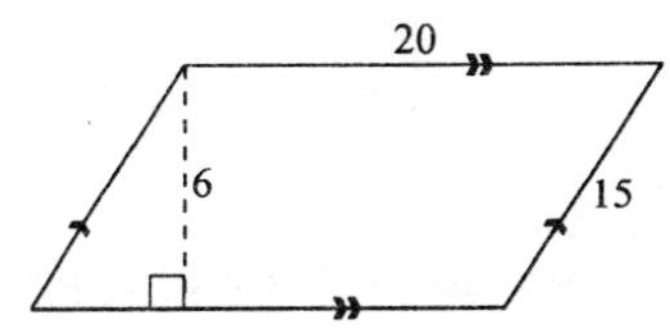

7. Find the base of a parallelogram if its area is 64 and the base is four times its altitude.

8. Find the base of a parallelogram if the area is 36 and its altitude is five more than its base.

9. Find the length of each leg of a right triangle if its area is 7 ½ and the length of one of its legs is one less than twice the length of its other leg.

10. The area of a rectangle is 72. If a square has the same area as the rectangle, find the length of each side of the square.

11. Find the altitude of a parallelogram that has an area of 64 and a base of 16.

12. If a parallelogram and a triangle have equal areas and equal bases, what is the ratio of their heights?

13. How is the area of a rectangle affected when:

a. The base is doubled while the altitude remains unchanged?

b. The altitude is doubled while the base remains unchanged?

c. The altitude and the base are both doubled?

14. How is the area of a square changed when:

a. The length of each side is doubled?

b. The length of each side is tripled?

15. Find the area of a circle with radius r and diameter d if:

a. $r = 25$ b. $r = \sqrt{3}$ c. $d = 36$ d. $r = x + 4$

16. Find the perimeter of each of the following:

a. A square with side of length 12

b. A rectangle with sides of lengths 8 and 10

c. A square with sides of lengths $5x - 2$ and $3x + 8$

d. A square with a side of length $3x$

17. Find the circumference of a circle with a diameter of 18.

18. Find the cost of carpet for a room that is 12 feet wide by 18 feet long if the carpet is $16 per square yard.

19.* △ ABC has a height of 6 and a base of 8.
△ PQR has a height of 6 and a base of 12.

Find the ratio of the base of △ ABC to the base of △ PQR. ____________
Find the ratio of the area of △ ABC to the area of △ PQR. ____________

△ ABC has a height of 10 and a base of 15.
△ PQR has a height of 10 and a base of 42.

Find the ratio of the base of △ ABC to the base of △ PQR. ____________
Find the ratio of the area of △ ABC to the area of △ PQR. ____________

△ ABC has a height of h and a base of b_1.
△ PQR has a height of h and a base of b_2.

Find the ratio of the base of △ ABC to the base of △ PQR. ____________
Find the ratio of the area of △ ABC to the area of △ PQR. ____________

Complete the following statement:

The ratio of the areas of two triangles with the same height is equal to the ratio of their ____________________.

20. Use the square below to fill in the blanks.

The area of square ABCD equals $(a + b)^2$

The area of square ABCD equals the sum of the following areas:

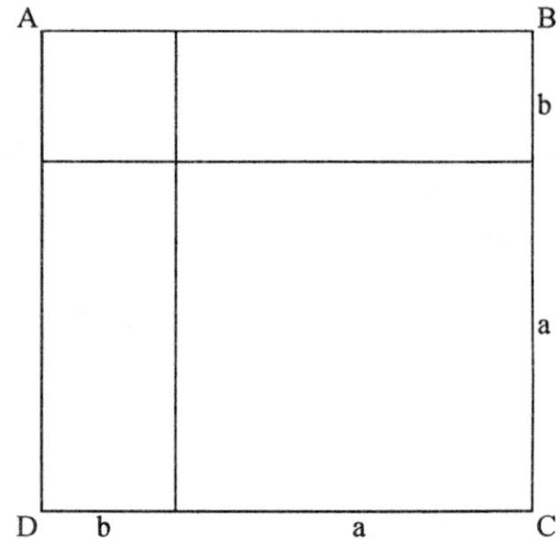

The area of the smallest square ____________________

The area of the two rectangles ____________________

The area of the square with sides of length a __________

$(a + b)^2 =$ ________ + ________ + ________

21. Find the area of the shaded part of each figure.

a. Circle O with radius of 4cm

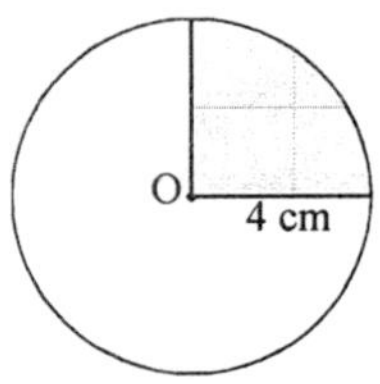

c. Circle O with radius of 8cm

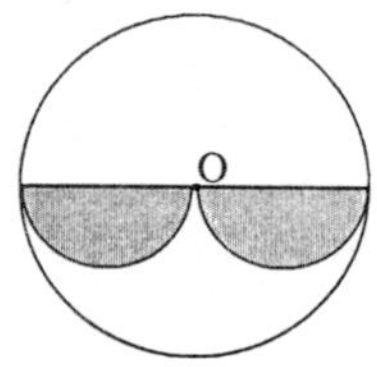

b. Circle O with radius 5
Circle O with radius 8

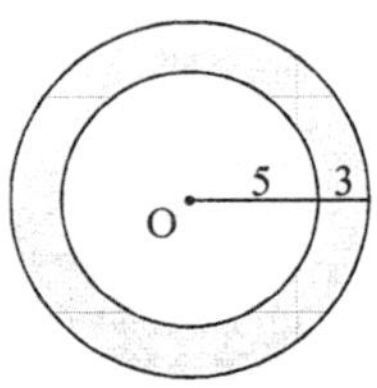

d. Square ABCD
AB = 10
Circle O with radius $5\sqrt{2}$

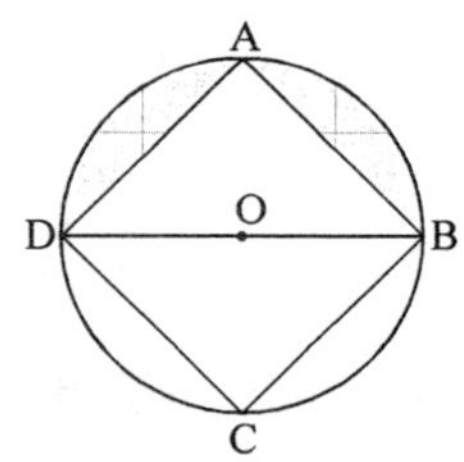

22* The center of a regular polygon is the common center of its inscribed and circumscribed circles. The radius of a regular polygon is the radius of its circumscribed circle, that is, it is a line segment joining its center to a vertex. The apothem of a regular polygon is the radius of its inscribed circle, that is, it is the perpendicular line segment from the center of the polygon to a side of the polygon. Can you write a formula for the area of a regular polygon when given the following information?

Regular polygon ABCDEF with n sides.
Side of length s
Apothem of length a
Perimeter p

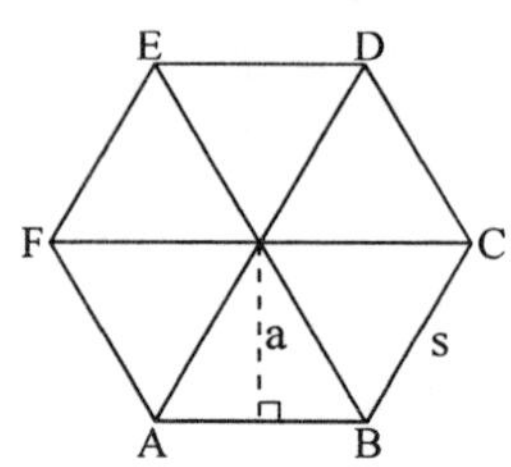

Can you write a formula for the area in terms of the perimeter and the apothem of the polygon?

23. The Garcia family is planning a rectangular garden plot. If they have 36 feet of fencing, what is the greatest area they can enclose, if they limit the dimensions of the plot to whole numbers?

24. Emily Baker wants to enclose a 48-square-foot rectangular garden plot. What is the least amount of fencing she will need if the dimensions of the plot must be limited to whole numbers?

Section 4.2 The Pythagorean Theorem

The **Pythagorean Theorem** is probably the most famous theorem in mathematics. Although the theorem bears the name of the Greek philosopher, Pythagoras, the ancient Egyptians and Babylonians are believed to have known of the theorem long before the birth of Pythagoras. According to the theorem, if squares are constructed on the three sides of a right triangle as shown in the figure below, the area of the square on the *hypotenuse* is equal to the sum of the areas of the squares on the two legs of the triangle. The Pythagorean Theorem is most often expressed in the following form:

$$\mathbf{c^2 = a^2 + b^2}$$

In the formula, c represents the hypotenuse of the triangle (the side opposite the right angle) while a and b represent the other two sides of the triangle which are called its *legs*. In short, the theorem indicates that *in a right triangle, the square of the hypotenuse is equal to the sum of the squares of its legs.*

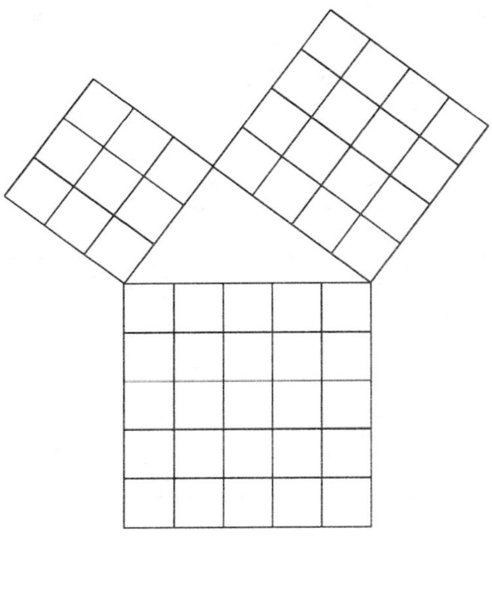

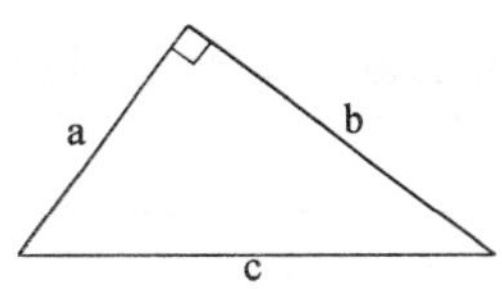

The Pythagorean Theorem may be used to find any side of a right triangle when two of its sides are given. For example, to find the length of the hypotenuse of a right triangle with legs of 5 and 12 units, we do the following:

EXAMPLE 1

$c^2 = a^2 + b^2$ Find the length of the hypotenuse

legs have lengths 5 and 12

$$\begin{aligned} c^2 &= 5^2 + 12^2 \\ c^2 &= 25 + 144 \\ c^2 &= 169 \\ c &= \pm\sqrt{169} \\ c &= 13 \end{aligned}$$

The hypotenuse is 13 units long.

The theorem also may be used to find a missing leg of a right triangle:

EXAMPLE 2

If a right triangle has a hypotenuse of 20 units and a leg of 12 units, find the other leg of the triangle.

Again $c^2 = a^2 + b^2$

$$\begin{aligned} 20^2 &= 12^2 + b^2 \\ 400 &= 144 + b^2 \\ b^2 &= 400 - 144 \\ b^2 &= 256 \\ b &= \pm\sqrt{256} \\ b &= 16 \end{aligned}$$

The other leg of the triangle is 16 units long.

Conversely, it is also true that a triangle must be a right triangle if the square of one side of the triangle is equal to the sum of the squares of the other two sides of the triangle.

EXAMPLE 3

Consider a triangle with sides of 8, 15, and 17 units:

$$\begin{aligned} 8 \cdot 8 &= 64 \\ 15 \cdot 15 &= 225 \\ 17 \cdot 17 &= 289 \end{aligned}$$

Since $64 + 225 = 289$, the triangle is a right triangle.

EXAMPLE 4

Consider now a triangle with sides of 8, 9 and 12 units:

$$\begin{aligned} 8 \cdot 8 &= 64 \\ 9 \cdot 9 &= 81 \\ 12 \cdot 12 &= 144 \end{aligned}$$

Since $64 + 81 \neq 144$, the triangle is not a right triangle.

Exercises 4.2

1. Find the length of the side marked x in each triangle:

a.

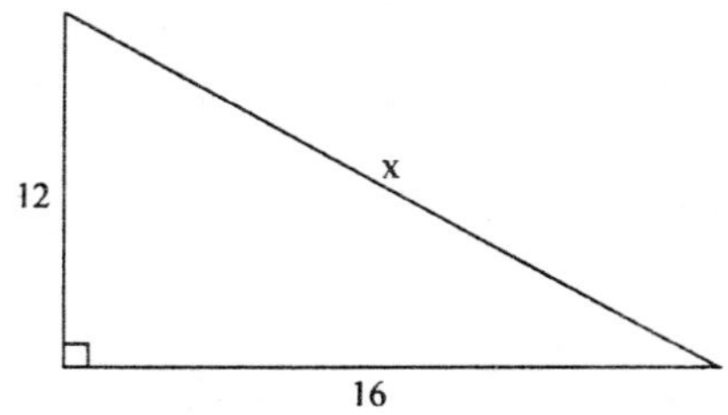

b.

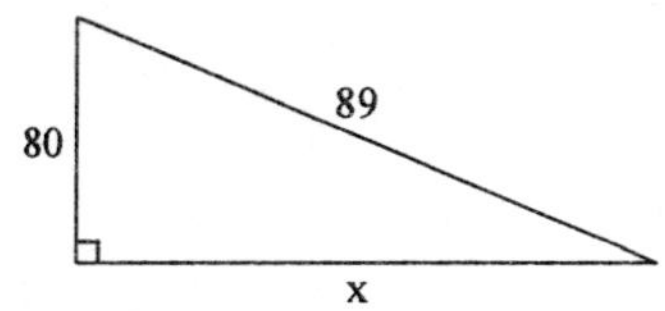

c.

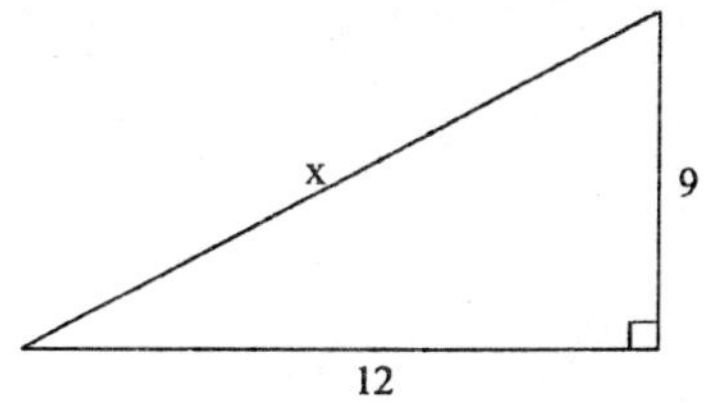

d.

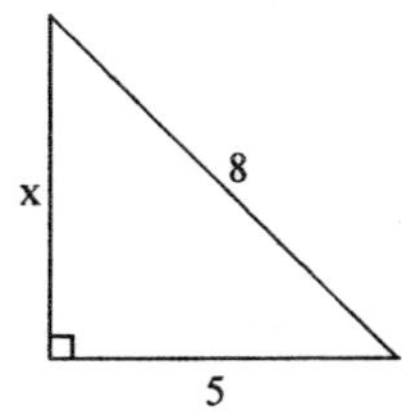

e.

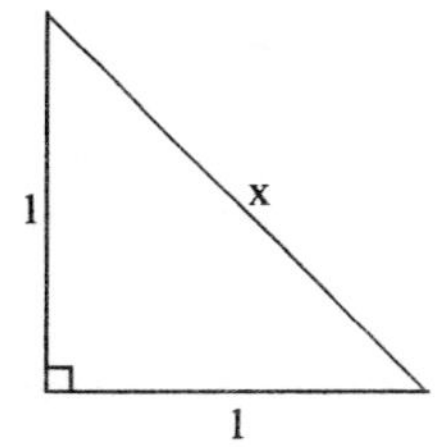

f.

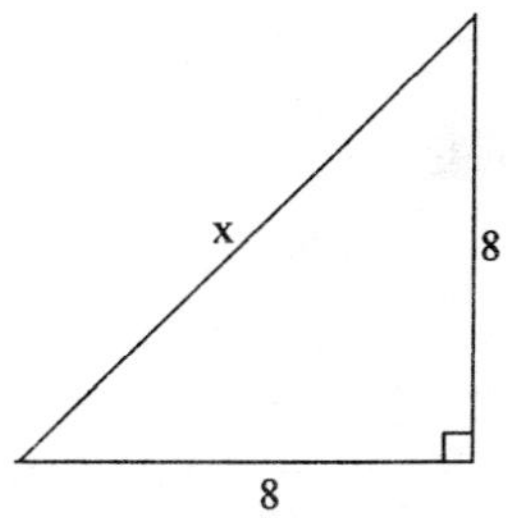

g.

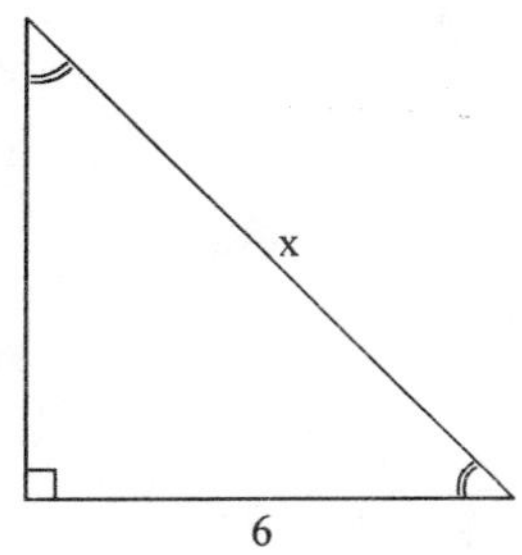

h.

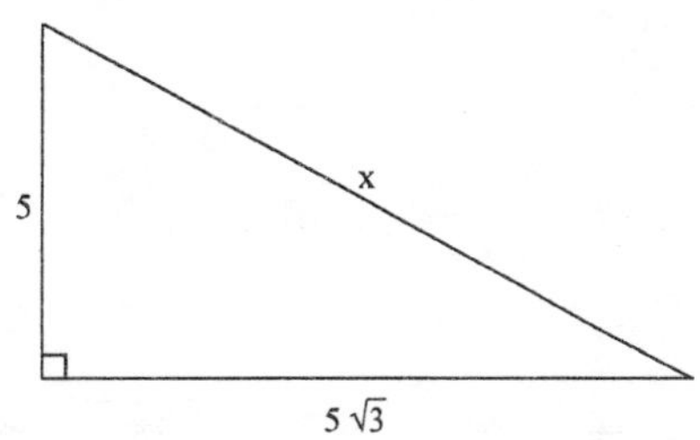

i.

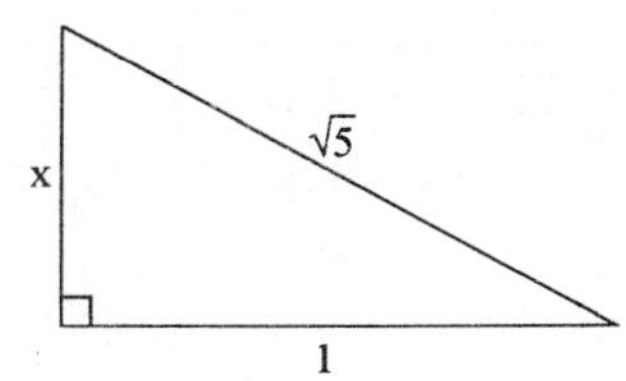

2. Which of the following sets of numbers could be the lengths of the sides of a right triangle?

a. 21, 28, 35

b. 24, 45, 57

c. $\sqrt{2}$, $\sqrt{3}$, $\sqrt{5}$

d. 16, 30, 34

e. 1/2, 3/4, $\sqrt{13}$ /4

3. The hypotenuse of a right triangle is 17 cm and one leg is 15 cm. Find the area of the triangle.

4. The hypotenuse of a right triangle is 9 units and one of its legs is 5 units long. Find the area of the triangle.

5. Find the length of each of the line segments in the following diagram. Which ones have rational lengths? Which ones have irrational lengths?

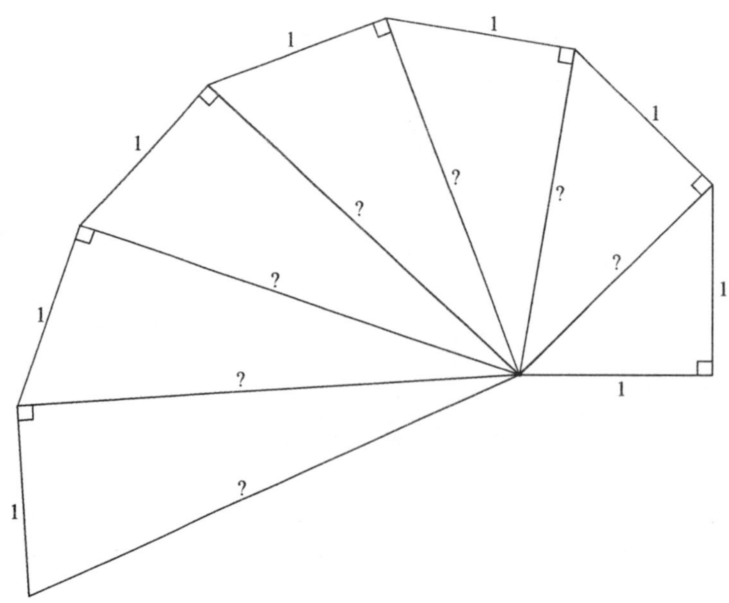

6. A ladder 25 ft long is placed against the vertical wall of a building. The foot of the ladder is 8 feet from the building. How high on the wall does the ladder reach? Round your answer to the nearest whole number.

7. A ladder 10 ft long is leaning against the vertical wall of a house. How high on the wall of the house will the ladder reach when the bottom of the ladder is 4 feet from the house? Round your answer to the nearest whole number.

8. A ladder 20 ft long is leaning against the vertical wall of a house. How high on the wall of the house will the ladder reach when the bottom of the ladder is 12 feet from the house? Round your answer to the nearest whole number.

Section 4.3 Surface Area

The **surface area** of a solid figure bounded by polygons is the sum of the area of its faces (faces are the polygons that form the figure). For example, a cube with edges of length 2cm has a surface area of 24 square centimeters. Notice that the cube has 6 faces, and each of these faces has an area of 4 square centimeters. Six times 4 square centimeters equals 24 square centimeters.

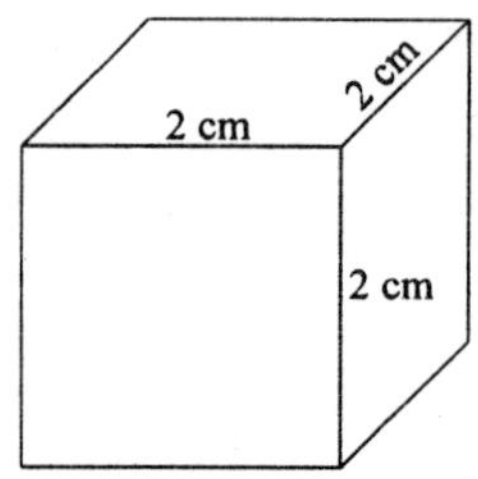

Here is a way to find the total surface area of the figure in the illustration below. If we cut the figure and flatten it out as shown, we see that its faces are polygons that we have studied; 2 triangles and 3 rectangles.

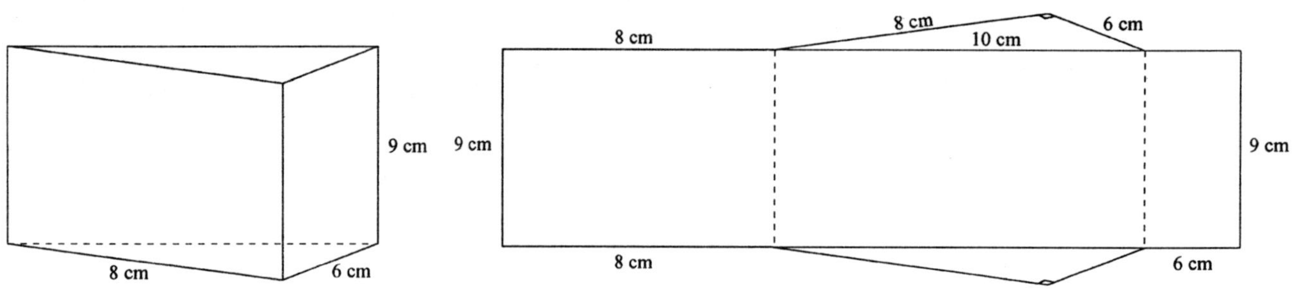

Because the two triangles are congruent right triangles, we use the Pythagorean Theorem to find the length of their hypotenuse. We determine that it has a length of 10cm. Since there are two of the triangles, we find the area of one of them, then multiply by two.

The area of one triangle is 1/2(6)(8) or 24 square centimeters.
The area of the two triangles is 2(24) or 48 square centimeters.

The other three faces are rectangles with the following areas:

$(10 \cdot 9)$ square centimeters
$(8 \cdot 9)$ square centimeters
$(6 \cdot 9)$ square centimeters

Therefore the total area of the side (lateral) faces of the figure is the sum of these areas which is 216 square centimeters.

The total surface area for the solid figure is then the sum of the areas of the triangles plus the sum of the areas of the rectangles.

The total surface area is $48\text{cm}^2 + 216\text{cm}^2$ or $264\ \text{cm}^2$.

In some cases, the surface area of a solid object with curved sides, like a cylinder, can be found in the same manner. You will be asked to find the surface area of such figures in your exercises.

Exercises 4.3

(Figures are not drawn to scale)

1. Find the surface area of each solid figure:

a.

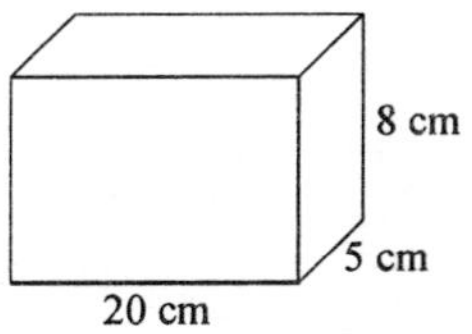

b.

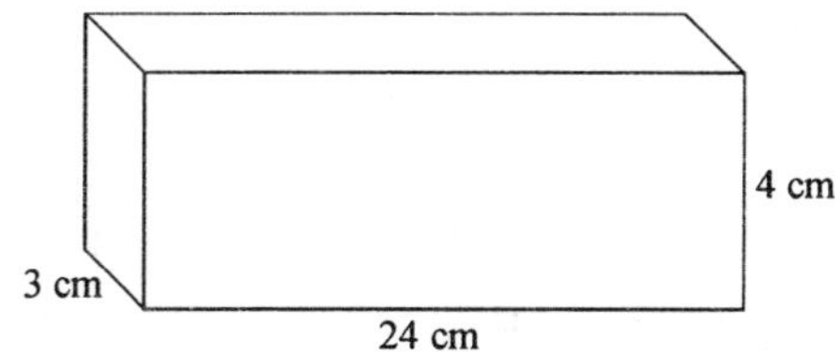

c.

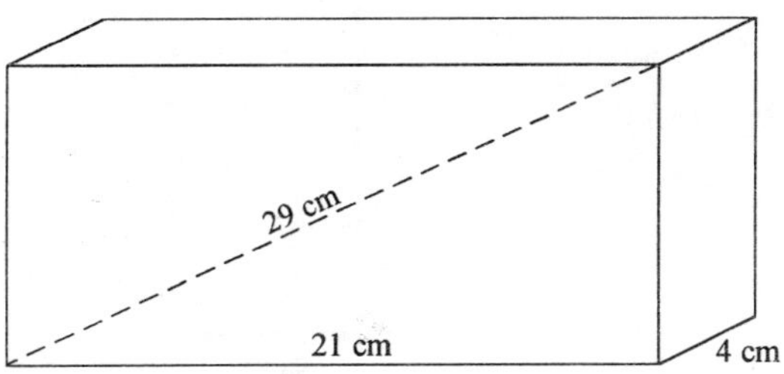

d.

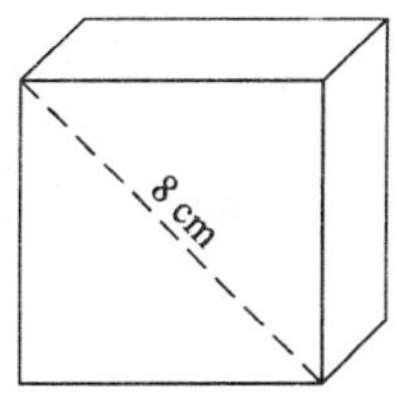

cube

e.

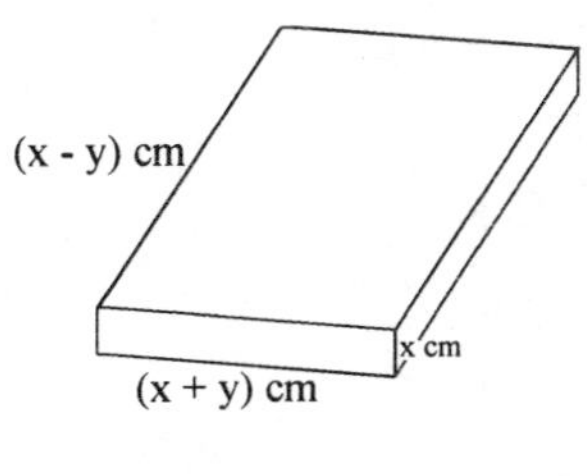

f.

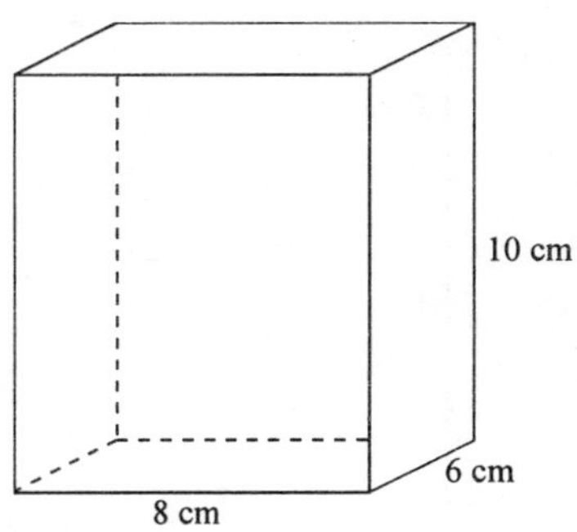

2. Find the surface area of each of the solids:

a.

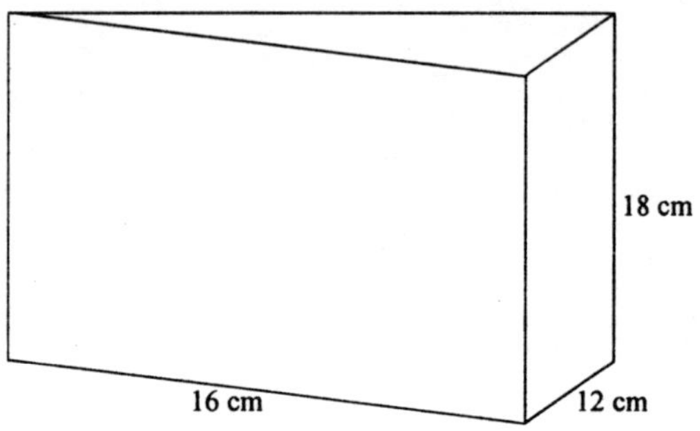

b.

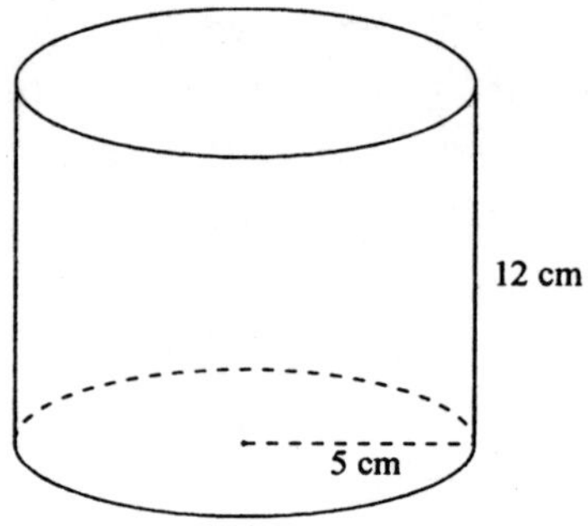

c.

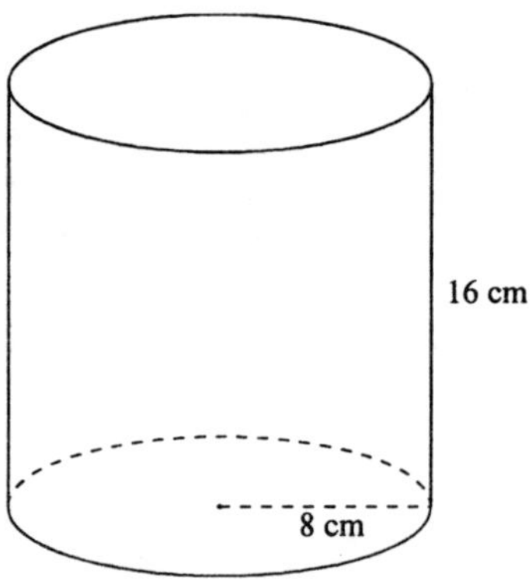

d.

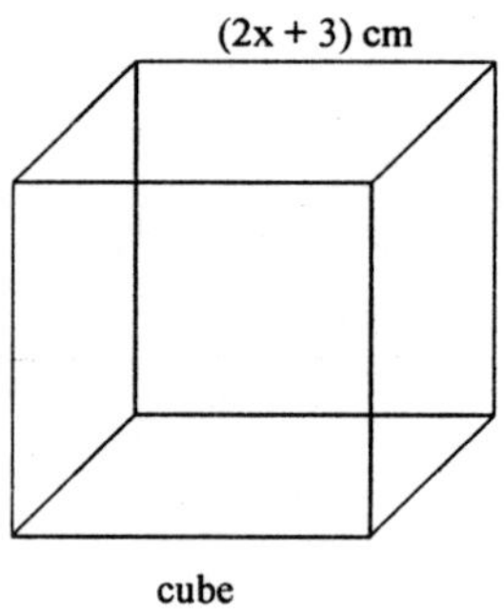

e.

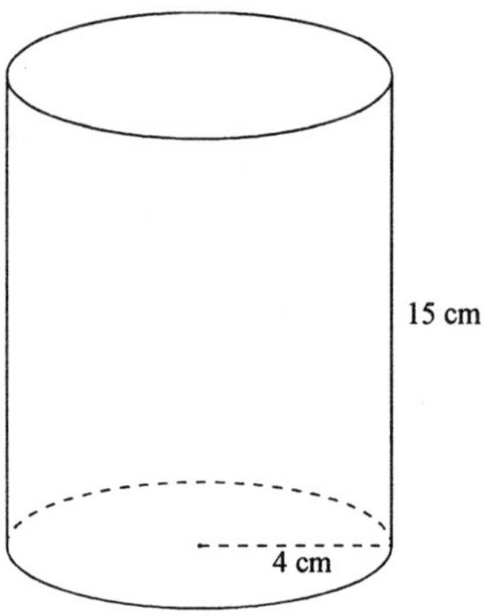

f.

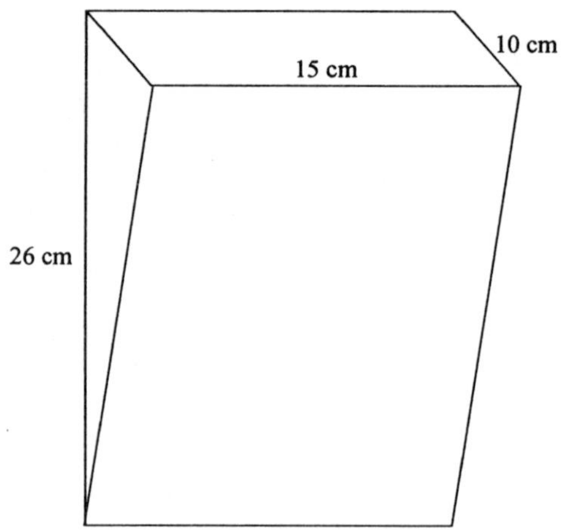

Chapter Four Highlights

Vocabulary

perimeter
circumference
pi
area
square units
parallelogram
bases of a parallelogram
height or altitude of a parallelogram
trapezoid
bases of a trapezoid
height of a trapezoid
base of a triangle
height of a triangle
right triangle
hypotenuse
legs of a right triangle

surface area

Conclusions

The perimeter of a rectangle with length L and width W is $P = 2L + 2W$
The perimeter P of a square with side of length S is $P = 4S$
Pi is the ratio of the circumference of a circle to its diameter.
The circumference C of a circle with radius r and diameter d is $C = 2\pi r$ or $C = \pi d$
The area A of a rectangle with width W and length L is $A = LW$
The area A of a square with side S is $A = S^2$
The area A of a parallelogram with height h and base b is $A = bh$
The area A of a trapezoid with height h and bases b_1 and b_2 is $A = 1/2\ h\ (b_1 + b_2)$
In a right triangle the square of the hypotenuse equals the sum of the squares of the legs of the triangle, $c^2 = a^2 + b^2$, where a and b represent the legs of the triangle and c represents the hypotenuse of the triangle.
A triangle must be a right triangle if the sum of the squares of two sides of the triangle equals the square of the remaining side.
The surface area of a solid figure bounded by polygons is the sum of the area of its faces (faces are the polygons that form the figure).

Chapter Five

Introduction to the Deductive Process

Objectives

Distinguish between inductive and deductive reasoning
List and explain the components of a deductive system
Write conditional statements in an if-then form
Form the converse, inverse and contrapositive of a conditional statement
Explain the relationship that a conditional statement has to its converse, inverse and contrapositive
Use Venn Diagrams to represent conditional statements
Determine the validity of a conclusion based upon a major and minor premise

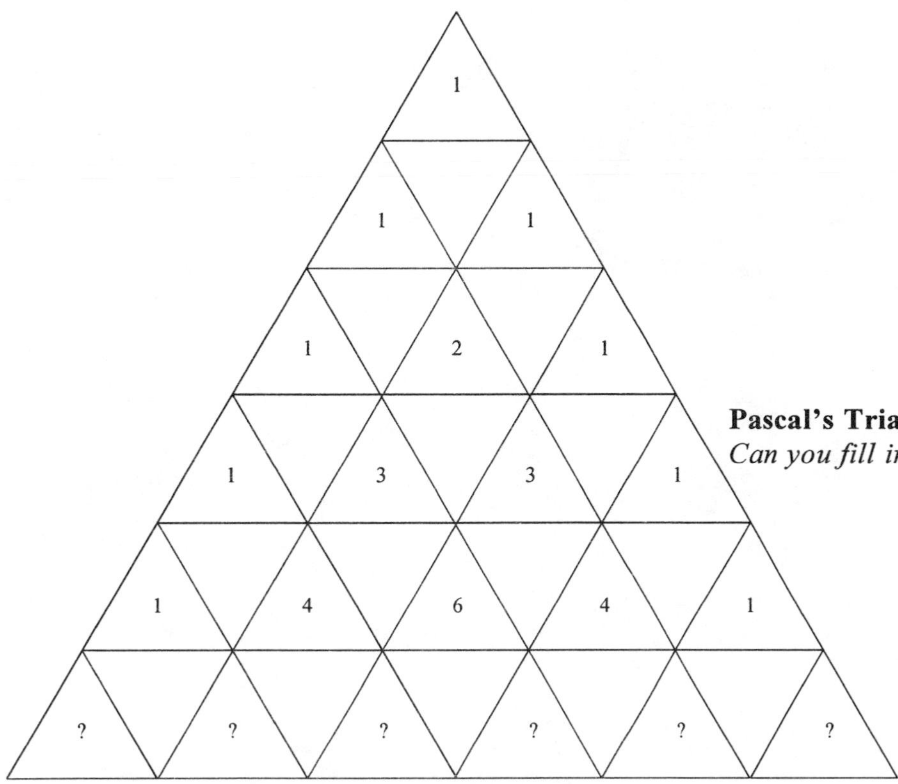

Pascal's Triangle
Can you fill in the next row?

Section 5.1 Making Conclusions

We have spent much of our time in this course making conclusions about geometric figures and relationships. Many of our conclusions were made primarily on the basis of **inductive reasoning** (the process of reaching a conclusion after observing a pattern). A little reflection, however, will convince us that this method can lead us to incorrect conclusions.

Consider the two line segments below. Without measuring, which line segment would you say is longer, or $\overline{AB}$ or $\overline{CD}$?

At first glance, many people will conclude that $\overline{CD}$ is the longer of the two segments. Actually both segments have the same length. This example and numerous other examples like it show us that our perception can be misleading.

Consider now the following exercise involving inductive reasoning. We want to find the number of regions formed by the line segments determined by connecting N points placed on a circle (Where N is a counting number).

We begin with N = 1

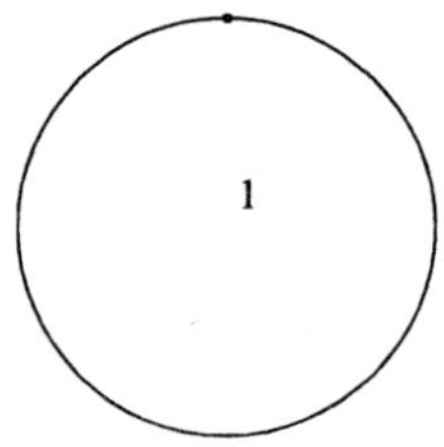

1 point 1 region

Next we try N = 2

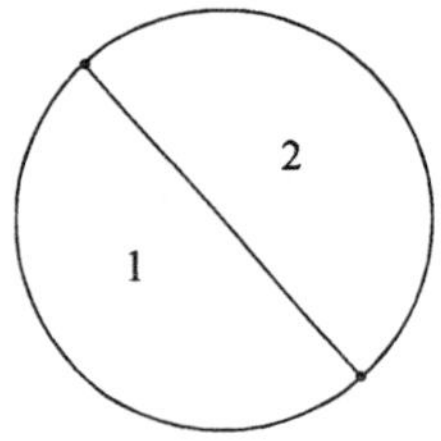

2 points 2 regions

Next we try N = 3

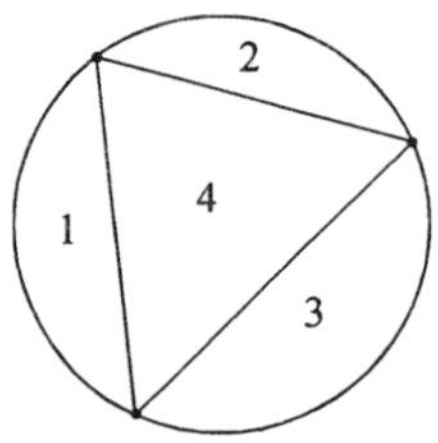

3 points 4 regions

Next N = 4

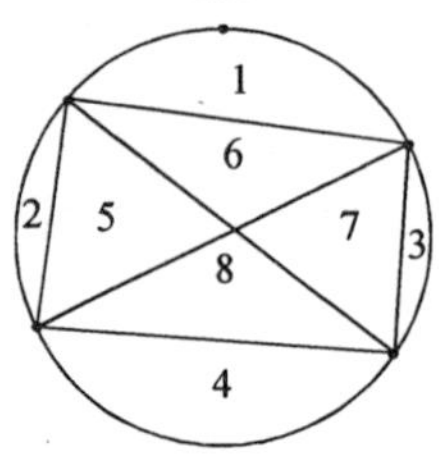

4 points 8 regions

Then N = 5

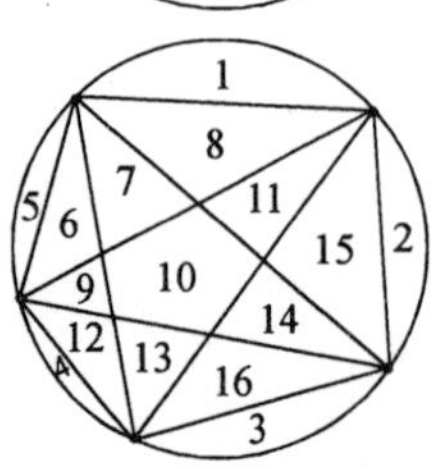

5 points 16 regions

Organizing these results in a chart, we observe a pattern:

Number of Points	1	2	3	4	5	6	
Number of Regions	1	2	4	8	16	?	
	2^0	2^1	2^2	2^3	2^4	?	

Can you use the information in the chart to make a prediction for 6 points?

A common conjecture is 2^5, or 32. Surprisingly, the correct answer is not 32! Try counting them! This example also indicates that making a conclusion on the basis of inductive reasoning can be risky business. Rest assured though that the conclusions we made earlier on the basis of inductive reasoning are perfectly valid!

How then, you may ask, are we to make conclusions about geometric figures and relationships if some of the methods we have used so far are not always reliable? Well, the mathematician uses a method that is called **deductive reasoning.** We have already used this type of reasoning in making some of our conclusions. Now we will investigate this way of making conclusions, and contrast it with inductive reasoning, the reasoning used by the scientist.

Inductive reasoning involves examining phenomena, observing a pattern, and then making a generalization. Deductive reasoning, on the other hand, involves accepting a generalization and then applying it to a specific phenomenon or situation. For example, the mathematician accepts the statement, "Two points determine a line." On the basis of this generalization, the mathematician concludes that if line l contains points A and B, no other line can contain those two points.

Of course, deductive reasoning also can lead to false conclusions. If a generalization is false, then any specific conclusion based upon it may also be false. Even with this potential for error, mathematicians have used the process of deductive reasoning to build up a powerful body of knowledge. Keep in mind that the generalizations accepted by mathematicians are not selected carelessly. They are believed to be "self-evident truths."

We must not conclude that inductive reasoning has no place in the world of mathematics. Quite the contrary, mathematicians use inductive reasoning in much the same way that we used it in the first part of the course to make conjectures.

Once conjectures are formulated, however, they must undergo the process of deductive reasoning before they are accepted into the body of knowledge called mathematics.

Exercises 5.1

1. Compare inductive and deductive reasoning.
2. What is the risk involved when inductive reasoning is used to make a conclusion?
3. What is the risk involved when deductive reasoning is used to make a conclusion?
4. Explain how theories about the position of the earth in the universe have changed throughout history. What type of reasoning was used to formulate these theories?
5. Explain how inductive reasoning is used to determine the probability that certain events will occur?
6. The mathematician does not use inductive reasoning to prove statements, however, this does not mean that the mathematician never uses this type of reasoning. Can you think of ways in which inductive reasoning may be used by the mathematician?

Section 5.2 The Components of a Deductive System

In this section, we will investigate the methods that are used in mathematics to make conclusions. We first observe that every system of study in mathematics is based upon the following components:

Undefined Terms
Definitions
Postulates (or Axioms)
Theorems

Undefined Terms: No attempt is made to define all the terms used in a course of study in mathematics. It is assumed that the meanings of some terms are so fundamental that they are clear to everyone. Undefined terms are necessary if circular definitions are to be avoided. Some of the undefined terms used in mathematics are point, line, plane, set and number.

Definitions: A **definition** is simply an agreement that a word or group of words may be used in place of another word or group of words. Definitions are generally used to simplify our language when we wish to refer to certain geometric figures and relationships.

Postulates: A **postulate** is a statement that we accept as being true without proof. It may be thought of as a self-evident truth, that is, a truth that we accept solely on the basis of our common experiences. An example of a postulate is, "Two points determine a line." Although it may be helpful to think of a postulate as a "self-evident truth," the actual acceptance of a postulate by mathematicians is independent of physical reality.

An **axiom** is also a statement that is accepted as true without proof. Sometimes an axiom is considered to be more general than a postulate, that is, it expresses a truth for all of mathematics in general; while a postulate is an accepted truth for a particular course of study in mathematics. For our purposes, we will simply refer to all self-evident truths as postulates.

Generally in mathematics, the number of postulates is kept to a minimum. Furthermore, all postulates that are accepted in a course of study in mathematics must be consistent, that is, no postulate can contradict another postulate.

Theorems: A **theorem** is a statement that must be proved before it is accepted as true. Once a theorem has been proved true, it may be used to prove other theorems.

Undefined terms, definitions, postulates and theorems form the foundation for every course of study in mathematics. A system of study that contains these components is called a **deductive system**. Basically such a system involves the use of relatively simple statements to derive more complex statements through a process called deductive logic. Although the components of a deductive system form every course of study in mathematics, traditionally, they are emphasized in the study of geometry. This is the case because Euclid, the Greek mathematician, for whom this course is named, organized much of the geometry known in his day (about 580 B.C.), into the first deductive system of study in mathematics.

Exercises 5.2

1. List the components of a deductive system.

2. What is a definition?

3. Why do we use undefined terms in mathematics?

4. How does a theorem differ from a postulate?

5. The following axioms are introduced in elementary algebra. Can you identify them?
   a. If a and b are real numbers, then $a + b = b + a$.

   b. If a and b are real numbers, then $a(b + c) = ab + ac$.

   c. If a is a real number, then $a \cdot 1 = a$.

6. Is an axiom a true statement?

7. How can axioms and postulates be compared to the rules for playing chess or baseball?

8. How can axioms and postulates be compared to the rules given in a Holy Book for a religion?

9. How can a group of postulates and axioms be compared to a country's constitution?

Section 5.3 Conditional Statements

A **deductive proof** is the process of using undefined terms, definitions, postulates and theorems that have been proved true to reach a conclusion. The statements that are crucial in this process are called conditional statements:

Conditional statement: A statement that can be written in an if-then form.
The following notation symbolizes the *format of a conditional statement*:

If A then B (Note: A and B represent clauses)

The A clause is called the **hypothesis** of the statement.

The B clause is called the **conclusion** of the statement.

We may also symbolize a conditional statement using the following expressions:

A → B	or	***A implies B***

EXAMPLE 1:

If *a figure is a square* (**hypothesis**) then *it is a polygon* (**conclusion**)

EXAMPLE 2:

I cry when it rains.

Rewrite in the if-then form: If *it rains* (**hypothesis**) then *I cry.* (**conclusion**)

EXAMPLE 3:

No square is a circle.

Rewrite: If *it is a square* (**hypothesis**) then *it is not a circle* (**conclusion**)

or Rewrite: If *it is a circle* (**hypothesis**) then *it is not a square.* (**conclusion**)

EXAMPLE 4:

A square is a rhombus.

Rewrite: If *it is a square* (**hypothesis**) then *it is a rhombus.* (**conclusion**)

EXAMPLE 5:

A quadrilateral is a rhombus if it has all equal sides.

Rewrite: If *it has all equal sides* (**hypothesis**)

then *a quadrilateral is a rhombus.* (**conclusion**)

We may also use circles called **Venn Diagrams** to represent conditional statements:

If A then B.

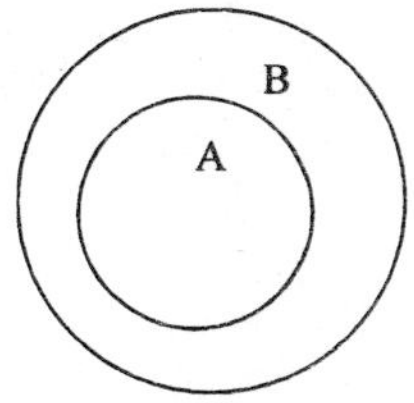

Objects that satisfy the condition specified by A must also satisfy the condition specified by B.

EXAMPLE:

Squares are polygons.

Rewrite: If it is a square then it is a polygon.

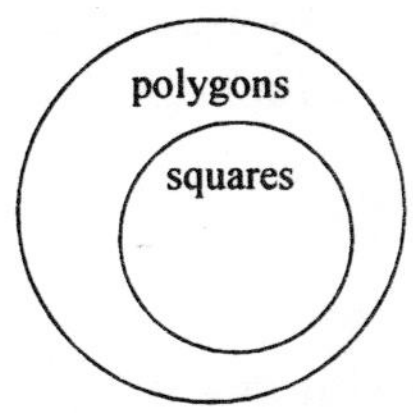

outside circle contains all polygons
inside circle contains all squares

Related Conditional Statements

Other statements that are related to a conditional statement are as follows:

Original Conditional Statement: If A then B

Example: If *it is a square* (A) then *it is a polygon.* (B)

1. **Converse:** The statement obtained by interchanging the hypothesis and the conclusion of the original conditional statement.

   If **B** then **A**

   If *it is a polygon* (**B**) then *it is a square.* (**A**)

2. **Inverse:** The statement obtained by negating the hypothesis and the conclusion of the original conditional statement.

   If **not A** then **not B**

   If *it is not a square* (**Not A**) then *it is not a polygon.* (**Not B**)

3. **Contrapositive**: The statement obtained by interchanging and negating the hypothesis and the conclusion of the original conditional statement.

   If **Not B** then **Not A**

   If *it is not a polygon* then *it is not a square.*
   Not B **Not A**

We should note the following:

a. *The original statement and its contrapositive are logically equivalent.* We may therefore replace a statement with its contrapositive any time we wish without changing the meaning of the original statement. By **logically equivalent** we mean that when one statement is true, the other statement is also true; and when one statement is false, the other statement is also false.
b. *The converse of a definition is always true.* A definition is faulty if its converse is not true. "A dog is an animal," is not a good definition for a dog, because the converse of the statement is not true.
c. *A statement and its converse can often be replaced by one statement.* Consider the following statement and its converse:
   1. A number is even if it is divisible by 2.
   2. A number is divisible by 2 if it is even.

   We may replace these two statements with the following statement:
   A number is even **if and only if** is it is divisible by 2.
   We may then replace the words, *"if and only if,"* with the *notation, "iff."*
   A number is even **iff** it is divisible by 2.
   In other words,

A statement containing the words, "if and only if," is composed of two statements: the statement as written and its converse.

Exercises 5.3

1. Write each statement in an if-then form (if necessary); then identify the hypothesis and conclusion of each:
   a. If two lines form a linear pair then they are supplementary.
   b. Vertical angles are equal.
   c. A square is a rectangle.
   d. No line has an endpoint.
   e. The base angles of an isosceles triangle are equal.
   f. Two lines are perpendicular if they form a right angle.
   g. No triangle has two right angles.
   h. If a quadrilateral is equiangular then it is a rectangle.
   i. A quadrilateral is a rhombus if it is equilateral.
   j. If two triangles are congruent, then they are similar.

2. Write the converse, inverse and contrapositive for each statement:

   a. If two angles form a linear pair then they are supplementary.

   b. If a triangle is equiangular then it is equilateral.

   c. An angle is acute if its measure is less than 90.

   d. A circle is not a polygon.

3. Write the converse, inverse and contrapositive of each statement; then indicate whether each is true or false.

   a. A banana is a fruit.

   b. If a figure is a rectangle then it is a polygon.

   c. If a number is even then it is divisible by 2.

   d. If a triangle is isosceles then it is equilateral.

4. "Iff" is an abbreviation for what phrase?

5. Write the two statements implied by the following statement: "A polygon is regular iff it is equiangular and equilateral."

6. Respond true or false:

   a. A statement and its converse are logically equivalent.
   b. The inverse and converse of a statement are logically equivalent.
   c. The contrapositive of a statement is false if the original statement is false.
   d. The converse of a statement is always false.
   e. The converse of a valid definition is always true.
   f. A postulate is a statement that is accepted as true without proof.
   g. An axiom is a statement that we prove.

7. Explain why the given statements are poor definitions for the italicized words.
   a. A *cat* is a mammal.
   b. A *bird* is an animal that flies.
   c. A *prime number* is an odd number greater than two.
   d. A *rational number* is a fraction.
   e. *Milk* is a white liquid.

8. Use Venn Diagrams to represent the given conditional statement.
   a. If a number is rational, then it is real.
   b. If a number is natural, then it is an integer.
   c. People who live in Nigeria live in Africa.
   d. People who live in the U.S.A. live in North America.

Section 5.4 Deductive Logic

Now we will examine the necessary conditions for making conclusions in a deductive system. Consider the following two statements:

1. If a quadrilateral is a rhombus then it is a polygon.
2. Quadrilateral PQRS is a rhombus.
3. Quadrilateral PQRS is a polygon.

If we accept the first two statements as being true, we must conclude that the third statement is also true. We say that the third statement is inescapable. Such an argument is called a **deductive argument.** *Symbolically a deductive argument can be represented in the following way:*

1. If A then B	**(major premise)**	**A and B are clauses**
2. A	**(minor premise)**	
3. B	**(deduction or conclusion)**	

In a deductive argument, if we accept the major premise, and the minor premise as an instance of the hypothesis of the major premise, then we must also accept the conclusion of the major premise as applied to the object or relationship expressed in the minor premise. In the following example, we see how such an argument is applied:

We accept:
1. If *a triangle is equiangular* then *it is isosceles.* (major premise)

We accept:
2. $\triangle$ **PQR** *is equiangular.* (minor premise)

We deduce:
3. $\triangle$ **PQR** *is isosceles.* (deduction)

Notice that the deduction is an application of the conclusion of the major premise to $\triangle$ **PQR.**

We may also make a deduction or conclusion when we are given the hypothesis of the contrapositive (of the major premise) as an instance of the minor premise. Symbolically we write:

1. If A then B	**(major premise)**
2. Not B	**(minor premise)**
3. Not A	**(conclusion or deduction)**

Notice that this argument is equivalent to the following argument:

1. If *not B* then *not A.* (contrapositive of major premise)
2. *Not B* (hypothesis of the contrapositive)
3. *Not A* (deduction)

In short, *we can make a valid deduction in an argument when the minor premise is an instance of either:*

1. *The hypothesis of the major premise,* or
2. *The hypothesis of the contrapositive of the major premise.*

EXAMPLE

We accept:

1. If a polygon is a square then it is a rhombus. (major premise)

We accept:

2. Polygon MNOP is not a rhombus. (minor premise)

We deduce:

3. Polygon MNOP is not a square. (deduction)

This deduction is an application of the conclusion of the contrapositive to the polygon referred to in the minor premise.

In each of the following examples, we determine whether a valid deduction may be made from the major and minor premises.

EXAMPLE 1

1. If *a polygon is a square* then *it is a rhombus.* (major premise)
2. *Polygon* **KLMN** *is a square.* (minor premise)

We can make a valid deduction. The minor premise is an instance of the hypothesis of the major premise.

3. *Polygon* **KLMN** *is a rhombus.*

EXAMPLE 2

1. If *a polygon is a square* then *it is a rhombus.* (major premise)
2. *Polygon* **KLMN** *is not a rhombus.* (minor premise)

We can make a valid deduction. The minor premise is an instance of the hypothesis of the contrapositive. We deduce:

3. *Polygon* **KLMN** *is not a square.*

EXAMPLE 3

1. If *a polygon is a square* then *it is a rhombus.* (major premise)
2. Polygon KLMN is a rhombus. (minor premise)

We can not make a valid conclusion. The minor premise is an instance of the hypothesis of the converse of the major premise. Making a deduction would result in what is called a converse error.

EXAMPLE 4

1. If *a polygon is a square* then *it is a rhombus.* (major premise)
2. Polygon KLMN is not a square. (minor premise)

We can not make a valid deduction. The minor premise is an instance of the hypothesis of the inverse of the major premise. Making a deduction would result in what is called an inverse error.

EXAMPLE 5

1. *This thing is an orange.* (minor premise)
2. If *it is an orange* then *it is a fruit.* (major premise)

We can make a valid deduction. The minor premise is an instance of the hypothesis of the major premise. Notice that the order in which the major and minor premises appear is not significant. We deduce: *This thing is a fruit.*

We make valid conclusions about geometric figures and relationships by using deductive reasoning to write proofs. These proofs generally consist of chains of deductive arguments. Here is an example of such a proof:

1. *∠ A and ∠ B form a linear pair* (*minor premise*)
2. *If two angles form a linear pair then they are supplementary* (*major premise*)
3. *∠ A and ∠ B are supplementary.* (*our deduction*)

This deduction now becomes the minor premise for the next argument:

4. *If two angles are supplementary then the sum of their measures is 180.* (*major premise*)
5. *The sum of the measures of ∠ A and ∠ B is 180, or* $\angle A + \angle B = 180$ (*our deduction*)

We have used deductive reasoning to make the following argument:

If A and B form a linear pair then $\angle A + \angle B = 180$.

Exercises 5.4

1. When possible, for each pair of statements, make a valid deduction. If no valid deduction is possible, write "no deduction."

   a. If two angles form a vertical pair then they have equal measures.
   b. ∠ A and ∠ B have equal measures.

   a. If two angles form a vertical pair then they have equal measures.
   b. ∠ A and ∠ B do not have equal measures.

   a. If two angles form a vertical pair then they have equal measures.
   b. ∠ A and ∠ B do not form a vertical pair.

   a. If two angles form a vertical pair then they have equal measures.
   b. ∠ A and ∠ B form a vertical pair.

2. Determine whether each argument is valid:

a. A square is a rectangle.
b. Polygon ABCD is a rectangle.
c. Polygon ABCD is a square. (conclusion)

a. No line has an endpoint.
b. This figure has an endpoint.
c. This figure is not a line. (conclusion)

a. If two angles form a linear pair then they are supplementary.
b. ∠ A and ∠ C form a linear pair.
c. $\angle A + \angle B = 180$ (conclusion)

Write a proof, that is easy to follow, of each italicized conditional statement by rearranging the set of statements that follow the italicized statement.

3. *If I eat vegetables, I will be rich.*

a. If I eat vegetables, I will be healthy.
b. If I have energy, I can work hard.
c. If I earn a lot of money, I will be rich.
d. If I am healthy, I will have energy.
e. If I work hard, I can earn a lot of money.

4. *If a number is used for counting, then it is real.*

a. If a number is natural, then it is an integer.
b. If a number is rational, then it is real.
c. If a number is not rational, then it is not an integer.
d. If a number is used for counting, then it is natural.

5. *If I fall in love, I will live forever.*

   a. If I fall in love, I will be happy.
   b. If I do not complain, I will be at peace.
   c. If I am happy, I will love life.
   d. If I love life, I will not complain.
   e. If I am at peace, I will live forever.

6. *If I study geometry everyday I will be happy.*

   a. If my test scores are not high, I do not understand geometry.
   b. If my test scores are high, I will be happy.
   c. If I study geometry everyday, then I will understand geometry.

Chapter 5 Highlights

Vocabulary

inductive reasoning
deductive reasoning
undefined terms
definitions
postulates
axioms
theorems
deductive proof
conditional statement
hypothesis
conclusion
Venn Diagrams
converse
inverse
contrapositive
logically equivalent
if and only if (iff)
deductive argument
major premise
minor premise
deduction or conclusion

Required Skills

Recognize notation for "A implies B" and "if and only if."
Distinguish between inductive and deductive reasoning
List and explain the components of a deductive system
Write conditional statements in an if-then form
Form the converse, inverse and contrapositive of a conditional statement
Explain the relationship that a conditional statement has to its converse, inverse and contrapositive
Use Venn Diagrams to represent conditional statements
Determine the validity of a conclusion based upon a major and minor premise

Conclusions

Deductive reasoning is used to make conclusions in mathematics.
Every course of study in mathematics is composed of undefined terms, definitions, postulates and theorems.
The original statement and its contrapositive are logically equivalent.
The converse of a definition must always be true.
We can make a valid deduction in an argument when the minor premise is either:

1. The hypothesis of the major premise, or
2. The hypothesis of the contrapositive of the major premise.

Chapter Six

Using the Deductive Process

Objectives

Recognize and use terms and notations related to deductive proofs
Use postulates of equality and congruence, existence postulates and definitions in deductive proofs
Understand and prove basic theorems about angles, isosceles triangles and equilateral triangles
Use theorems to prove other theorems in deductive proofs

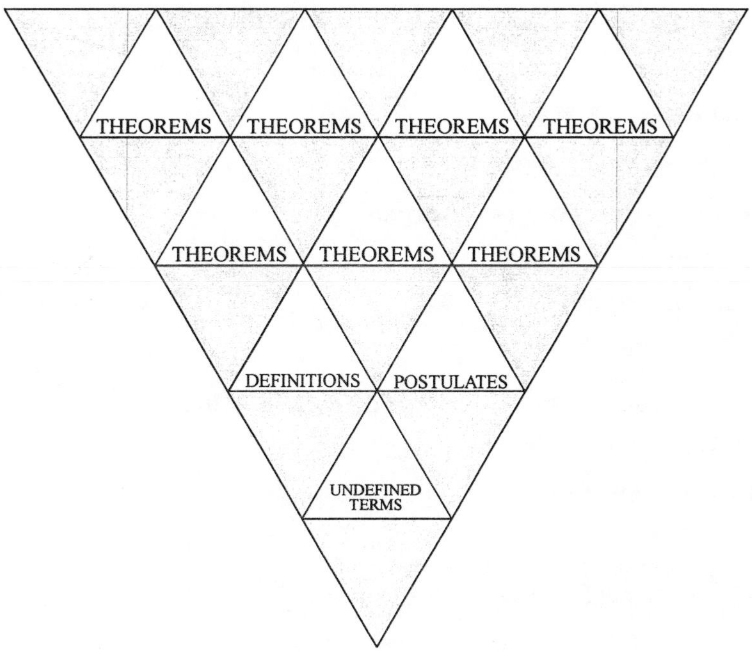

Section 6.1 Introduction to Deductive Proofs

In the last chapter, we learned how to use major and minor premises to make deductions. In this chapter, we use those procedures to form *arguments* about geometric relationships which we will call "proofs." As we proceed, we must keep in mind that every conclusion we make must be supported with a deductive argument. In other words, whenever we make a conclusion from deduced or given information, that conclusion and the information must be related by some major premise. For our purposes, major premises will be definitions, postulates or axioms, and theorems that already have been proved using the deductive process. We refer to these major premises as "reasons" when they are used to make a deduction, or to support statements in proofs.

At this time, it may be a good idea to review some of the postulates that we will use in this course. These postulates have been separated into two groups. The first group contains general postulates that you probably have encountered in other courses. The second group contains postulates that deal with geometric concepts. *All variables in these postulates represent positive real numbers.*

General Postulates

1. **A number may be substituted for its equal in any expression.**

   If $a = b$ then a may be replaced by b, and b may be replaced by a in any expression.

   Substitution Postulate of Equality

   Example: If $x = y$ and $x + 2y = 4$, then we may replace the x in the second expression by y obtaining $y + 2y = 4$

2. **Any number equals itself. $a = a$**

   Reflexive Postulate of Equality

3. **The members of an equation may be interchanged.**

   If $a = b$ then $b = a$

   Symmetric Postulate of Equality

   Example: If $x + y = 180$ then $180 = x + y$

4. **Numbers equal to the same number are equal to each other.**

   If $a = b$ and $b = c$ then $a = c$

   Transitive Postulate of Equality

   Example: If $\angle A + \angle B = 180$ and $\angle C + \angle B = 180$ then $\angle A + \angle B = \angle C + \angle B$

5. **If equal numbers are added to equal numbers, the sums are equal.**

   If $a = b$ and $c = d$ then $a + c = b + d$

   Addition Postulate of Equality

   Example:

If	$x + y$	=	2
and	$x - y$	=	6
then	$2x$	=	8

6. **If equal numbers are subtracted from equal numbers, the differences are equal.**

If $a = b$ and $c = d$ then $a - c = b - d$

Subtraction Postulate of Equality

Example: If $\angle A + \angle B = 180$
and $\angle B = 90$
then $\angle A = 90$

7. **If equal numbers are multiplied by equal numbers, the products are equal.**

If $a = b$ and $c = d$ then $ac = bd$

Multiplication Postulate of Equality

Example: If $1/2 \angle A = 45$ and $2 = 2$
then $\angle A = 90$

8. **If equal numbers are divided by nonzero equal numbers, the quotients are equal.**

If $a = b$ and $c = d$ where c and d are not zero, then $a/c = b/d$

Division Postulate of Equality

Example: If $2 \angle A + 2 \angle B = 180$ and $2 = 2$
then $\angle A + \angle B = 90$

Before a postulate is applied and its conclusion reached, we must show that the requirements of all parts of its hypothesis have been met. Notice that postulates 5, 6, 7, and 8 have hypotheses that require two equations (if $a = b$ and $c = d$). In the examples for postulates 7 and 8, the Reflexive Postulate was used to generate a second equation needed to satisfy the two part hypotheses of these postulates.

Other general postulates that we will find useful in some of our proofs deal with powers, exponents and inequalities. Applications of some of these postulates will be seen later as we proceed with our work.

9. **Like powers of equal numbers are equal.**

If $a = b$ then $a^n = b^n$

Powers Postulate of Equality

Example: if $x = y$ then $x^2 = y^2$

10. **Like roots of equal numbers are equal.**

If $a = b$ then $\sqrt{a} = \sqrt{b}$

Roots Postulate of Equality

Example: If $x = y$ then $\sqrt{x} = \sqrt{y}$

11. **If $a + b = c$, then $c > a$ and $c > b$**

Whole Greater Than Part Postulate

This postulate allows us to make the transition from an equality to an inequality.

Example:

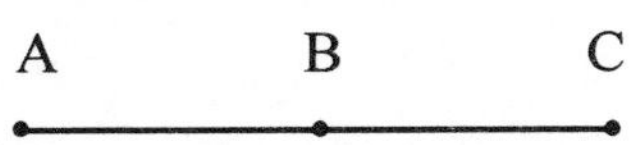

Since $AB + BC = AC$, we may write $AC > AB$ or $AC > BC$

12. **If a and b are real numbers, then exactly one of the following statements must be true:**

$a = b$ or $a > b$ or $a < b$

Trichotomy Postulate of Inequality

13. **If $a > b$ and $b > c$ then $a > c$**
    Transitive Postulate of Inequality
14. **If $a > b$ then $a + c > b + c$**
    Addition Postulate of Inequality
15. **If $a > b$ and $c > 0$ then $ac > bc$**
    If $a > b$ and $c < 0$ then $ac < bc$
    Multiplication Postulate of Inequality

Now let us look at some geometric postulates that we will be using in this chapter and other chapters. You are already familiar with these postulates because we studied them in the first part of the course. We now refer to these statements as postulates.

1. **If points A, B and C are collinear with B between A and C, then $AB + BC = AC$.**
   Whole Equals Sum of Parts Postulate
2. **If $\overrightarrow{BD}$ is between $\overrightarrow{BA}$ and $\overrightarrow{BC}$, then $\angle ABD + \angle DBC = \angle ABC$.**
   Whole Equals Sum of Parts Postulate
3. **If two angles form a linear pair, then they are supplementary.**
   Linear Pair Postulate
4. **A line contains an infinite number of points.**
   Line Postulate

In this chapter our objective is to learn how the deductive system is used in mathematics. Therefore, for our purposes, we need not be unduly concerned with the level of rigor required by mathematicians. The following guidelines that we will use in our work will keep the amount of rigor to a minimum and facilitate our understanding as we explore the world of proofs.

Guidelines For Our Proofs

1. *We are allowed to use only the following as reasons in our proofs:*
   a. *Definitions*
   b. *Postulates*
   c. *Theorems that have been proved using the deductive process*
   d. *Information that is given*
2. *When we perform an algebraic computation in our proofs, we agree to use the Substitution Postulate of Equality as the reason for the computation.*

   Example: If we have the statement $\angle A + \angle A = 180$
   We may conclude $2\angle A = 180$
   We write the *Substitution Postulate of Equality* as the reason for the conclusion.
3. *We will use "definition" as a reason for stating any of the following conditions:*
   a. *Betweenness of points and rays*
   b. *Collinearity of points*
   c. *Points, lines, line segments, rays, angles, polygons, circles, arcs, etc.*
   d. *Parts of geometric figures such as the vertex of an angle, the side of a polygon, the opposite side of a polygon, etc.*
   e. *Vertical angles, adjacent angles, linear pairs, angles formed by transversals, intersecting lines or their point of intersection, etc.*

More precisely, we accept from the figure information gathered from the relative location of points or sets of points. We never accept from the figure information that deals with measurement. For example, in the following figure, we may conclude that A, B and C are collinear points with B between A and C. We may not conclude that B is the midpoint of $\overline{AC}$.

In *some* of our proofs, we will choose not to state a condition before making deductions that require that the condition be met. For example, we need not state that B is between A and C before making the conclusion that AB + BC = AC.

4. *Unless we indicate otherwise, each proof will consist of the following parts:*
   a. *A figure*
   b. *The information that is given; including the figure*
   c. *The statement that is to be proved*
   d. *An argument (proof) in column form with each statement in the proof supported by an appropriate reason*

   Most of our proofs will be written in column form because this format is effective for organizing ideas. Mathematicians, however, do not limit themselves to this format. They sometimes use a more natural discussion type format.

Well, the best way to learn how the deductive process works is to use it. We begin by listing some very simple proofs. You should study these proofs carefully, then try to do them all by yourself on a separate sheet of paper. Some of the proofs require that you supply missing reasons. Be sure that you take the time to supply those reasons. Going through this process will prepare you gradually for the more challenging proofs that you will be expected to do later on.

As you study the proofs, you will notice that the theorems (statements that we must prove) are not always expressed in accurate and precise language. For example, the word "equal" is used in many theorems in place of "equal lengths" or "equal measures." Try not to let this bother you. You will find that in higher math courses, authors generally do not distinguish between these concepts because they know from context the concept intended. Of course, if you wish, you may rewrite the statements of such theorems using precise language. Generally, the procedure is to rewrite the theorem in terms of a geometric figure that illustrates the condition posed by the theorem. The hypothesis of the theorem is then referred to as the "given information" and the conclusion of the theorem is referred to as the "information to be proved."

THEOREM: Supplements of equal angles are equal.
(Here of course, the word "equal" means "equal in measure.")

GIVEN:
$\measuredangle$ A and $\measuredangle$ C are supplementary.
$\measuredangle$ B and $\measuredangle$ D are supplementary.
$\angle A = \angle B$

PROVE:
$\angle C = \angle D$

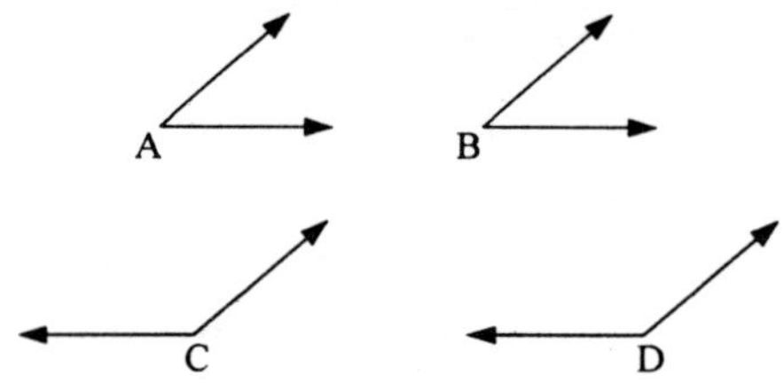

STATEMENTS	REASONS
1. $\measuredangle$ A and $\measuredangle$ C are supplementary. $\measuredangle$ B and $\measuredangle$ D are supplementary.	1. Given
2. $\angle A + \angle C = 180$; $\angle B + \angle D = 180$	2. Definition of supplementary angles
3. $\angle A + \angle C = \angle B + \angle D$	3. Transitive Postulate of Equality; or Substitution Postulate of Equality
4. $\angle A = \angle B$	4. Given
5. $\therefore \angle C = \angle D$ (The symbol $\therefore$ means *therefore*)	5. Subtraction Postulate of Equality

Now for another theorem that is related closely to this one:

THEOREM: Supplements of the same angle are equal.

GIVEN:
$\measuredangle$ A and $\measuredangle$ C are supplementary.
$\measuredangle$ B and $\measuredangle$ C are supplementary.

PROVE:
$\angle A = \angle B$

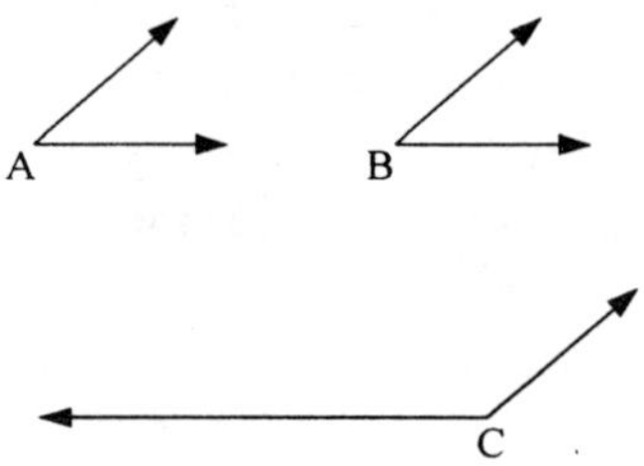

STATEMENTS	REASONS
1. $\measuredangle$ A and $\measuredangle$ C are supplementary. $\measuredangle$ B and $\measuredangle$ C are supplementary.	1. Given
2. $\angle A + \angle C = 180$; $\angle B + \angle C = 180$	2. Definition of supplementary angles
3. $\angle A + \angle C = \angle B + \angle C$	3. Transitive Postulate of Equality; or Substitution Postulate of Equality
4. $\angle C = \angle C$	4. Reflexive Postulate of Equality
5. $\therefore \angle A = \angle B$	5. Subtraction Postulate of Equality

You should be able to prove the following two theorems with no difficulty. The proofs are very similar to the proofs that we have just studied.

THEOREM: Complements of equal angles are equal.
THEOREM: Complements of the same angle are equal.

Now we will use one of the theorems that we have just proved in the proof of the following theorem:

THEOREM: Vertical angles are equal.

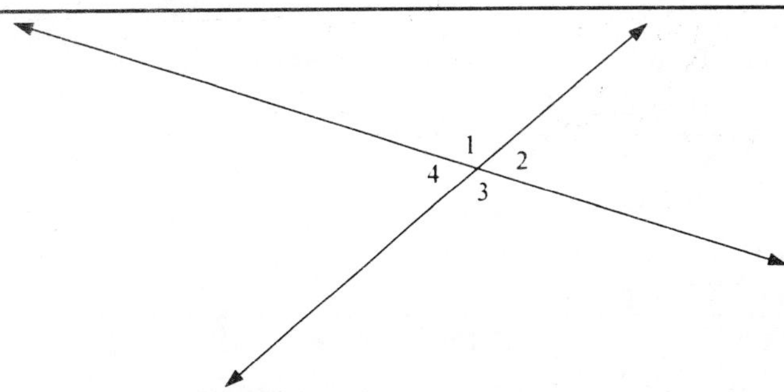

GIVEN:
The figure with vertical angles ∡ 1 and ∡ 3

PROVE:
$\angle 1 = \angle 3$

STATEMENTS	REASONS
1. In the figure ∡ 1 and ∡ 3 are vertical angles.	1. Given
2. ∡ 1 and ∡ 2 form a linear pair. ∡ 2 and ∡ 3 form a linear pair.	2. Definition of linear pair
3. ∡ 1 and ∡ 2 are supplementary. ∡ 2 and ∡ 3 are supplementary.	3. Linear Pair Postulate
4. $\therefore \angle 1 = \angle 3$	4. Theorem: Supplements of the same angle are equal.

The proof of our next theorem is rather simple, however, the theorem itself will be quite useful in the proofs of other theorems.

THEOREM: All right angles are equal.

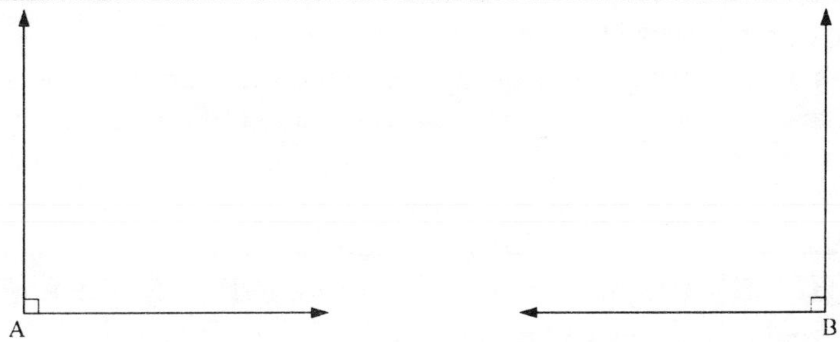

GIVEN:
∡ A is a right angle.
∡ B is a right angle.

PROVE:
$\angle A = \angle B$

STATEMENTS	REASONS
1. ∡ A is a right angle. ∡ B is a right angle.	1. Given
2. $\angle A = 90$; $\angle B = 90$	2. Definition of right angle
3. $\angle A = \angle B$	3. Transitive Postulate of Equality

The next two theorems will provide us with more practice in using the deductive process. Be sure to study them carefully.

THEOREM: If two angles form a linear pair and are equal, then each is a right angle.

GIVEN:

∠ 1 and ∠ 2 form a linear pair.

$\angle 1 = \angle 2$

PROVE:

∠ 1 is a right angle; ∠ 2 is a right angle.

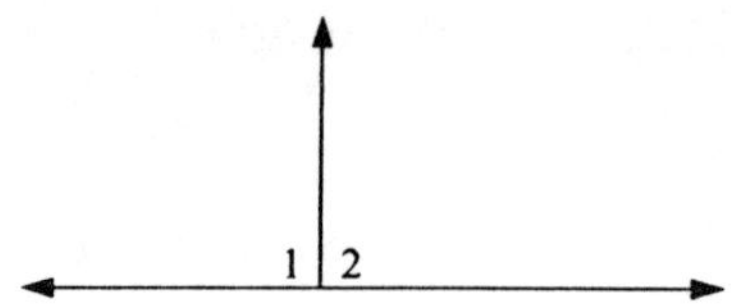

STATEMENTS	REASONS
1. ∠ 1 and ∠ 2 form a linear pair.	1. Given
2. ∠ 1 and ∠ 2 are supplementary.	2. Linear Pair Postulate
3. $\angle 1 + \angle 2 = 180$	3. Definition of Supplementary Angles
4. $\angle 1 = \angle 2$	4. Given
5. $\angle 1 + \angle 1 = 180$	5. Substitution Postulate of Equality
6. $2 \angle 1 = 180$	6. Substitution Postulate of Equality
7. $\angle 1 = 90$	7. Division Postulate of Equality
8. $\angle 2 = 90$	8. Substitution Postulate of Equality
9. ∠ 1 is a right angle; ∠ 2 is a right angle.	9. Definition of Right Angle

Notice that in step 7 of this proof, the Division Postulate of Equality was used to support the division of the equation members in step 6 by the number 2. Technically the Reflexive Postulate also should be used to satisfy the two part requirements of the hypothesis of the Division Postulate of Equality. We do not want our proofs to become too bulky, so we agree that in cases where we multiply or divide both sides of an equation by the same number, add the same number to both sides of an equation, or subtract the same number from both sides of an equation, we will not have to use the Reflexive Postulate of Equality. We agree that it is acceptable to write only the Addition, Subtraction, Multiplication or Division Postulate as a reason.

THEOREM:The bisector of an angle divides the angle into two angles, each 1/2 as large as the original angle.

GIVEN:

$\overrightarrow{BD}$ is the bisector of ∠ ABC.

PROVE:

$\angle 1 = 1/2 \angle ABC$

$\angle 2 = 1/2 \angle ABC$

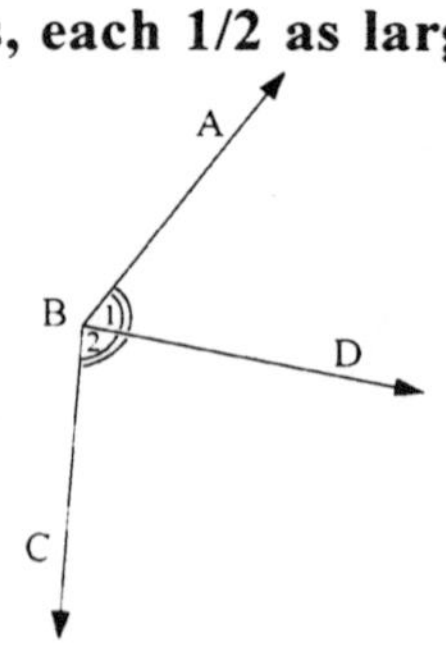

STATEMENTS	REASONS
1. $\overrightarrow{BD}$ is the bisector of ∠ ABC.	1. Given
2. $\angle 1 = \angle 2$	2. Definition of Angle Bisector
3. $\angle 1 + \angle 2 = \angle ABC$	3. Whole Equals Sum of Parts Postulate
4. $\angle 1 + \angle 1 = \angle ABC$	4. Substitution Postulate of Equality
5. $2 \angle 1 = \angle ABC$	5. Substitution Postulate of Equality
6. $\angle 1 = 1/2 \angle ABC$	6. Multiplication Postulate of Equality
7. $\therefore \angle 2 = 1/2 \angle ABC$	7. Substitution Postulate of Equality

Why don't you try your hand at the next theorem. Its proof is very similar to the the proof of the theorem that we just completed.

THEOREM: The bisector of a line segment divides the segment into two segments, each 1/2 as large as the original line segment.

Exercises 6.1

(Figures may not be drawn to scale.)

1. **GIVEN:** The figure with PR = SQ
PROVE: PS = RQ
Supply the reasons in the following proof:

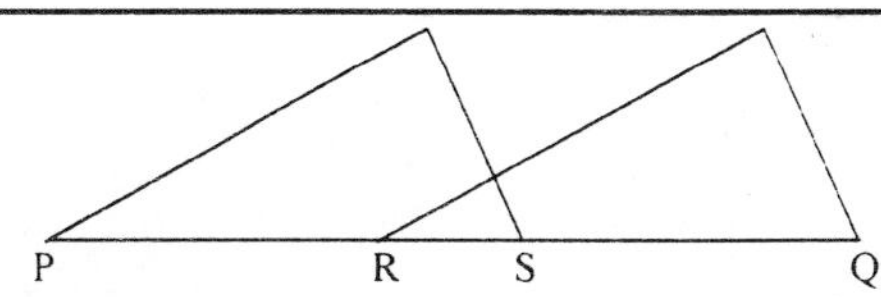

STATEMENTS	REASONS
1. The figure with PR = SQ	1. ?
2. RS = RS	2. ?
3. PR + RS = QS + RS	3. ?
4. PR + RS = PS	4. ?
5. SQ + RS = RQ	5. ?
6. PS = RQ	6. ?

2. **GIVEN:** The figure with ∠ 1 = ∠ 3
PROVE: ∠ ABE = ∠ DBC

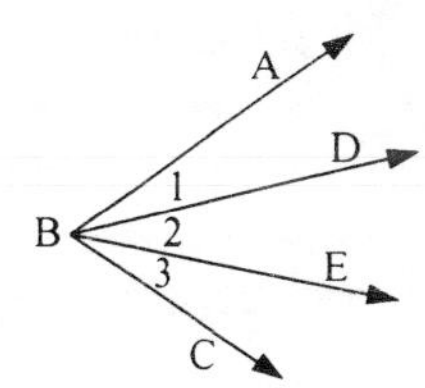

3. Supply the reasons in the following proof:

GIVEN: The figure with ∠ ABE = ∠ DBC
PROVE: ∠ 1 = ∠ 3

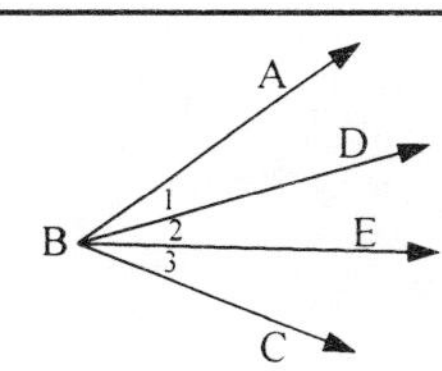

STATEMENTS	REASONS
1. The figure with ∠ ABE = ∠ DBC	1. ?
2. ∠ 1 + ∠ 2 = ∠ ABE	2. ?
3. ∠ 2 + ∠ 3 = ∠ DBC	3. ?
4. ∠ 1 + ∠ 2 = ∠ 2 + ∠ 3	4. ?
5. ∠ 2 = ∠ 2	5. ?
6. ∠ 1 = ∠ 3	6. ?

4. **GIVEN:** The figure with PS = RQ
PROVE: PR = SQ

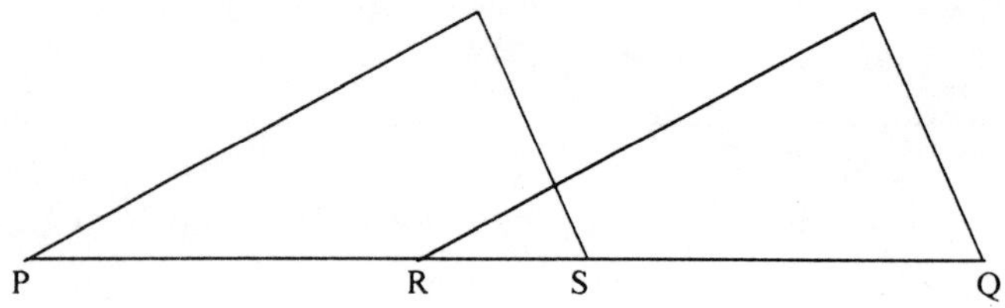

5. **GIVEN:** The figure with ∠ 1 = ∠ 3; ∠ 2 = ∠ 4
PROVE: ∠ RPQ = ∠ RQP

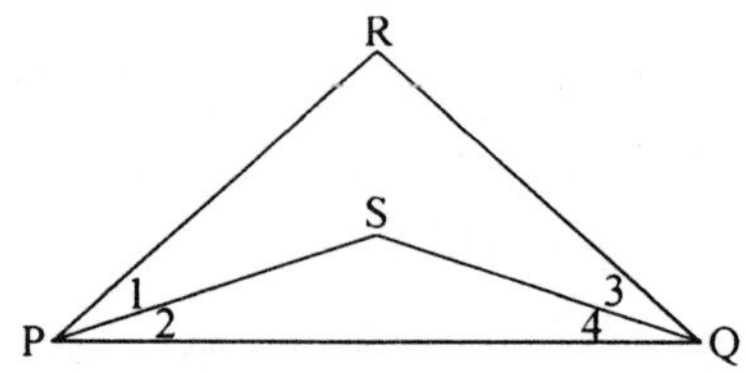

6. Supply the reasons in the following proof:

GIVEN: The figure with ∠ 1 = ∠ 4
PROVE: ∠ 2 = ∠ 3

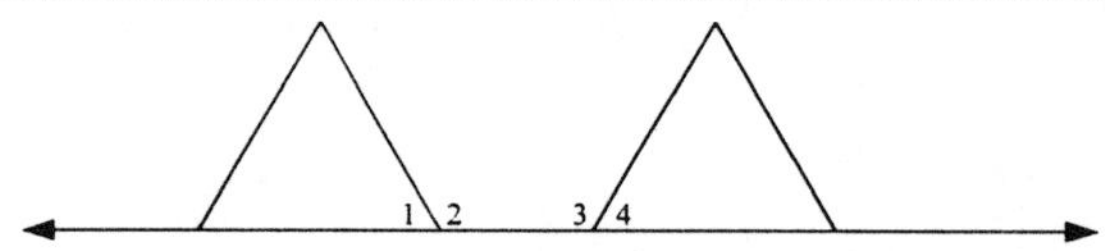

STATEMENTS	**REASONS**
1. The figure with ∠ 1 = ∠ 4	1. ?
2. ∠ 1 and ∠ 2 form a linear pair. ∠ 3 and ∠ 4 form a linear pair.	2. ?
3. ∠ 1 and ∠ 2 are supplementary. ∠ 3 and ∠ 4 are supplementary.	3. ?
4. ∠ 2 = ∠ 3	4. ?

7. **GIVEN:** The figure with ∠ 2 = ∠ 4
PROVE: ∠ 1 = ∠ 3

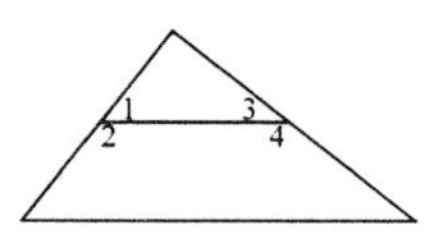

8. **GIVEN:** The figure with ∠ 1 = ∠ 3 ; ∠ 2 = ∠ 4
PROVE: ∠ ABC = ∠ ACB

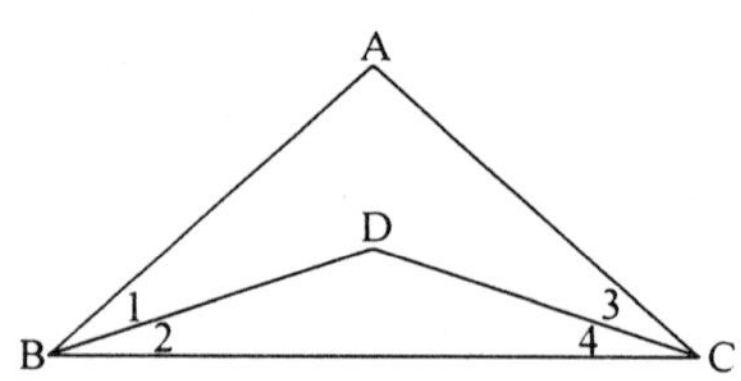

9. **GIVEN:** The figure with ∠ 3 = ∠ 5
PROVE: ∠ 1 = ∠ 6

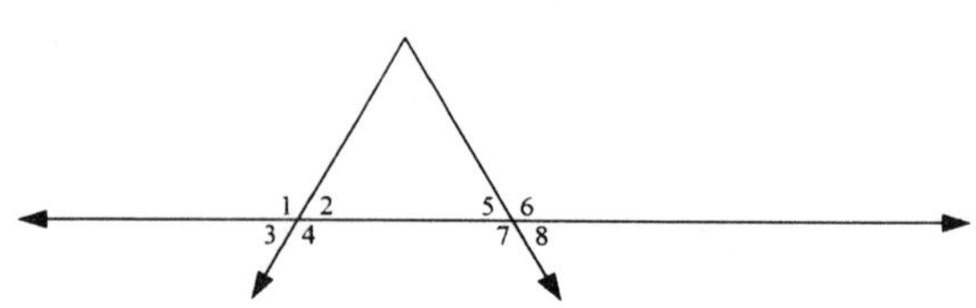

10. **GIVEN:** The figure with AD = AE ; DB = EC
    PROVE: AB = AC

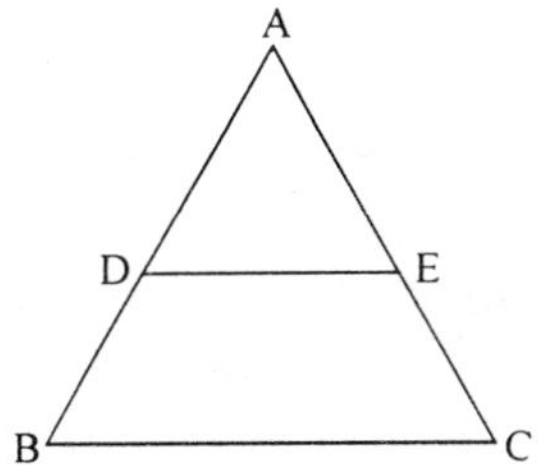

11. **GIVEN:** The figure with $\angle 1 = \angle 3$
    PROVE: $\angle 2$ and $\angle 3$ are supplementary

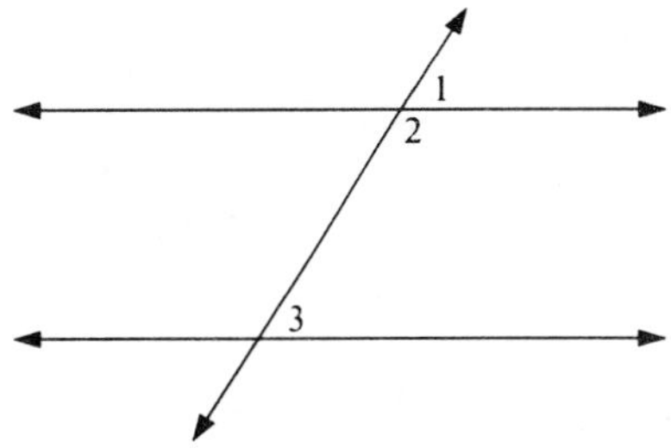

12. **GIVEN:** The figure with $\overleftrightarrow{DB} \perp \overleftrightarrow{BC}$; $\angle 1 = \angle 3$
    PROVE: $\angle 2 = \angle 4$

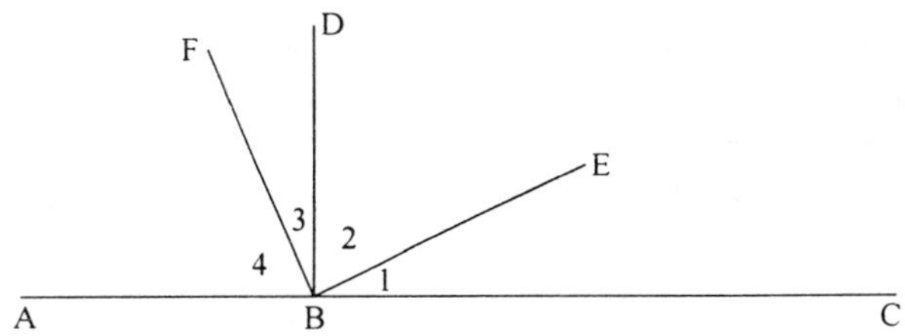

Section 6.2 Congruent Triangle Proofs

Before beginning this section, you may want to review section 2 of chapter 2. As you recall from that chapter, congruent triangles are triangles that have the same shape and the same size. In this section we will use the deductive process to show that certain triangles are congruent or that corresponding parts of congruent triangles have the same measure or length. These are the postulates that we will use in our work. Of course you are already familiar with them.

1. **Two triangles are congruent if three sides of one triangle have the same lengths, respectively, as three sides of the other triangle.**

   Side Side Side Postulate **SSS Postulate**

2. **Two triangles are congruent if two sides and the included angle of one triangle have the same lengths and measure respectively, as two sides and the included angle of the other triangle.**

   Side Angle Side Postulate **SAS Postulate**

3. **Two triangles are congruent if two angles and the included side of one triangle have the same measures and length, respectively, as two angles and the included side of the other triangle.**

   Angle Side Angle Postulate **ASA Postulate**

Often in our proofs, we will find it necessary to use the definition of congruent triangles as a reason for concluding that corresponding parts of congruent triangles have equal lengths and measures. To make our work easier, we will agree that the **abbreviation CPCTE** (short for corresponding parts of congruent triangles are "equal") can be used in place of the definition.

Keep in mind that the converse of every definition is true. Therefore every definition can be used in two forms. Consider this restatement of the definition for congruent triangles:

Two triangles are congruent if and only if their corresponding parts are "equal."

This definition encompasses each of the following statements:

If all corresponding parts of two triangles are "equal," then the triangles are congruent.

If two triangles are congruent, then all of their corresponding parts are "equal."

The first statement is used to conclude that two triangles are congruent, after the equality of their corresponding lengths and measures has been established; while the second statement is used to conclude that parts of triangles have the same measure or length, after it has been established that the two triangles are congruent.

The SSS, SAS and ASA Postulates can be used to prove that two triangles are congruent. Carefully study the following proofs:

GIVEN:
The figure with $\angle A = \angle E$; $AC = CE$

PROVE:
$\triangle ACB \cong \triangle ECD$

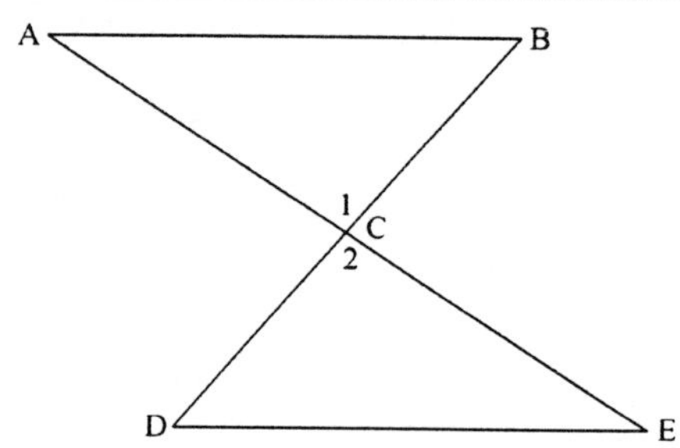

STATEMENTS	REASONS
1. The figure with $\angle A = \angle E$; $AC = CE$	1. Given
2. $\angle 1$ and $\angle 2$ are vertical angles.	2. Definition of vertical angles
3. $\angle 1 = \angle 2$	3. Theorem: Vertical angles are equal
4. $\triangle ACB \cong \triangle ECD$	4. ASA Postulate (Steps 1 and 3)

GIVEN:
The figure with $\overleftrightarrow{AC} \perp \overleftrightarrow{BD}$; $BC = CD$

PROVE:
$\triangle ACB \cong \triangle ACD$

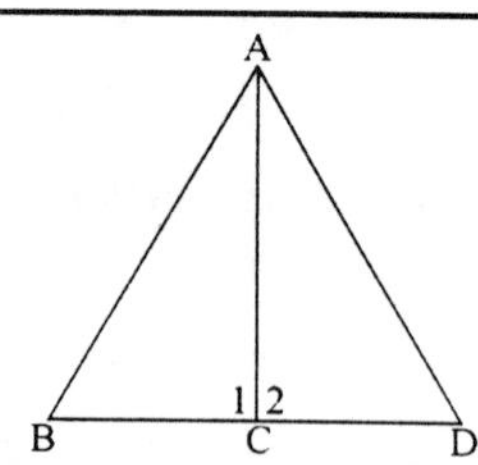

STATEMENTS	REASONS
1. The figure with $\overleftrightarrow{AC} \perp \overleftrightarrow{BD}$; $BC = CD$	1. Given
2. $\angle 1$ and $\angle 2$ are right angles.	2. Definition of Perpendicular Lines
3. $\angle 1 = \angle 2$.	3. Theorem: All right angles have equal measures
4. $AC = AC$	4. Reflexive Postulate of Equality
5. $\therefore$ $\triangle ACB \cong \triangle ACD$	5. SAS (Steps 1, 3, 4)

GIVEN:
The figure with Q, the midpoint of $\overline{TV}$ and $\overline{PR}$

PROVE:
∠ TPQ = ∠ VRQ

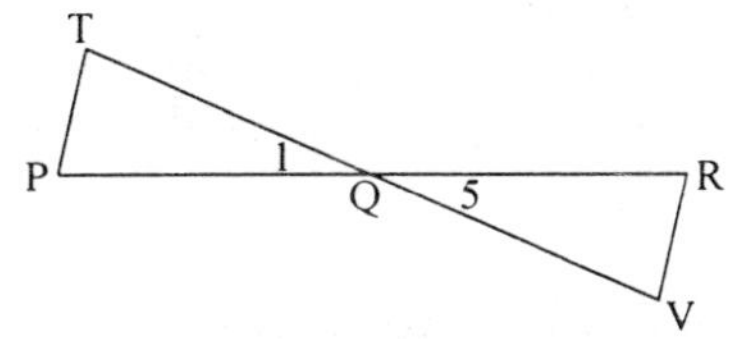

STATEMENTS	REASONS
1. Q is the midpoint of TV and PR in the figure.	1. Given
2. TQ = QV, PQ = QR	2. Definition of Midpoint
3. ∡ 1 and ∡ 5 are vertical angles.	3. Definition of Vertical Angles
4. ∠ 1 = ∠ 5	4. Theorem: Vertical Angles Are Equal
5. △ TQP ≅ △ VQR	5. SAS Postulate (Steps 2, 4)
6. ∴ ∠ TPQ = ∠ VRQ	6. CPCTE

In the next proof, △ DBC and △ ECB are overlapping triangles. You may find it helpful to take the triangles apart as illustrated:

GIVEN:
The figure with ∠ 1 = ∠ 2 ; BE = CD

PROVE:
△ DBC ≅ △ ECB

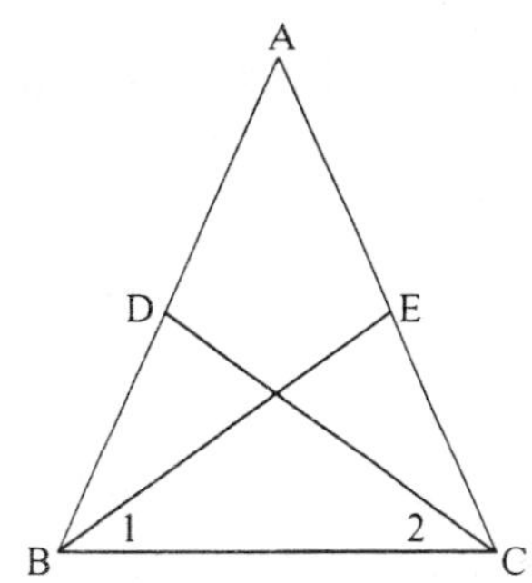

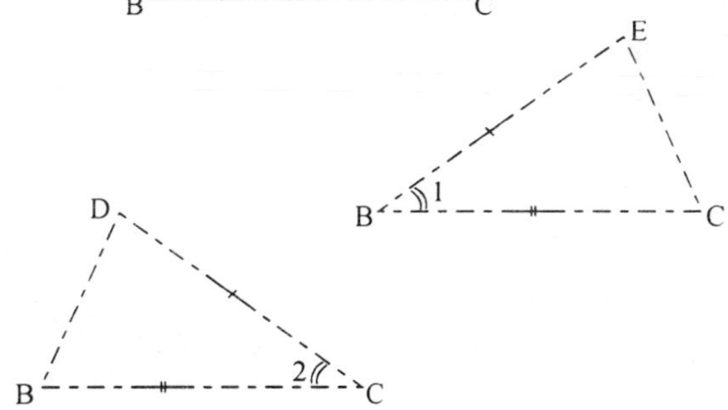

STATEMENTS	REASONS
1. ∠ 1 = ∠ 2 and BE = CD in the figure	1. Given
2. BC = BC	2. Reflexive Postulate of Equality
3. △ DBC ≅ △ ECB	3. SAS Postulate of Equality (Steps 1, 2)

Exercises 6.2

(Figures are not drawn to scale)

1. **GIVEN:** The figure with ∠ N = ∠ P
∠ M = ∠ Q
MN = PQ

PROVE: △ MON ≅ △ QOP

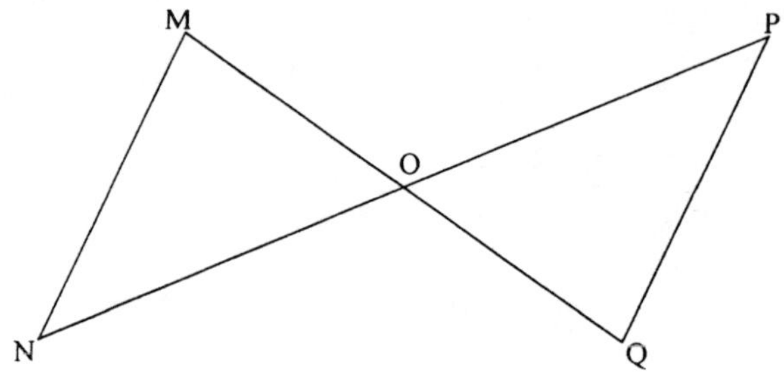

STATEMENTS	REASONS

2. **GIVEN:** The figure with ∠ A = ∠ R
∠ C = ∠ Q
AC = RQ

PROVE: ∠ B = ∠ P

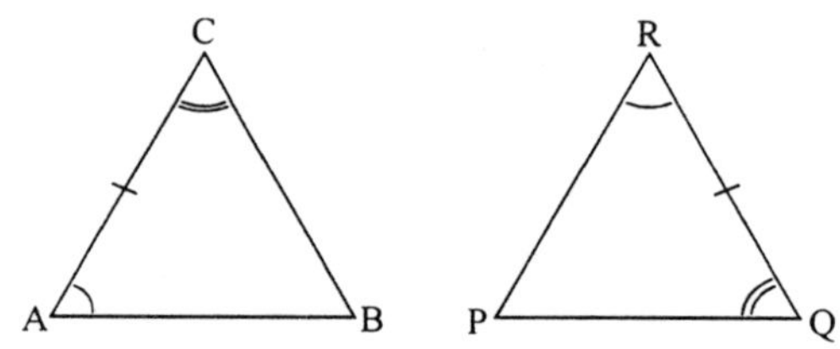

STATEMENTS	REASONS

3. **GIVEN:** The figure with LM = MO
KM = MP

PROVE: KL = OP

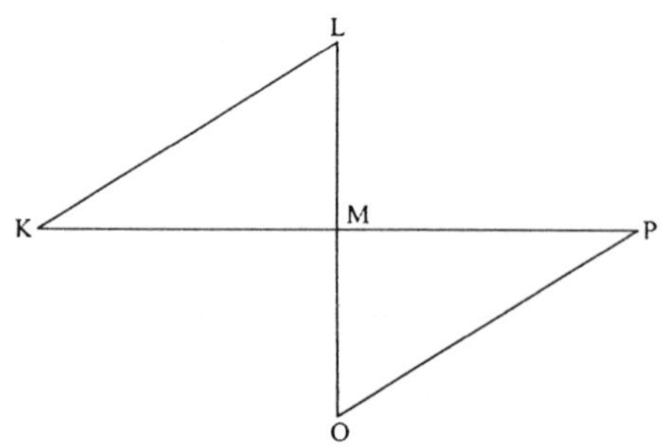

STATEMENTS	REASONS

4. **GIVEN:** The figure with ∠ ABD = ∠ CDB
∠ CBD = ∠ ADB

PROVE: ∠ DAB = ∠ BCD

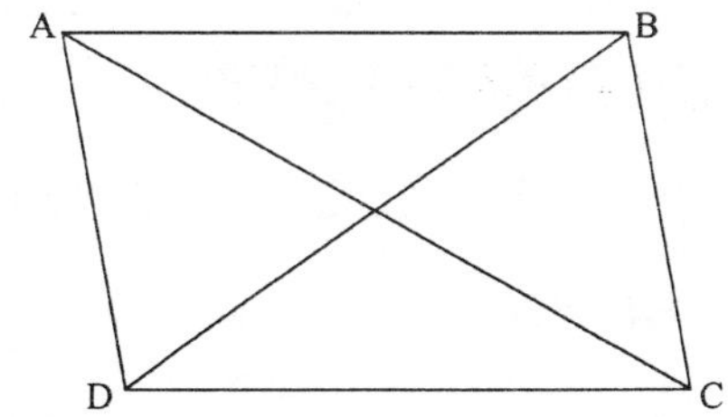

STATEMENTS	REASONS

5. **GIVEN:** The figure with ∠ 1 = ∠ 2
∠ 3 = ∠ 4
AB = CD

PROVE: P bisects $\overline{AC}$ and $\overline{BD}$

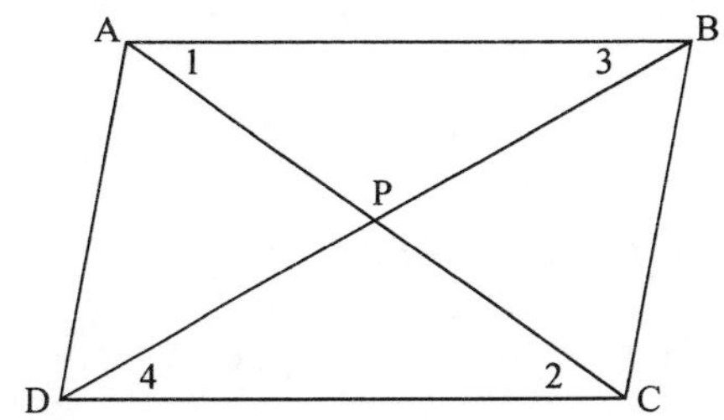

STATEMENTS	REASONS

6. **GIVEN:** The figure with AD = CD
B is the midpoint of $\overline{AC}$.

PROVE: △ DBA ≅ △ DBC

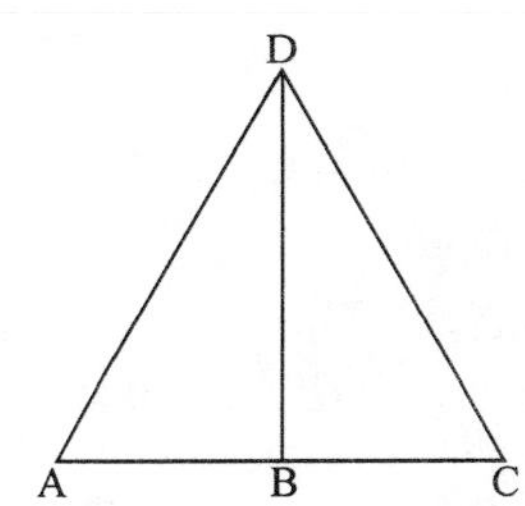

STATEMENTS	REASONS

7. **GIVEN:** The figure; Point O bisects $\overline{PT}$ and $\overline{RQ}$.

PROVE: $\triangle$ QOP $\cong$ $\triangle$ ROT

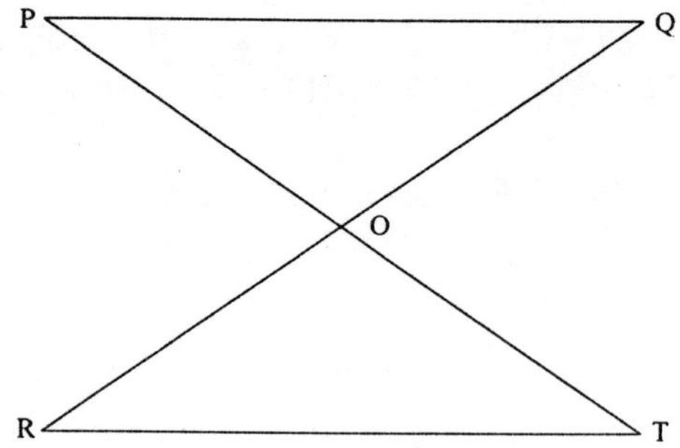

STATEMENTS	REASONS

8. **GIVEN:** The figure; Point O bisects $\overline{PN}$.
$\angle$ N = $\angle$ P
MN = PQ

PROVE: Point O bisects $\overline{MQ}$

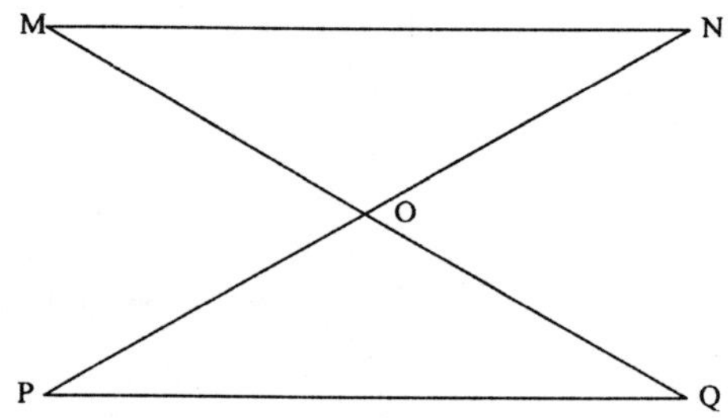

STATEMENTS	REASONS

9. **GIVEN:** The figure with $\overleftrightarrow{DB} \perp \overleftrightarrow{AC}$
$\overrightarrow{DB}$ bisects $\angle$ ADC.

PROVE: B bisects $\overline{AC}$.

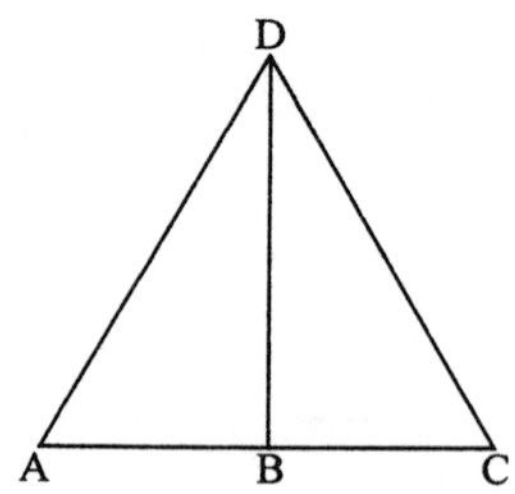

STATEMENTS	REASONS

10. **GIVEN:** The figure with AB = CD
AC = BD

PROVE: ∠ DBC = ∠ ACB

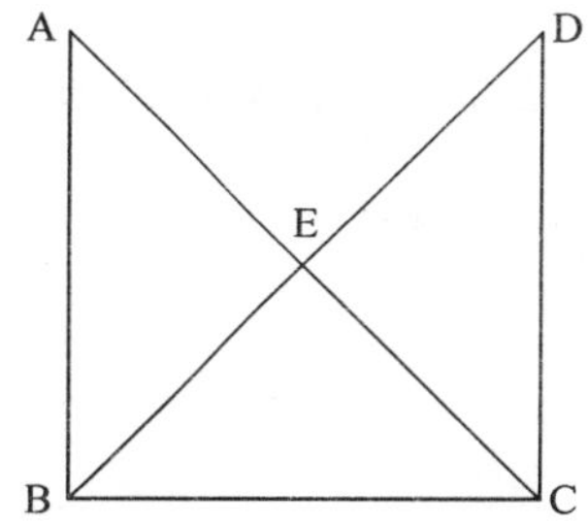

STATEMENTS	REASONS

11. **GIVEN:** The figure with AB = CD
BC = AD

PROVE: ∠ ACB = ∠ CAD

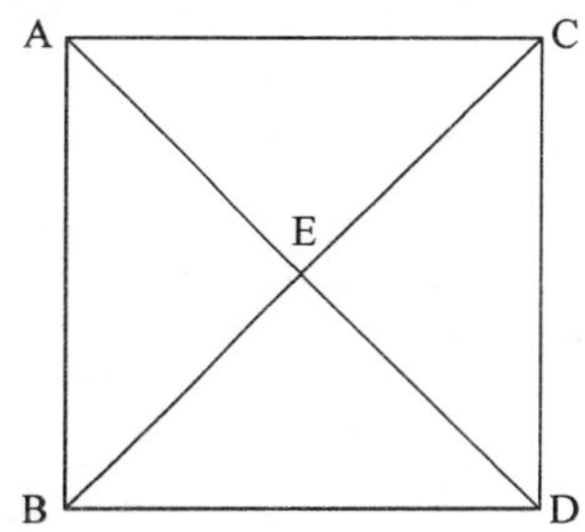

STATEMENTS	REASONS

12. **GIVEN:** The figure with ∠ 1 = ∠ 2
PT = QU
PQ = RS

PROVE: △ TRP ≅ △ USQ

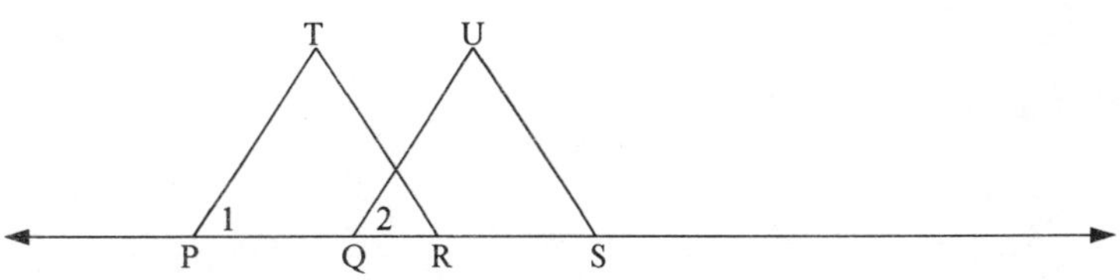

STATEMENTS	REASONS

13. **GIVEN:** The figure with AC = BD
∠ ECA = ∠ FDB
∠ 1 = ∠ 3

PROVE: △ ECA ≅ △ FDB

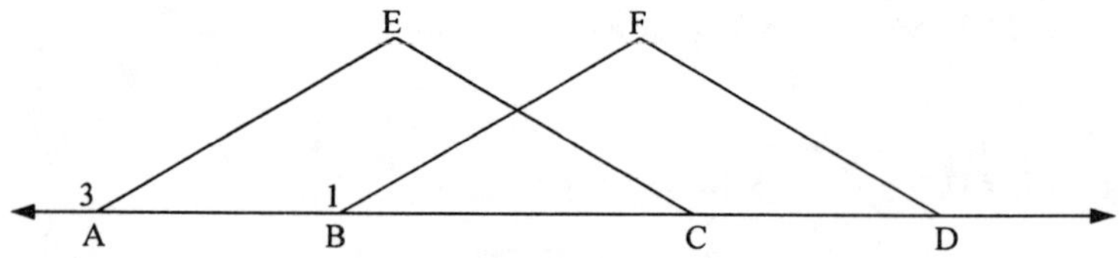

STATEMENTS	REASONS

14. **GIVEN:** The figure with QM = RN
MN = OP
∠ 1 = ∠ 3

PROVE: QO = RP

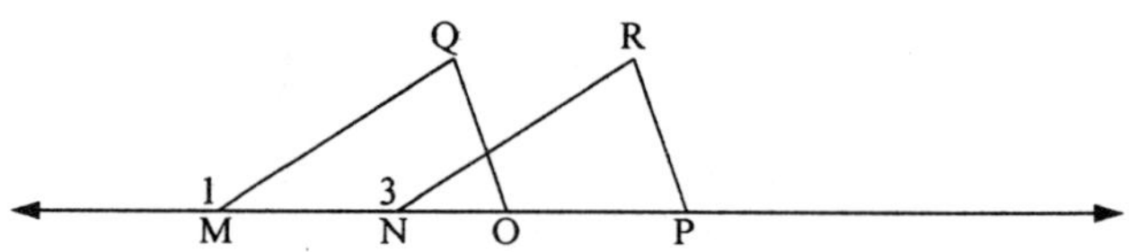

STATEMENTS	REASONS

15.***PROVE:** ***THEOREM:*** **Two triangles congruent to a third triangle are congruent to each other.**

16.* A relation is called an *equivalence relation* if the following postulates are true for the relation.

The letters a and b represent elements in the set affected by the relation.

R represents the relation.

Reflexive Postulate	a **R** a
Symmetric Postulate	If a **R** b then b **R** a
Transitive Postulate	If a **R** b and b **R** c then a **R** c

The *equality* relation is an equivalence relation because for all real numbers a and b the following postulates hold. (In this case **R** represents the relation " = .")

Reflexive Postulate of Equality	a = a
Symmetric Postulate of Equality	If a = b then b = a
Transitive Postulate of Equality	If a = b and b = c then a = c

a. Is the congruence relation for triangles an equivalence relation?

b. Is the similarity relation for triangles an equivalence relation?

c. Is the inequality ($<$ or $>$) relation an equivalence relation?

Section 6.3 Isosceles Triangle Proofs

In this chapter, we will prove some very useful theorems about isosceles triangles, so before continuing you may want to review the information on isosceles triangles given in section 1 of chapter 2. We will also introduce the idea of an auxiliary set of points. An **auxiliary set of points** is a line, a line segment, a ray or simply one point that is not illustrated in the original figure. For example, in the figure below, we see an illustration of triangle BAD.

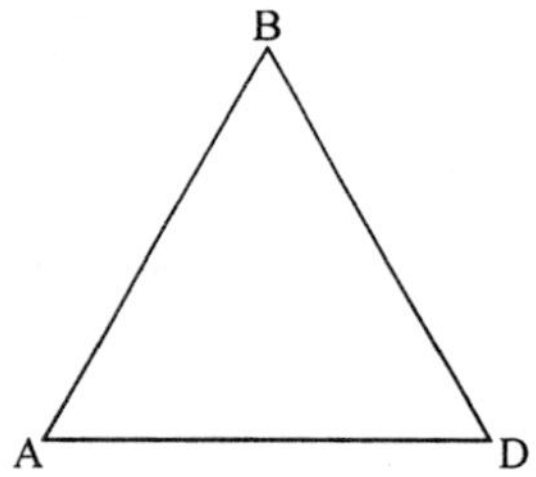

Line segment $\overline{BC}$, indicated with a broken line segment, also exists although it is not shown in the original figure. We call $\overline{BC}$ an auxiliary set of points,

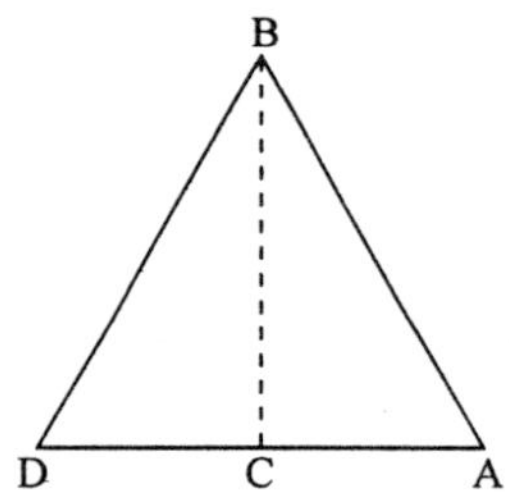

As you recall from the first part of this course, space contains all points, and these points exist in space regardless of our choice to acknowledge or not to acknowledge their presence. In some of our proofs, we may find it useful to introduce points that are not shown. We must, however, be careful when introducing such points. We must be sure that they actually exist and possibly are unique. For example, it would be foolish to introduce the midpoint of a line, since a line does not have a midpoint. Postulates that guarantee the existence or the uniqueness of a set of points may be called existence postulates. In this chapter, we will use the following existence postulate:

An angle has exactly one bisector, that is, a ray with the vertex of the angle as its endpoint which separates the angle into two angles of equal measure.
The Angle Bisector Postulate

In the proof of the next theorem, we make use of an auxiliary line segment. Notice that the *auxiliary segment is represented by a broken line segment.* This theorem is a very important theorem in mathematics. We will call it the *Isosceles Triangle Theorem.*

We will call this theorem **THE ISOSCELES TRIANGLE THEOREM**

THEOREM:
If two sides of a triangle have equal lengths, then the angles opposite those sides have equal measures, or the base angles of an isosceles triangle are equal.

GIVEN:
△ ABC with AC = BC

PROVE:
∠ A = ∠ B

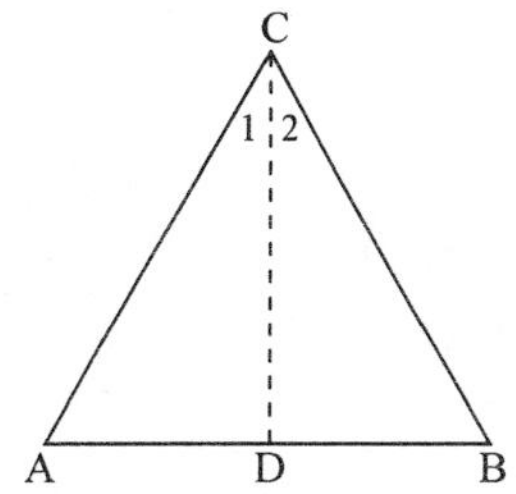

STATEMENTS	REASONS
1. △ ABC with AC = BC	1. Given
2. Construct the angle bisector of Angle ACB. Call it $\overrightarrow{CD}$.	2. Postulate: Every angle has exactly one angle bisector.
3. ∠ 1 = ∠ 2	3. Definition of angle bisector
4. CD = CD	4. Reflexive Postulate of Equality
5. △ ACD ≅ △ BCD	5. SAS (Steps 2, 3, 4)
6. ∴ ∠ A = ∠ B	6. CPCTE

Following are two theorems that are related to the Isosceles Triangle Theorem. You should have no difficulty proving these theorems.

THEOREM:
The bisector of the vertex angle of an isosceles triangle separates the triangle into two congruent triangles.

THEOREM:
The bisector of the vertex angle of an isosceles triangle is the perpendicular bisector of the base of the triangle.

The converse of the Isosceles Triangle Theorem is also true.

THEOREM:

If two angles of a triangle have equal measures, then the sides opposite those angles have equal lengths.

GIVEN: $\triangle$ ABC in the figure with $\angle A = \angle B$

PROVE: AC = BC

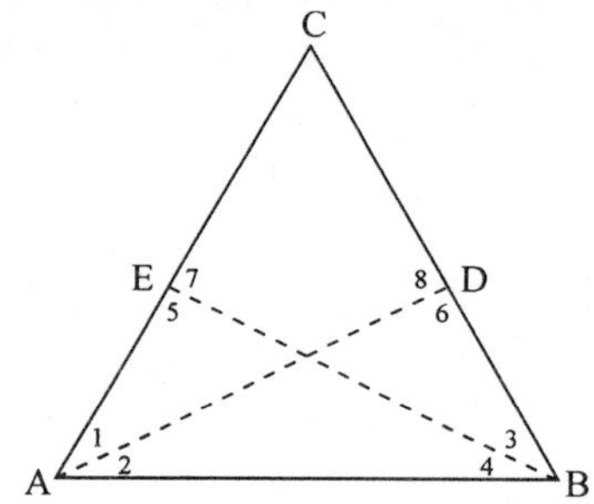

1. $\triangle$ ABC in the figure with $\angle A = \angle B$	1. ?
2. Construct $\overrightarrow{AD}$ and $\overrightarrow{BE}$, the bisectors of $\measuredangle$ A and $\measuredangle$ B, respectively.	2. Every angle has exactly one bisector: Postulate
3. $\angle 1 = \frac{1}{2} \angle CAB$; $\angle 3 = \frac{1}{2} \angle CBA$ $\angle 2 = \frac{1}{2} \angle CAB$; $\angle 4 = \frac{1}{2} \angle CBA$	3. Theorem: The bisector of an angle separates the angle into two angles each ½ the original angle.
4. $\frac{1}{2} \angle CAB = \frac{1}{2} \angle CBA$	4. Multiplication Postulate of Equality (Step 1)
5. $\angle 2 = \angle 4$ and $\angle 1 = \angle 3$	5. Substitution Postulate of Equality
6. AB = AB	6. Reflexive Postulate of Equality
7. $\triangle EAB \cong \triangle DBA$	7. ASA (Steps 1,5,6)
8. EB = DA; $\angle 5 = \angle 6$	8. CPCTE
9. $\measuredangle$ 5 and $\measuredangle$ 7 form a linear pair. $\measuredangle$ 6 and $\measuredangle$ 8 form a linear pair.	9. Definition of Linear Pair
10. $\measuredangle$ 5 and $\measuredangle$ 7 are supplementary. $\measuredangle$ 6 and $\measuredangle$ 8 are supplementary.	10. Linear Pair Postulate
11. $\angle 7 = \angle 8$	11. Supplements of equal angles are equal Theorem
12. $\triangle CAD \cong \triangle CBE$	12. ASA Postulate (Steps 5,8,11)
13. AC = BC QED	13. CPCTE

QED or Q.E.D. (quod erat demonstrandum) is Latin for "which was to be proved." This phrase may be used when you finish a proof.

A theorem that is an immediate consequence of another theorem is sometimes called a **corollary.** The following theorems may be called corollaries of the Isosceles Triangle Theorem and its converse. See if you can prove them.

COROLLARY: If a triangle is equilateral, then it is equiangular.

COROLLARY: If a triangle is equiangular, then it is equilateral.

Exercises 6.3

(*Figures are not drawn to scale*)

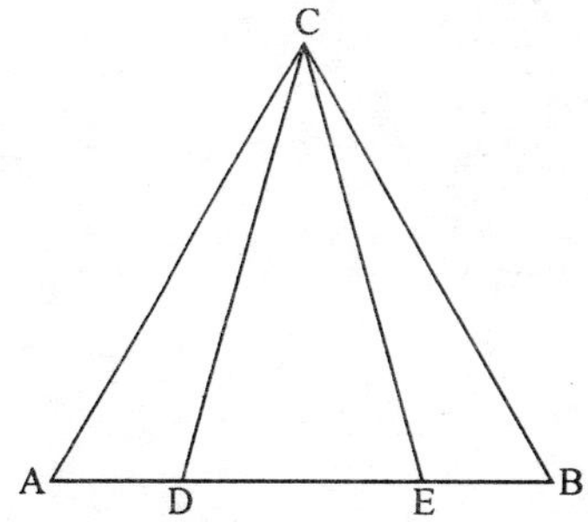

(figure for exercises 1 through 4)

1. **GIVEN:** In △ ABC, AC = BC; AD = BE

PROVE: △ CAD ≅ △ CBE

STATEMENTS	REASONS

2. **GIVEN:** In △ ABC, AC = BC
AD = BE

PROVE: △ CAE ≅ △ CBD

STATEMENTS	REASONS

3. **GIVEN:** In △ ABC, ∠ A = ∠ B
AE = DB

PROVE: △ ACE ≅ △ BCD

STATEMENTS	REASONS

4. **GIVEN:** In △ ABC, ∠ A = ∠ B
AE = BD

PROVE: △ ACD ≅ △ BCE

STATEMENTS	REASONS

5. **GIVEN:** △ PST with PT = ST and
PQ = RS

PROVE: ∠ 1 = ∠ 2

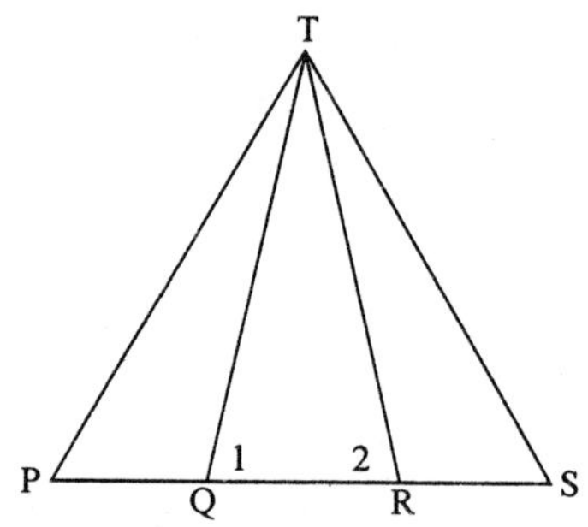

STATEMENTS	REASONS

6. **GIVEN:** AC = BC; AD = BD

PROVE: ∠ CAD = ∠ CBD

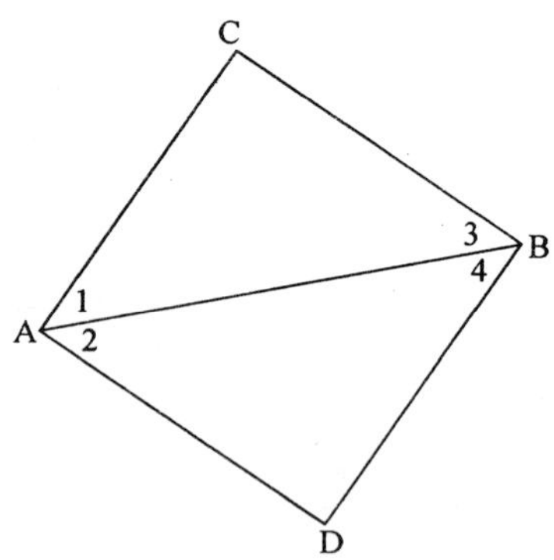

STATEMENTS	REASONS

7. **GIVEN:** $\triangle$ PST with PT = ST and PQ = RS

PROVE: $\angle 1 = \angle 2$

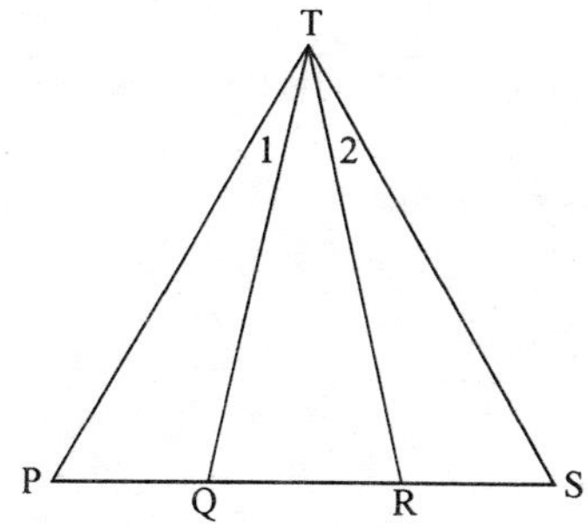

STATEMENTS	REASONS

8. **GIVEN:** $\triangle$ ABD with $\angle A = \angle 1$
$\triangle$ BDC with $\angle 2 = \angle C$

PROVE: AB = CD

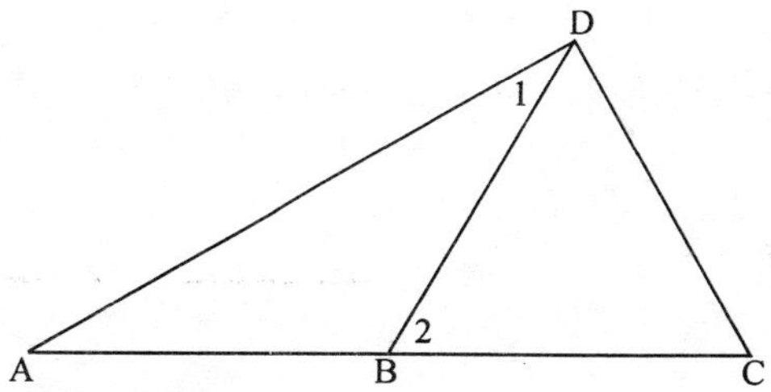

STATEMENTS	REASONS

9. **GIVEN:** The figure with RT = ST and RP = SQ

PROVE: $\triangle SRP \cong \triangle RSQ$

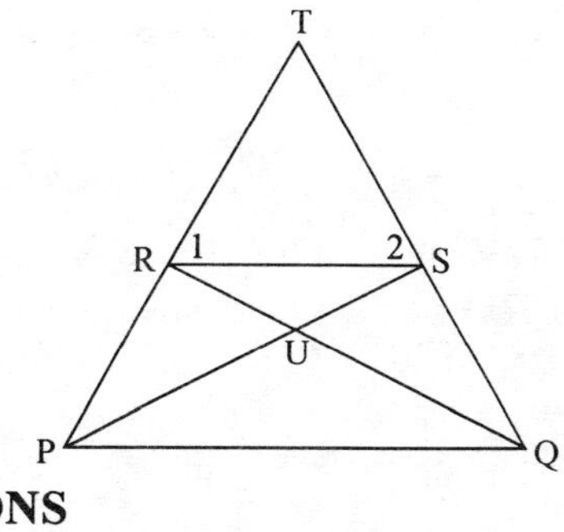

STATEMENTS	REASONS

10. **GIVEN:** The figure with ∠1 = ∠2 and ∠SRU = ∠RSU

PROVE: RQ = SP

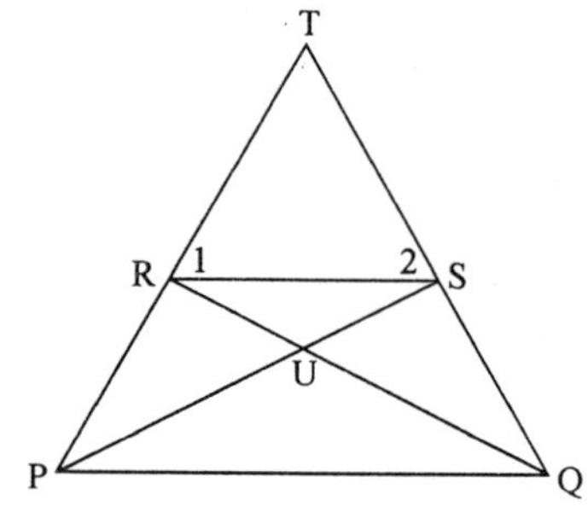

STATEMENTS	REASONS

11. **GIVEN:** The figure with RU = SU and ∠1 = ∠2

PROVE: △ PUQ is isosceles.

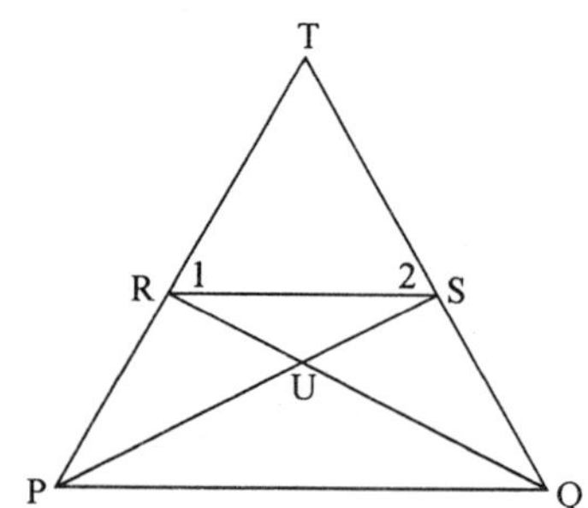

STATEMENTS	REASONS

12. **GIVEN:** The figure with PT = QT and RT = ST

PROVE: RU = US

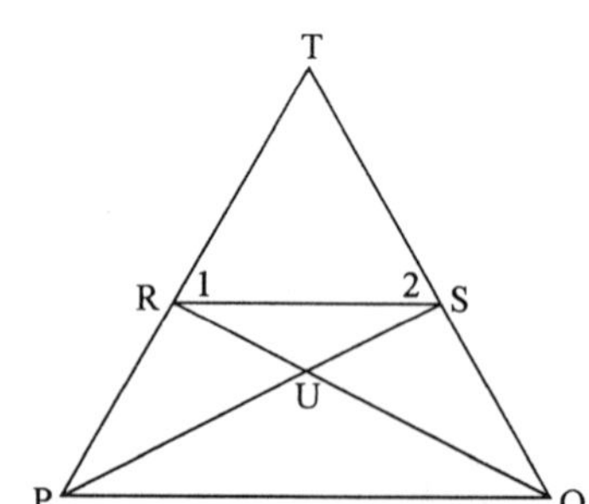

STATEMENTS	REASONS

13. Prove each theorem:

THEOREM:
The bisector of the vertex angle of an isosceles triangle separates the triangle into two congruent triangles.

THEOREM:
The bisector of the vertex angle of an isosceles triangle is the perpendicular bisector of the base of the triangle.

Chapter Six Highlights

Do You Know These Terms?

supplementary angles
complementary angles
vertical angles
angles that form a linear pair
right angle
angle bisector
midpoint of a line segment
bisector of a line segment
perpendicular lines
CPCTE
QED
auxiliary set of points
corollary
proof (or argument)
equivalence relation*

*(Introduced in an exercise)

Do You Know and Understand These Postulates?

Transitive Postulate of Equality
Reflexive Postulate of Equality
Substitution Postulate of Equality
Subtraction Postulate of Equality
Symmetric Postulate of Equality
Addition Postulate of Equality
Multiplication Postulate of Equality
Division Postulate of Equality
Linear Pair Postulate
Whole Equals Sum of Parts Postulate
Line Postulate
Side Side Side Postulate
Side Angle Side Postulate
Angle Side Angle Postulate
Angle Bisector Postulate

Do You Know and Understand These Theorems?

Supplements of Equal Angles are Equal.
Supplements of the Same Angle are Equal.
Complements of Equal Angles are Equal.
Complements of the Same Angles are Equal.
Vertical Angles are Equal.
All Right Angles are Equal.
If Two Angles Form a Linear Pair and are Equal, Then Each is a Right Angle.
The Bisector of an Angle Divides the Angle into Two Angles, Each ½ as Large as the Original Angle.
The Bisector of a Line Segment Divides the Segment into Two Line Segments, Each ½ as Large as the Original Line Segment.
The Isosceles Triangle Theorem.
Converse of the Isosceles Triangle Theorem.
The Bisector of the Vertex Angle of an Isosceles Triangle separates the Triangle into Two Congruent Triangles.
The Bisector of the Vertex Angle of an Isosceles Triangle is the Perpendicular Bisector of the Base of the Triangle.
If a Triangle is Equilateral then it is Equiangular.
If a Triangle is Equiangular then it is Equilateral.
Two Triangles Congruent to a Third Triangle are Congruent to Each Other.

Chapter Seven

Proofs Involving Transversals and Inequalities

Objectives

- State the Parallel Postulate and explain its significance
- State and prove the Transversal Theorems
- State and prove Inequality Theorems for triangles
- Prove that the sum of the measures of angles in a triangle is 180
- List the methods for proving that coplanar lines are parallel
- Apply theorems to exercises involving geometric figures

Does This Proof Convince You That 2 = 1

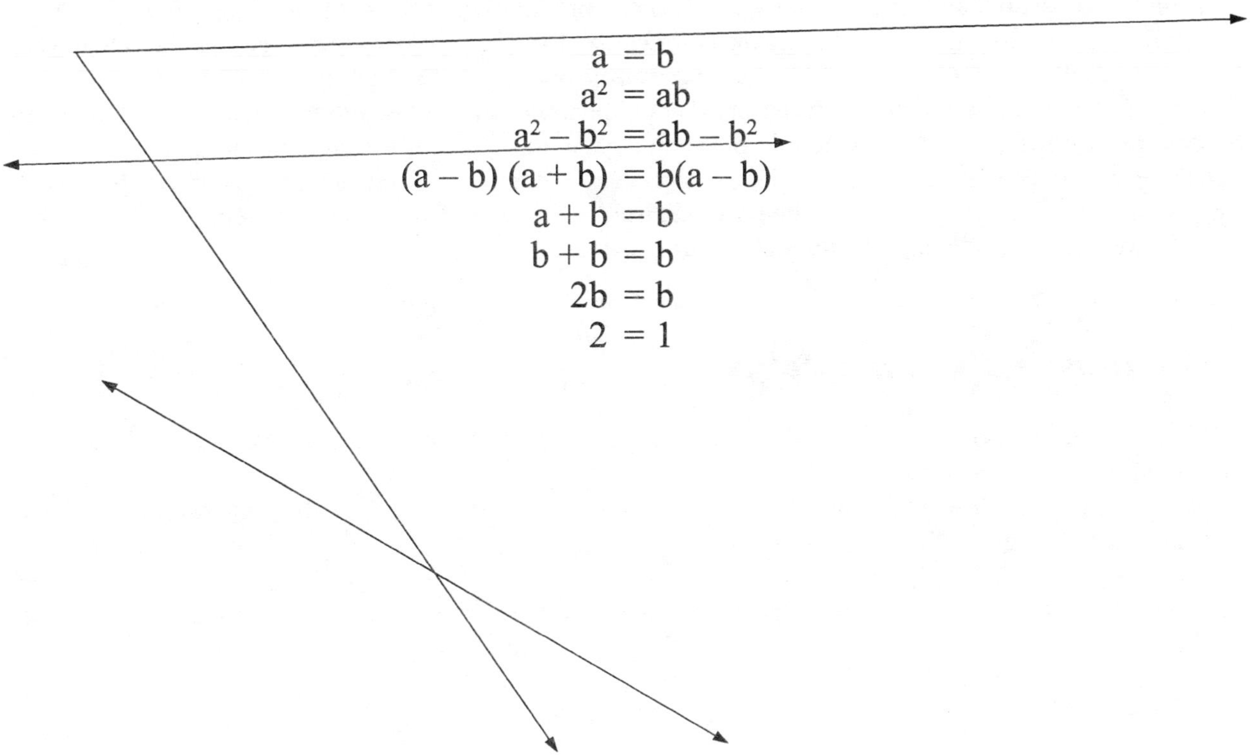

$$a = b$$
$$a^2 = ab$$
$$a^2 - b^2 = ab - b^2$$
$$(a - b)(a + b) = b(a - b)$$
$$a + b = b$$
$$b + b = b$$
$$2b = b$$
$$2 = 1$$

Section 7.1 Transversals and Inequalities

We recall from section 3 of chapter 1 that special relationships exist between angles that are formed when two parallel lines are cut by a transversal. In this chapter, we will use the deductive process to prove that these relationships are true. Some of the postulates used in this chapter are existence and uniqueness postulates. As you recall, they allow us to enter into our proofs sets of points that may not be shown as a part of the given figure. We will take time now to discuss these and other postulates before attempting our proofs.

The Transversal Postulate

If two parallel lines are cut by a transversal, then corresponding angles have equal measures.
The Transversal Postulate is often treated as a theorem, however, it is more convenient for our purposes to treat it as a postulate.

The Parallel Postulate is the postulate that distinguishes Euclidean Geometry from other geometries. The postulate is stated in several ways, however we will use the following version in our studies.

The Parallel Postulate and Theorem

Given a line and a point not on the line, there is exactly one line through the point that is parallel to the given line.

The Parallel Postulate

Given a line and a point not on the line, there can be no more than one line through the point that is parallel to the given line.
The first statement differs from the Parallel Postulate in that it claims not only that the line through the point is unique, but also that the line actually does exist. For many centuries mathematicians believed that it was possible to prove both the *existence* and the *uniqueness* of such a line using the other postulates that were included in Euclid's *Elements.* They were able to prove the existence of such a line on the basis of the other postulates, but realized in the nineteenth century that uniqueness could not be proved. This realization led to the discovery of other geometries which do not include Euclid's Parallel Postulate as an assumption. Although the Parallel Postulate deals only with the uniqueness of the line, for convenience in our proofs, we will use the statement that implies both uniqueness and existence. Whenever we use this statement as a reason in our proofs, we will write "Parallel Postulate" as our reason.

The next group of postulates are likewise existence and uniqueness postulates.

The Perpendicular Postulates

Given a line and a point not on the line, there can be no more than one line through the point that is perpendicular to the given line.
In a plane, given a line and a point on the line, there is exactly one line through the point that is perpendicular to the given line.

The next postulate is also an existence postulate. It allows us to draw a line when we are given two points on the line.

The Unique Line Postulate

If A and B are two points, then there is exactly one line that contains points A and B
OR
Two points determine a line.

This postulate guarantees us that when we have two points, there is exactly one line that will contain those two points. Therefore in our proofs, we may refer to "the line through two points" although the line may not be illustrated in the original figure. Likewise we may use this postulate to refer to "the line segment joining two points" although the line segment may not be illustrated in the original figure.

The Unique Point Postulate

If AB is a ray, then there is a unique point P on ray AB such that AP equals any specified length.

The Unique Point Postulate is an existence postulate. It allows us to mark off a line segment of any length on a ray.

The Line Postulate

A line contains an infinite number of points.

The Line Postulate allows us to introduce one or more points on a line. It is also an existence postulate for our work.

Whole Greater Than Part Postulate

If $a + b = c$, then $c > a$ and $c > b$

With the Whole Greater Than Part Postulate we can make a transition from an equality to an inequality. It will be used primarily in cases where we have a point between two other points on a line segment and in cases where we have a ray between two other rays in a plane.

Consider the following:

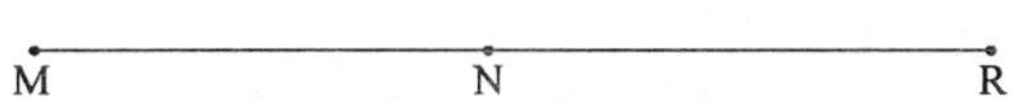

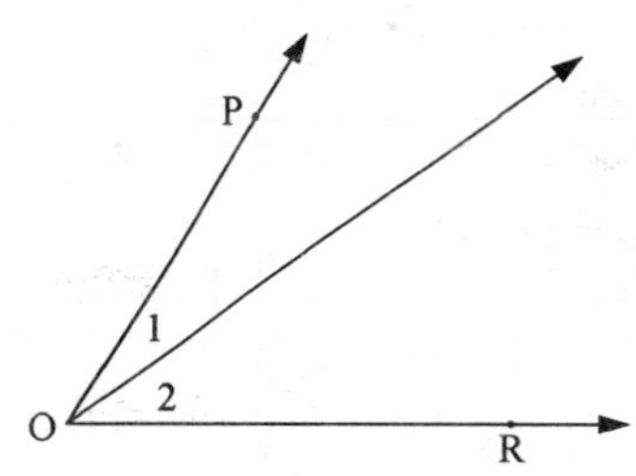

In the first illustration, we see that MN + NR = MR. We may use the Whole Greater Than Part Postulate to conclude that MR > MN and MR > NR.

In the second illustration we see that $\angle 1 + \angle 2 = \angle POR$. We may use the Whole Greater Than Part Postulate to conclude that $\angle POR > \angle 1$ and $\angle POR > \angle 2$

The next three postulates should not present much difficulty.

Trichotomy Postulate

If a and b are real numbers, then exactly one of the following statements must be true: a = b or a > b or a < b

Transitive Postulate of Inequality

If a > b and b > c then a > c

Addition Postulate of Inequality

If a > b then a + c > b + c

In the two proofs that follow, we will use the deductive process to prove that when two parallel lines are cut by a transversal, alternate interior angles have equal measures, and same side interior angles are supplementary.

We will call this theorem the:
ALTERNATE INTERIOR ANGLES THEOREM

THEOREM:

If two parallel lines are cut by a transversal, then alternate interior angles have equal measures.

GIVEN:

The figure with $\ell_1 \parallel \ell_2$

PROVE:

$\angle 1 = \angle 2$

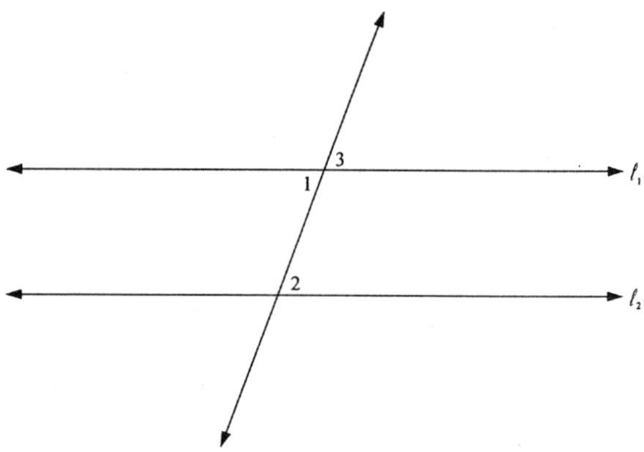

STATEMENTS	REASONS
1. $\ell_1 \parallel \ell_2$	1. Given
2. $\angle 2 = \angle 3$	2. Transversal Postulate: If two parallel lines are cut by a transversal, then the corresponding angles have equal measures.
3. $\angle 1 = \angle 3$	3. Theorem: Vertical angles have the same measure
4. $\angle 1 = \angle 2$	4. Transitive Postulate of Equality

We will call this theorem the:
SAME SIDE INTERIOR ANGLES THEOREM

THEOREM:
If two parallel lines are cut by a transversal, then same side interior angles are supplementary.

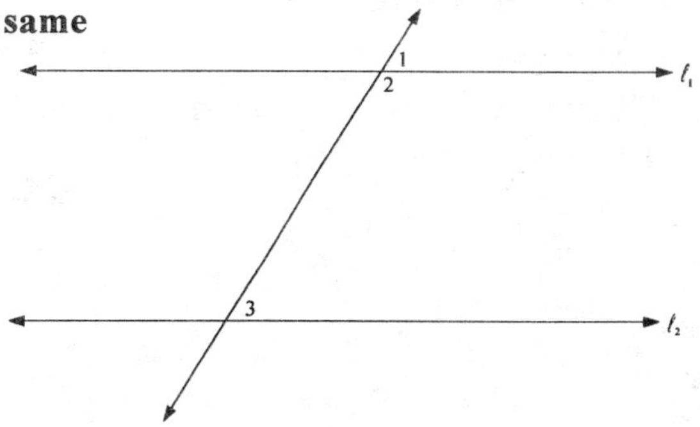

GIVEN:
The figure with $\ell_1 \parallel \ell_2$

PROVE:
$\measuredangle$ 2 and $\measuredangle$ 3 are supplementary.

Why don't you try to fill in the reasons in this proof?

STATEMENTS	REASONS
1. $\measuredangle$ 1 and $\measuredangle$ 2 form a linear pair.	1. Definition of Linear Pair
2. $\measuredangle$ 1 and $\measuredangle$ 2 are supplementary.	2. ?
3. $\angle 1 + \angle 2 = 180$	3. ?
4. $\ell_1 \parallel \ell_2$	4. Given
5. $\angle 1 = \angle 3$	5. Transversal Postulate
6. $\angle 3 + \angle 2 = 180$	6. Substitution Postulate of Equality
7. $\measuredangle$ 2 and $\measuredangle$ 3 are supplementary.	7. ?

With the Parallel Postulate, we can prove that the sum of the measures of the angles of a triangle is 180.

THEOREM:
The sum of the measures of the angles of a triangle is 180.

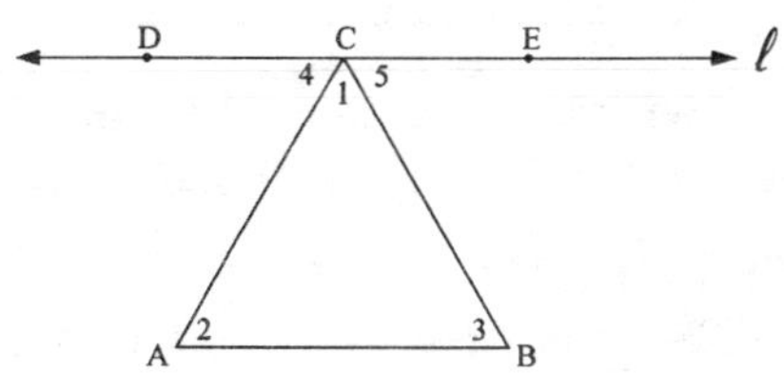

GIVEN:
The figure with $\triangle$ ABC

PROVE:
$\angle 1 + \angle 2 + \angle 3 = 180$

STATEMENTS	REASONS
1. Let ℓ be the line through C parallel to line AB.	1. Parallel Postulate
2. $\measuredangle$ DCA and $\measuredangle$ ACE form a linear pair.	2. Definition of Linear Pair
3. $\measuredangle$ DCA and $\measuredangle$ ACE are supplementary.	3. Linear Pair Postulate
4. $\angle DCA + \angle ACE = 180$	4. Definition of Supplementary Angles
5. $\angle 1 + \angle 5 = \angle ACE$	5. Whole Equals Sum Of Parts Postulate
6. $\angle 1 + \angle 5 + \angle 4 = 180$ (Recall $\measuredangle$ 4 is another way to name $\measuredangle$ DCA.)	6. Substitution Postulate of Equality
7. $\angle 4 = \angle 2$; $\angle 5 = \angle 3$	7. Alternate Interior Angles Theorem: If parallel lines are cut by a transversal, alternate interior angles have equal measures.
8. $\angle 1 + \angle 3 + \angle 2 = 180$	8. Substitution Postulate of Equality

Now we will prove other theorems that are based upon the theorem that we have just proved. See if you can fill in the missing information in these proofs.

We will call this theorem the:
EXTERIOR ANGLE THEOREM

THEOREM:
The measure of an exterior angle of a triangle equals the sum of the measures of its two remote interior angles.

GIVEN:
The figure with $\triangle$ ABC
exterior angle $\measuredangle$ 4
interior angles $\measuredangle$ 1 , $\measuredangle$ 2 and $\measuredangle$ 3
remote interior angles $\measuredangle$ 1 and $\measuredangle$ 2

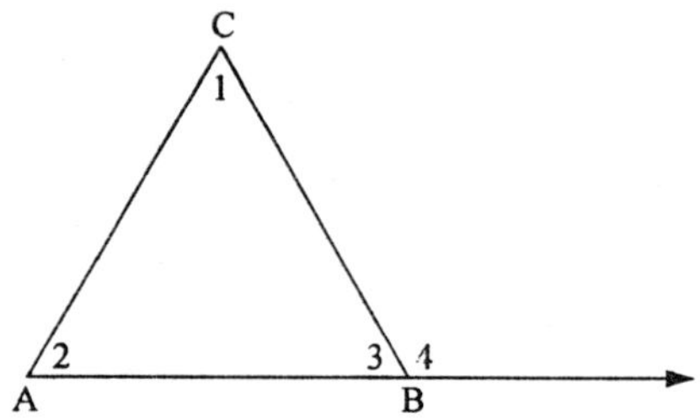

PROVE:
$\angle 4 = \angle 1 + \angle 2$

STATEMENTS	REASONS
1. The figure with $\triangle$ ABC exterior angle $\measuredangle$ 4 interior angles $\measuredangle$ 1 , $\measuredangle$ 2 and $\measuredangle$ 3 remote interior angles $\measuredangle$ 1 and $\measuredangle$ 2	1. ?
2. $\angle 1 + \angle 2 + \angle 3 = 180$	2. ?
3. $\measuredangle$ 3 and $\measuredangle$ 4 form a linear pair.	3. Definition of Linear Pair
4. $\measuredangle$ 3 and $\measuredangle$ 4 are supplementary.	4. ?
5. $\angle 3 + \angle 4 = 180$	5. ?
6. $\angle 1 + \angle 2 + \angle 3 = \angle 3 + \angle 4$	6. ?
7. $\angle 3 = \angle 3$	7. ?
8. $\angle 1 + \angle 2 = \angle 4$	8. Subtraction Postulate of Equality

We will call this theorem the:
REMOTE INTERIOR ANGLE THEOREM

THEOREM:
The measure of an exterior angle of a triangle is greater than the measure of each of its remote interior angles.

GIVEN:
The figure with $\triangle$ ABC
exterior angle $\measuredangle$ 4
interior angles $\measuredangle$ 1, $\measuredangle$ 2 and $\measuredangle$ 3
remote interior angles $\measuredangle$ 1 , $\measuredangle$ 2

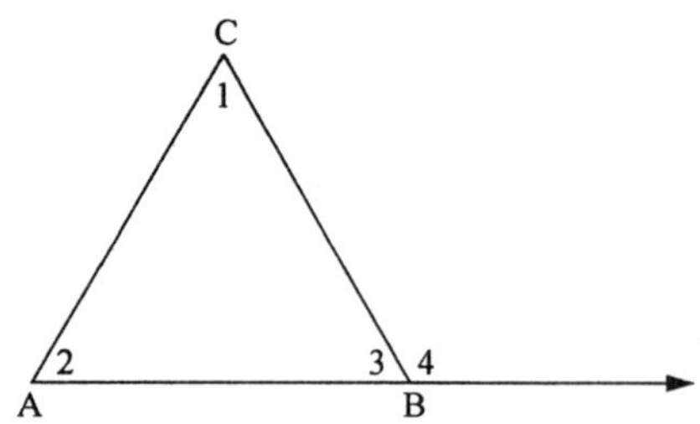

PROVE: $\angle 4 > \angle 1$; $\angle 4 > \angle 2$

STATEMENTS	REASONS
1. The figure with $\triangle$ ABC exterior angle $\measuredangle$ 4 interior angles $\measuredangle$ 1 , $\measuredangle$ 2 and $\measuredangle$ 3 remote interior angles $\measuredangle$ 1 , $\measuredangle$ 2	1. ?
2. $\angle 1 + \angle 2 = \angle 4$	2. ?
3. $\angle 4 > \angle 1$; $\angle 4 > \angle 2$	3. Whole Greater Than Part Postulate

THEOREM:

If the measures of two angles of a triangle equal, respectively, the measures of two angles of another triangle, then the two remaining angles have equal measures.

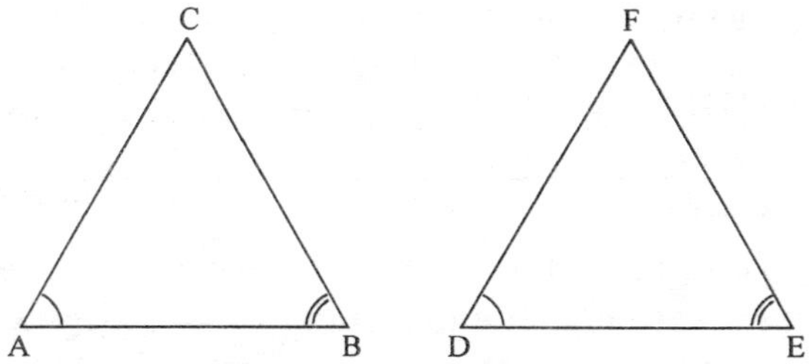

GIVEN:

△ ABC and △ DEF with ∠ A = ∠ D
and ∠ B = ∠ E

PROVE:

∠ C = ∠ F

STATEMENTS	REASONS
1. △ ABC and △ DEF with ∠ A = ∠ D and ∠ B = ∠ E	1. ?
2. ∠ A + ∠ B = ∠ D + ∠ E	2. Addition Postulate of Equality
3. ∠ A + ∠ B + ∠ C = 180 ∠ D + ∠ E + ∠ F = 180	3. ?
4. ∠ A + ∠ B + ∠ C = ∠ D + ∠ E + ∠ F	4. ?
5. ∠ C = ∠ F	5. Subtraction Postulate of Equality

You will be asked to prove these theorems as exercises:

THEOREM:

In a right triangle, the two angles that are not right angles are complementary.

THEOREM:

In a right triangle, the two angles that are not right angles are acute.

THEOREM:

The acute angles of a right triangle are complementary.

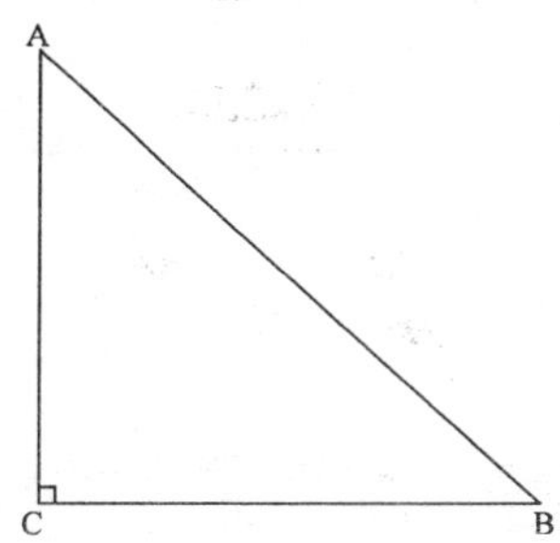

GIVEN:

Right triangle ACB with
acute angles ∡ A and ∡ B and right angle, ∡ C

PROVE:

∡ A and ∡ B are complementary.

STATEMENTS	REASONS
1. Right triangle ACB with acute angles ∡ A and ∡ B and right angle, ∡ C	1. ?
2. ∠ C = 90	2. ?
3. ∠ A + ∠ B + ∠ C = 180	3. ?
4. ∠ A + ∠ B = 90	4. Subtraction Postulate of Equality
5. ∡ A and ∡ B are complementary.	5. ?

THEOREM:
Each angle of an equiangular triangle has a measure of 60.

You will be asked to prove this theorem as one of your exercises.

The next two theorems give us two more methods for proving that two triangles are congruent. The first theorem applies to any type of triangles; while the second theorem applies only to right triangles.

THEOREM:
If the measures of two angles and the length of the side opposite one of these angles equal, respectively, the measures of two angles and the length of the corresponding side in another triangle, then the two triangles are congruent. AAS THEOREM

GIVEN:
△ ABC ≅ △ DEF with ∠ A = ∠ D
∠ B = ∠ E and BC = EF

PROVE:
△ ABC ≅ △ DEF

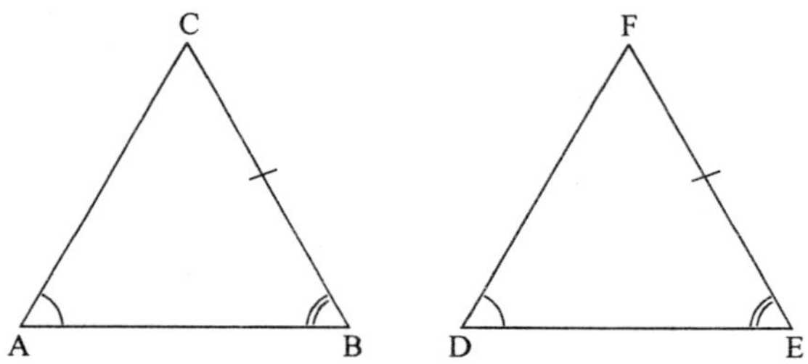

STATEMENTS	REASONS
1. △ ABC ≅ △ DEF with ∠ A = ∠ D ∠ B = ∠ E	1. ?
2. ∠ C = ∠ F	2. ?
3. BC = EF	3. ?
4. △ ABC ≅ △ DEF	4. ASA Postulate (Steps 1, 2, 3)

THEOREM:

If the lengths of the hypotenuse and a leg of one triangle equal, respectively, the lengths of the hypotenuse and a leg of another triangle, then the triangles are congruent.
HL THEOREM

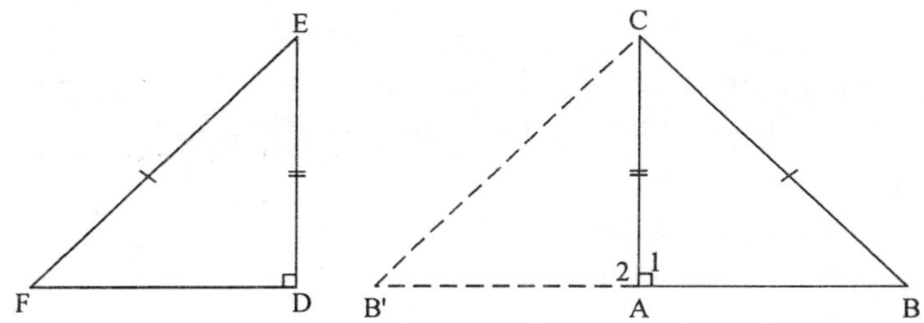

GIVEN:

△ CAB with right angle, ∡ A
△ EDF with right angle, ∡ D
BC = FE and AC = DE

PROVE:

△ CAB ≅ △ EDF

STATEMENTS	REASONS
1. △ CAB with right angle, ∡ A △ EDF with right angle, ∡ D	1. ?
2. Locate point B' on the ray opposite ray AB so that AB' = DF.	2. Unique Point Postulate
3. Construct line segment B'C.	3. Postulate: Two points determine a line
4. $\overleftrightarrow{CA} \perp \overleftrightarrow{BB'}$	4. Definition of perpendicular lines
5. ∡ 2 is a right angle.	5. Definition of perpendicular lines
6. ∠ 2 = ∠ 1 = ∠ D	6. Theorem: All right angles have equal measures
7. DE = AC	7. Given
8. △ CAB' ≅ △ EDF	8. SAS Postulate (Steps 2, 6, 7)
9. ∠ B' = ∠ F	9. CPCTE
10. FE = B'C	10. CPCTE
11. FE = BC	11. Given
12. B'C = BC	12. Transitive Postulate of Equality
13. ∠ B' = ∠ B in △ B'CB	13. Isosceles Triangle Theorem
14. ∠ B = ∠ F	14. Transitive Postulate of Equality (Steps 9, 13)
15. △ CAB ≅ △ EDF	15. AAS Theorem (Steps 6, 7, 14)

We may also prove that right triangles are congruent by proving that the two triangles have hypotenuses with equal lengths and a pair of acute angles with equal measures. You will be asked to prove this theorem as one of your exercises.

THEOREM:

If the length of the hypotenuse and the measure of an acute angle of one triangle equal, respectively, the length of the hypotenuse and the measure of an acute angle of another triangle, then the two triangles are congruent. HA THEOREM

The next three theorems that we will prove involve inequalities. Carefully study the proofs of these theorems.

We will call this theorem the:
GREATER SIDE THEOREM

THEOREM:
Given two sides of a triangle and the angles opposite those sides, then the angle that has the greater measure is opposite the side that has the greater length.

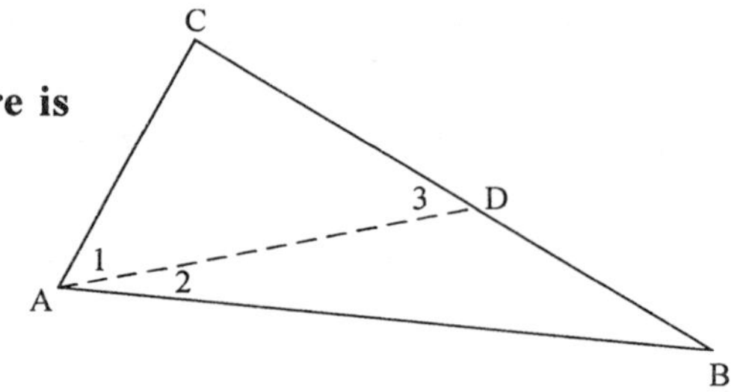

GIVEN:
The figure with BC > AC

PROVE:
$\angle$ CAB > $\angle$ B

STATEMENTS	REASONS
1. Locate point D on ray CB so that CD = AC.	1. Unique Point Postulate
2. $\angle$ 1 = $\angle$ 3 in $\triangle$ ACD	2. ?
3. $\angle$ CAB = $\angle$ 1 + $\angle$ 2	3. Whole Equals Sum of Parts Postulate
4. $\angle$ CAB > $\angle$ 1	4. Whole Greater Than Parts Postulate
5. $\angle$ CAB > $\angle$ 3	5. Substitution Postulate of Equality
6. $\angle$ 3 > $\angle$ B	6. Remote Interior Angle Theorem
7. $\angle$ CAB > $\angle$ B	7. Transitive Postulate of Equality

You will recognize the next theorem as the converse of the theorem that we just proved. Instead of proving the theorem directly, we will prove that its contrapositive is true. A proof using this technique is called an indirect proof. In an indirect proof we suppose that the statement we want to prove true is false, and then show that such a supposition leads to a contradiction of our given information. When we use the indirect method of proof, it is not always convenient to use a two-column format. In this proof we choose to use a paragraph, or discussion format.

We will call this theorem the:
GREATER ANGLE THEOREM

THEOREM:
Given two angles of a triangle and the sides opposite those angles, the side with the greater length is opposite the angle with the greater measure.

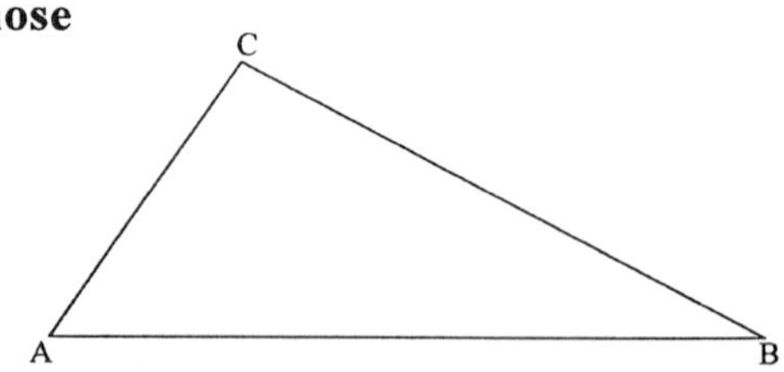

GIVEN:
In $\triangle$ ABC, $\angle$ A > $\angle$ B

PROVE: BC > AC

By the Trichotomy Postulate, exactly one of the following conditions must be true for BC and AC:

1. BC = AC   2. BC < AC   3. BC > AC

Let us suppose that #1 is true, that is, BC = AC. Then we must conclude that $\angle$ A = $\angle$ B (Isosceles Triangle Theorem). This conclusion contradicts our given information, $\angle$ A > $\angle$ B. The Trichotomy Postulate states that both relationships cannot be true.
Let us suppose that #2 is true, that is, BC < AC. Then we must conclude that $\angle$ A < $\angle$ B (Greater Side Theorem). This conclusion also contradicts our given information. The Trichotomy Postulate states that both relationships cannot be true.
We have no choice but to conclude that #3 is true, that is, BC > AC.
We have proved the contrapositive of the theorem: If BC $\ngtr$ AC, then in $\triangle$ ABC, $\angle$ A $\ngtr$ $\angle$ B.

Now we can prove that the hypotenuse is the longest side in a right triangle,

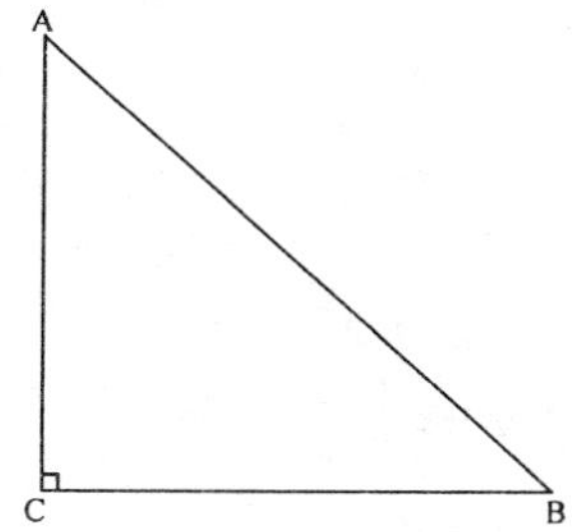

THEOREM:

In a right triangle, the hypotenuse is the longest side.

GIVEN:

△ ACB with right angle ACB and acute angles, ∠ A and ∠ B

PROVE:

AB > AC and AB > BC

STATEMENTS	**REASONS**
1. △ ACB with right angle ACB and acute angles A and B	1. Given
2. ∠ A and ∠ B are complementary.	2. Theorem: In a right triangle the acute angles are complementary.
3. ∠ A + ∠ B = 90	3. ?
4. 90 > ∠ A and 90 > ∠ B	4. Whole Greater Than Part Postulate
5. AB > BC and AB > AC	5. ?

The next theorem is used frequently in advanced courses in mathematics.

This theorem is called the:

TRIANGULAR INEQUALITY THEOREM

THEOREM:

The sum of the lengths of two sides of a triangle is always greater than the length of the remaining side of the triangle.

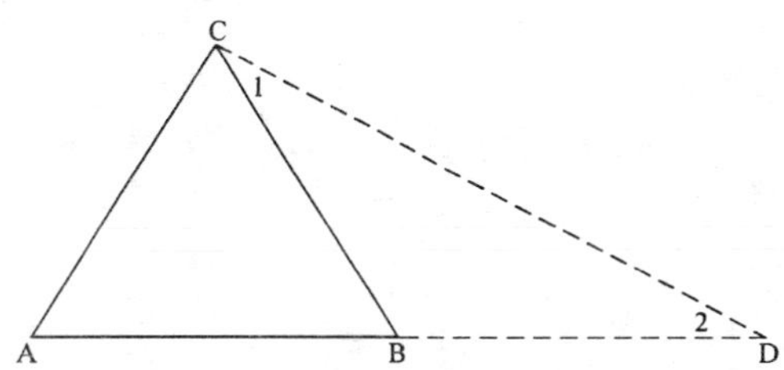

GIVEN:

△ ABC

PROVE:

AB + BC > AC
AC + CB > AB
AC + AB > BC

NOTE: We will prove only one of these relationships (AB + BC > AC) since the proofs of the other two inequalities are similar.

STATEMENTS	**REASONS**
1. Locate point D on the ray opposite ray BA so that BD = BC.	1. Unique Point Postulate
2. Construct line segment DC.	2. ?
3. ∠ 1 = ∠ 2 in △ CBD	3. ?
4. ∠ ACD = ∠ ACB + ∠ 1	4. ?
5. ∠ ACD > ∠ 1	5. ?
6. ∠ ACD > ∠ 2	6. ?
7. In △ ACD, AD > AC	7. Greater Angle Theorem
8. AB + BD = AD	8. ?
9. AB + BC = AD	9. Substitution Postulate of Equality (Steps 1, 8)
10. AB + BC > AC	10. Substitution Postulate of Equality (Steps 7 , 9)

The following theorem may be useful in some of the proofs that we will do later.

We will call this theorem the:
ADDITION THEOREM OF INEQUALITY

THEOREM:
If a > b and c > d, then a + c > b + d.

GIVEN:
a > b and c > d

PROVE:
a + c > b + d

STATEMENTS	REASONS
1. a > b	1. Given
2. a + c > b + c	2. Addition Postulate of Inequality
3. c > d	3. Given
4. b + c > b + d	4. Addition Postulate of Inequality
5. a + c > b + d	5. Transitive Postulate of Inequality

Exercises Section 7.1

(*Figures are not drawn to scale*)

1.* Prove the following theorem:

THEOREM
Each angle of an equiangular triangle has a measure of 60.

2.* Prove the following theorem:

THEOREM
If the length of the hypotenuse and the measure of an acute angle of one triangle equal, respectively, the length of the hypotenuse and the measure of an acute angle of another triangle, then the two triangles are congruent.

3. Name the sides of the triangle in order of decreasing lengths.

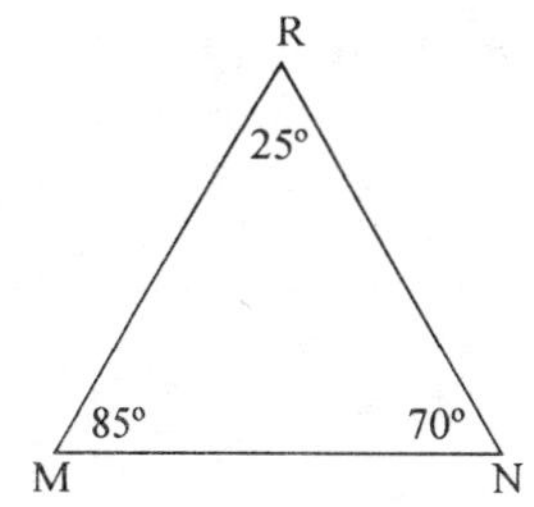

4. In △ KLM, ∠ K > ∠ M. Which of the following statements can be true for ML and KL?

a. ML = KL   b. ML > KL   c. ML < KL

5. In △ RST, SR < RT. Which of the following statements can be true for ∠ T and ∠ S?

a. ∠ T = ∠ S   b. ∠ T > ∠ S   c. ∠ T < ∠ S

6. In the figure, which segment is longest?

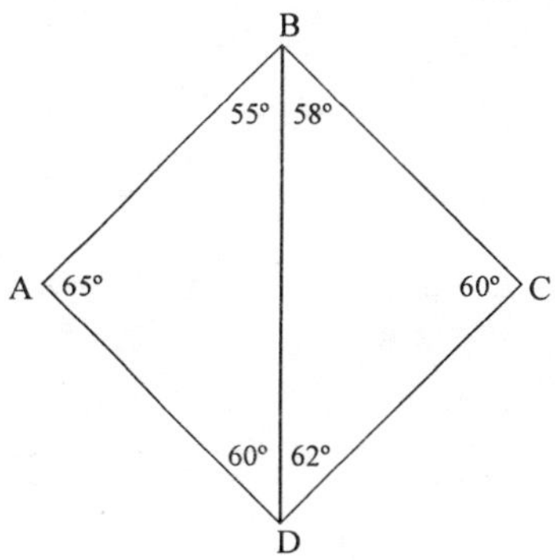

7. In the figure, which segment is shortest?

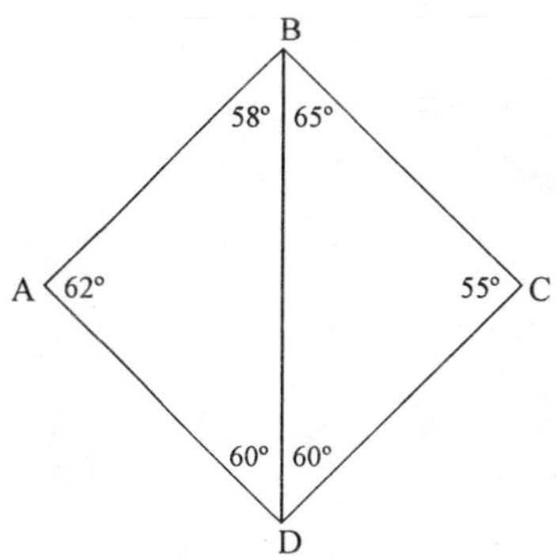

8. Which of the following sets of numbers can be the lengths of sides of a triangle?

a. 6, 6, 6   b. 15, 23, 42   c. 7, 8, 9   d. 6, 6, 12

9. Determine the acceptable value(s) for x, if the following variable expressions represent the lengths of sides of a triangle:

a. x $\quad x + 2$ $\quad x + 5$

b. x $\quad x + 1$ $\quad x + 2$

c. $x - 1$ $\quad x - 3$ $\quad x - 4$

d. $x + 8$ $\quad x - 6$ $\quad x + 5$

e. $3x - 2$ $\quad 3x + 2$ $\quad 3x + 5$

10. The following expressions represent the measures of angles of a triangle. Determine the measure of each angle for each triangle:

a. $2x + 5$ $3x + 8$ $4x + 2$

b. $3x$ $10x$ $7x$

11. **Given:** The figure

Prove: PR > QR

Hint: The first step of your proof should be PQ + QR = PR

STATEMENTS	REASONS

12. **Given:** The figure

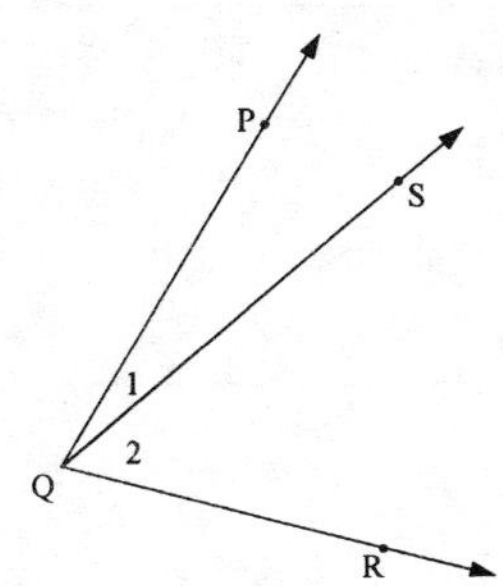

Prove: $\angle$ PQR > $\angle$ 2

STATEMENTS	REASONS

13. **Given:** The figure with AB > DE

Prove: AD > BE

1. AB > DE	1. ?
2. AB + BD > DE + BD	2. Addition Postulate of Inequality
3. AB + BD = AD DE + BD = BE	3. ?
4. AD > BE	4. Substitution Postulate of Equality

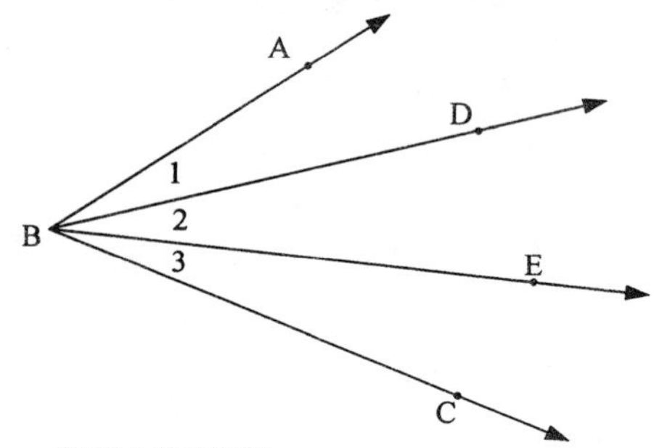

14. **Given:** The figure with ∠ 1 > ∠ 3

Prove: ∠ ABE > ∠ DBC

STATEMENTS	REASONS

15. **Given:** The figure with AB > DB and BC > BE

Prove: AC > DE

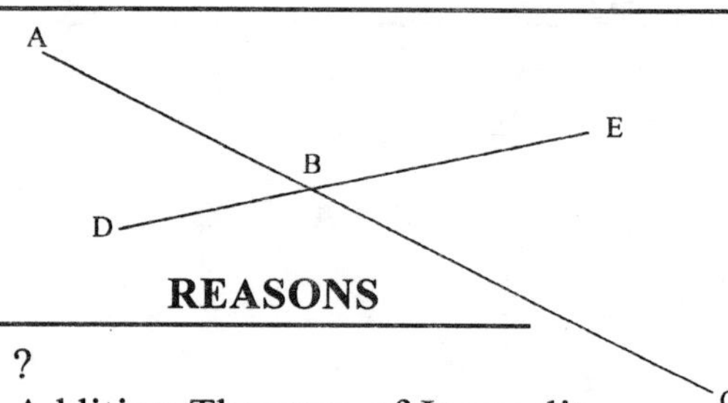

STATEMENTS	REASONS
1. AB > DB; BC > BE	1. ?
2. AB + BC > DB + BE	2. Addition Theorem of Inequality
3. AB + BC = AC DB + BE = DE	3. ?
4. AC > DE	4. Substitution Postulate of Equality

16. **Given:** The figure with ∠ 1 > ∠ 3; ∠ 2 > ∠ 4

Prove: ∠ ABC > ∠ ACB

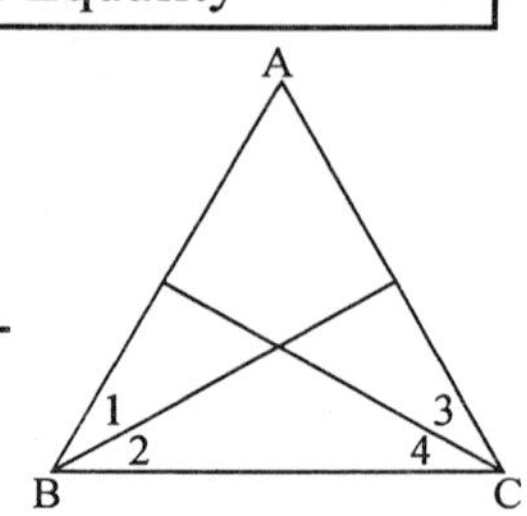

STATEMENTS	REASONS

17. In the figure, prove that $\angle 1 > \angle 3$.

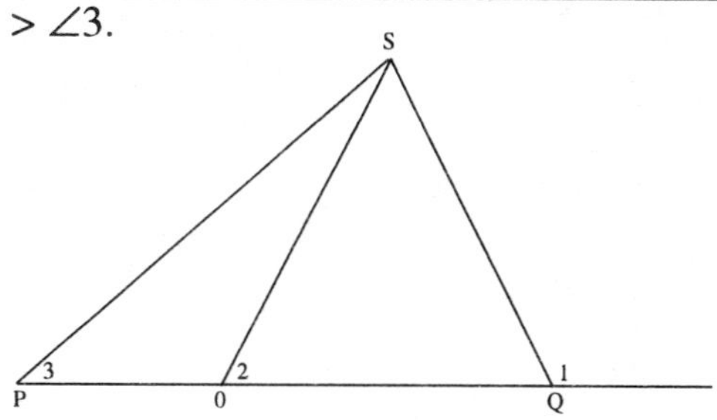

STATEMENTS	REASONS

18. **Given:** The figure

Prove: $\angle 1 > \angle P$

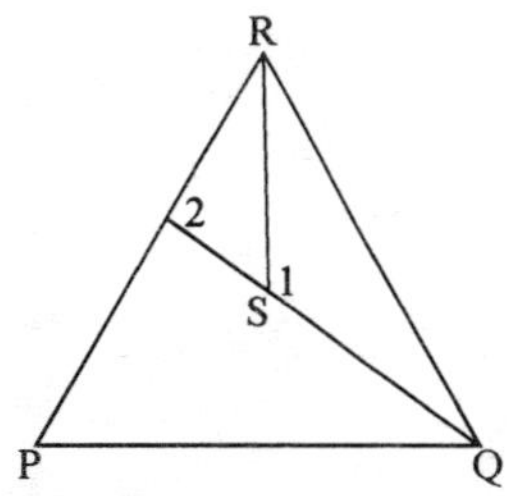

STATEMENTS	REASONS

19. **Given:** The figure with $\angle 1 > \angle 2$

Prove: AC > AD by filling in the missing reasons in the proof:

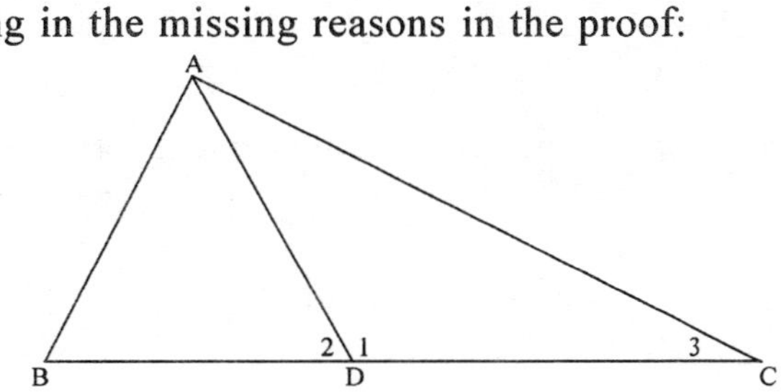

STATEMENTS	REASONS
1. $\angle 1 > \angle 2$	1. ?
2. $\angle 2 > \angle 3$	2. ?
3. $\angle 1 > \angle 3$	3. ?
4. AC > AD	4. Greater Angle Theorem

20. **Given:** The figure with $\angle 1 > \angle 2$

Prove: PS > QS

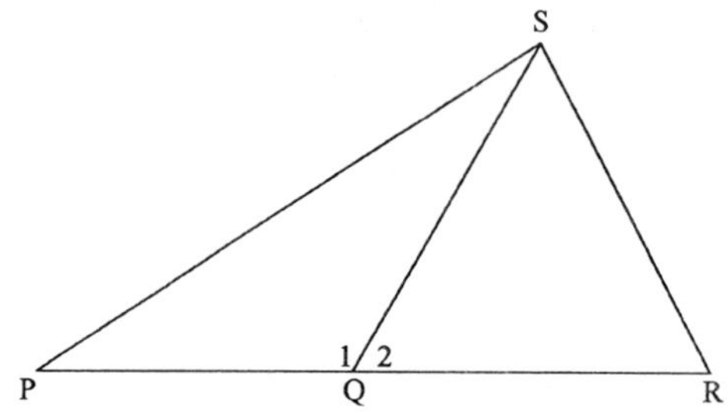

STATEMENTS	**REASONS**

21. **Given:** $\angle R > \angle S$; $\angle Q > \angle P$

Prove: SP > RQ

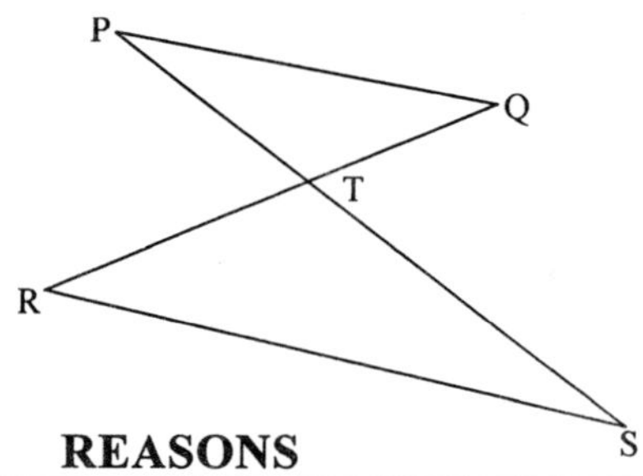

STATEMENTS	**REASONS**

22. Can you prove that $a^2 + 5$ is always a positive number when a represents a real number?

Section 7.2 Proving that Lines are Parallel and Perpendicular Lines

Now we will prove that the converses of the transversal theorems are also true. As you will see, these theorems may be used to prove that two lines in a plane are parallel if certain relationships exist between the angles formed by these lines and a transversal. We will use the indirect method to prove the first theorem.

We will call this theorem the:
CORRESPONDING ANGLES/PARALLEL LINES THEOREM

THEOREM:
If two lines are cut by a transversal so that corresponding angles have equal measures, then the two lines are parallel.

GIVEN:
ℓ_1 and ℓ_2 cut by transversal t with
$\angle 1 = \angle 2$
ℓ_1, ℓ_2 and t are coplanar.

PROVE:
$\ell_1 \parallel \ell_2$

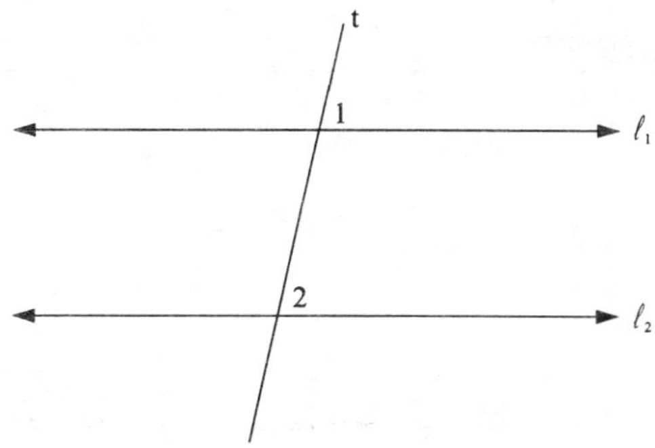

We begin our proof with the following supposition:

Supposition: ℓ_1 is not parallel to ℓ_2, that is, ℓ_1 and ℓ_2 intersect in a point that we will call P as shown in the figure.

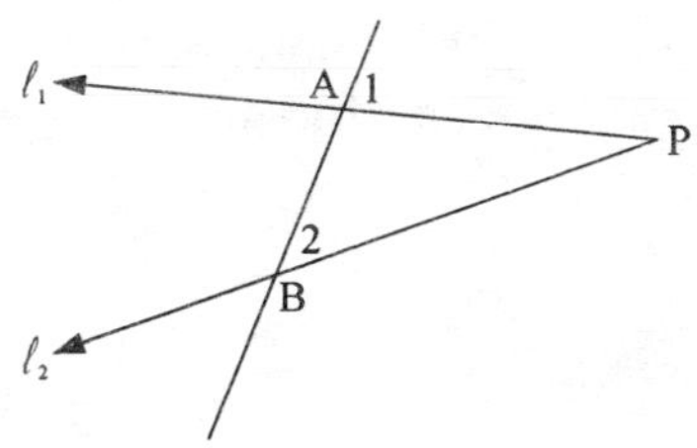

In $\triangle$ ABP, $\angle$ 1 is an exterior angle with $\angle$ 2 as one of its remote interior angles. $\angle$ 1 must then be greater than $\angle$ 2. Recall that the measure of an exterior angle of a triangle must be greater than the measure of either of its remote interior angles. This conclusion contradicts our given information that $\angle 1 = \angle 2$. By the Trichotomy Postulate, both relationships cannot be true. We are forced to reject our supposition and conclude that ℓ_1 must be parallel to ℓ_2

For the next three proofs, we will use our usual two column format. You should fill in the missing steps and reasons in these proofs.

We will call this theorem the:
ALTERNATE INTERIOR ANGLE/PARALLEL LINES THEOREM

THEOREM:
If two lines are cut by a transversal so that alternate interior angles have equal measures, then the lines are parallel.

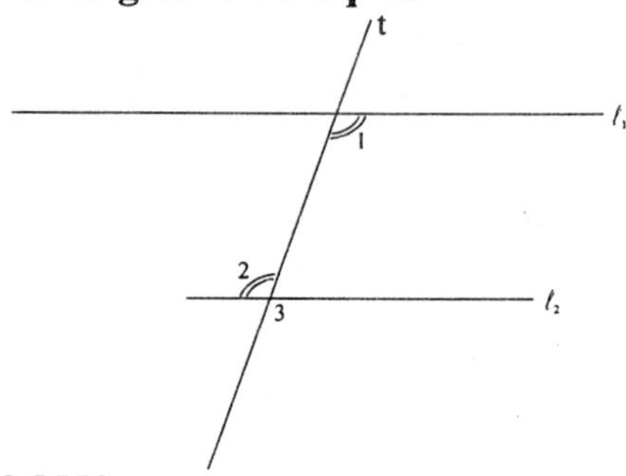

GIVEN:
The figure with $\angle 1 = \angle 2$
ℓ_1, ℓ_2 and t are coplanar.

PROVE:
$\ell_1 \parallel \ell_2$

STATEMENTS	REASONS
1. In the figure, $\angle 3$ and $\angle 2$ are vertical angles.	1. ?
2. ?	2. ?
3. $\angle 1 = \angle 2$	3. ?
4. $\angle 1 = \angle 3$	4. ?
5. $\ell_1 \parallel \ell_2$	5. Corresponding Angles/Parallel Lines Theorem

We will call this theorem the:
SAME SIDE INTERIOR ANGLES/PARALLEL LINES THEOREM

THEOREM:
If two lines are cut by a transversal so that same side interior angles are supplementary, then the lines are parallel.

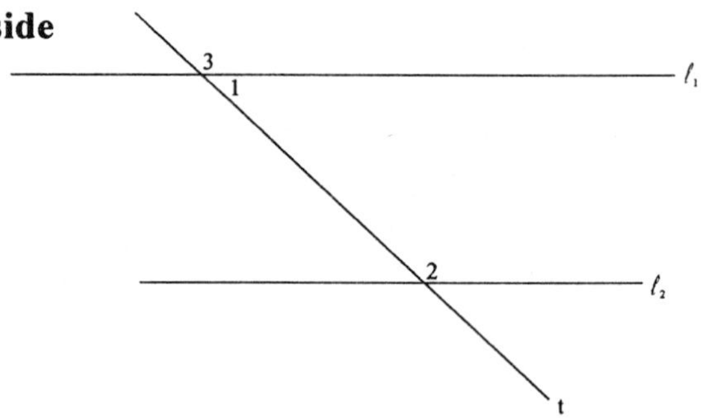

GIVEN:
The figure with $\angle 1$ supplementary to $\angle 2$

PROVE:
$\ell_1 \parallel \ell_2$

STATEMENTS	REASONS
1. In the figure, $\angle 1$ and $\angle 3$ form a linear pair.	1. Definition of Linear Pair
2. $\angle 1$ and $\angle 3$ are supplementary.	2. ?
3. $\angle 1$ and $\angle 2$ are supplementary.	3. ?
4. $\angle 2 = \angle 3$	4. ?
5. $\ell_1 \parallel \ell_2$	5. ?

THEOREM:

In a plane, if two lines are perpendicular to the same line, then the two lines are parallel.

GIVEN:

ℓ_1, ℓ_2 and t are coplanar lines with $\ell_1 \perp t$ and $\ell_2 \perp t$

PROVE:

$\ell_1 \parallel \ell_2$

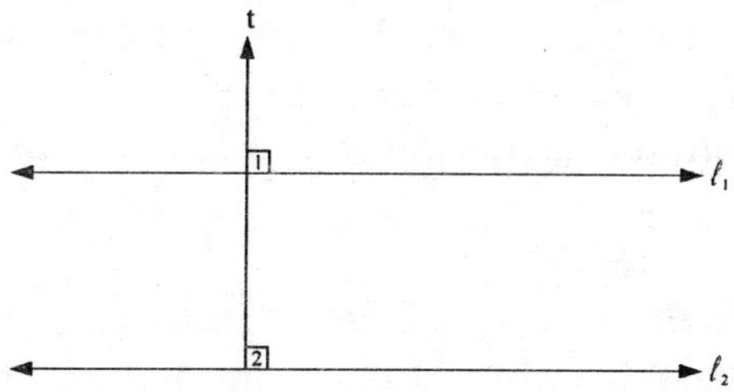

STATEMENTS	REASONS
1. $\ell_1 \perp t$, $\ell_2 \perp t$	1. Given
2. $\measuredangle$ 1 and $\measuredangle$ 2 are right angles	2. ?
3. $\angle 1 = \angle 2$	3. ?
4. $\ell_1 \parallel \ell_2$	4. ?

If you recall, we defined the distance from a point not on a line to the line to be the length of the perpendicular line segment joining the point to the line. This definition is a good one because this length is the shortest distance from the point to the line. Now, we will prove that this is so.

THEOREM:

The shortest distance from a point not on a line to the line is the length of the perpendicular line segment joining the point to the line.

GIVEN:

Line ℓ with $\overline{PA} \perp \ell$, Point B is a point on ℓ different from point A

PROVE:

PB > PA

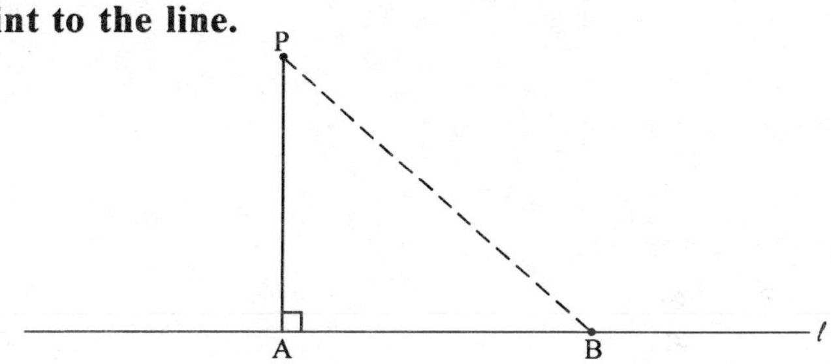

STATEMENTS	REASONS
1. Line ℓ with $\overline{PA} \perp \ell$, Point B is a point on ℓ different from point A	1. Given
2. Construct $\overline{PB}$.	2. ?
3. $\measuredangle$ PAB is a right angle.	3. ?
4. $\triangle$ PAB is a right triangle.	4. ?
5. PB > PA	5. THEOREM: In a right triangle, the hypotenuse is the longest side.

Exercises 7.2

1.* A line that is perpendicular to a line segment and also bisects the segment is called the perpendicular bisector of the segment. Prove that each point on the perpendicular bisector of a line segment is equidistant from the endpoints of the line segment.

THEOREM:
Each point on the perpendicular bisector of a line segment is equidistant from the endpoints of the line segment.

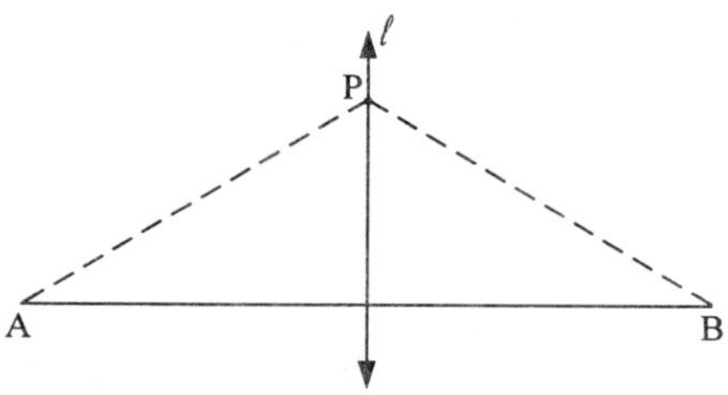

GIVEN:
Line segment with perpendicular bisector, line ℓ
P represents any point on line ℓ.

PROVE:
PA = PB

STATEMENTS	REASONS

2. Why are lines m and n parallel?

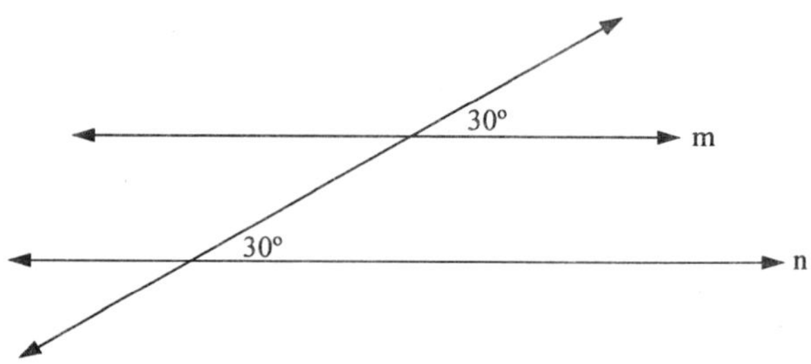

3. Why are lines m and n parallel?

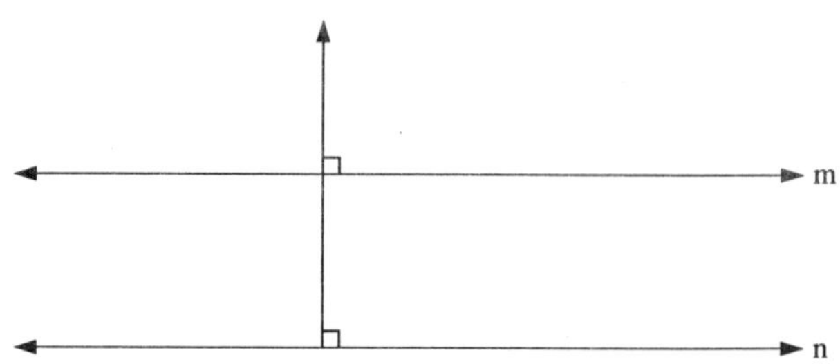

4. Why are lines m and n parallel?

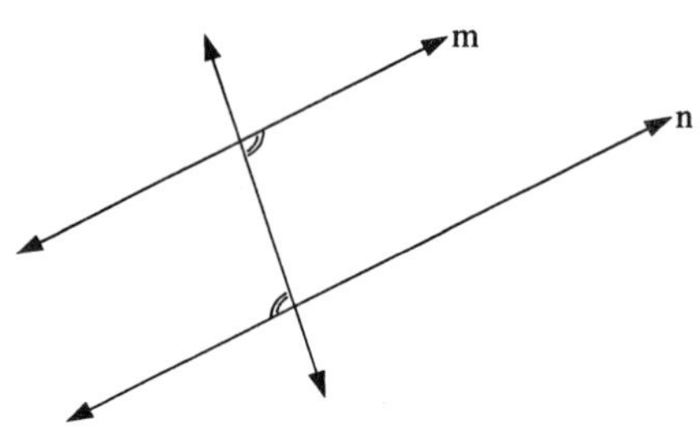

Chapter 7 Highlights

Can You Define the Following?

*The Perpendicular Bisector of a Line Segment
Distance from a Point Not on a Line to the Line

Do You Know and Understand these Postulates?

The Transversal Postulate
If two parallel lines are cut by a transversal, then corresponding angles have equal measures.

The Parallel Postulate *And Theorem*
Given a line and a point not on the line, there is exactly one line through the point that is parallel to the given line.

The Parallel Postulate
Given a line and a point not on the line, there can be no more than one line through the point that is parallel to the given line.

The Perpendicular Postulates
Given a line and a point not on the line, there can be no more than one line through the point that is perpendicular to the given line.
In a plane, given a line and a point on the line, there is exactly one line through the point that is perpendicular to the given line.

The Unique Line Postulate
If A and B are two points, then there is exactly one line that contains points A and B
OR
Two points determine a line.

The Unique Point Postulate
If AB is a ray then there is a unique point P on ray AB such that AP equals any specified length.

The Line Postulate
A line contains an infinite number of points.

Whole Greater Than Part Postulate
If $a + b = c$, then $c > a$ and $c > b$

Trichotomy Postulate
If a and b are real numbers, then exactly one of the following statements must be true:
$a = b$ or $a > b$ or $a < b$

Transitive Postulate of Inequality
If $a > b$ and $b > c$ then $a > c$

Addition Postulate of Inequality
If $a > b$ then $a + c > b + c$

Do You Know and Understand these Theorems?

Alternate Interior Angles Theorem

If two parallel lines are cut by a transversal, then alternate interior angles have equal measures.

Same Side Interior Angles Theorem

If two parallel lines are cut by a transversal, then same side interior angles are supplementary.

The sum of the measures of the angles of a triangle is 180.

Exterior Angle Theorem

The measure of an exterior angle of a triangle equals the sum of the measures of its two remote interior angles.

Remote Interior Angle Theorem

The measure of an exterior angle of a triangle is greater than the measure of each of its remote interior angles.

If the measures of two angles of a triangle equal, respectively, the measures of two angles of another triangle, then the two remaining angles have equal measures.

In a right triangle, the two angles that are not right angles are complementary.

In a right triangle, the two angles that are not right angles are acute.

The acute angles of a right triangle are complementary.

Each angle of an equiangular triangle has a measure of 60.

If the measures of two angles and the length of the side opposite one of these angles equal, respectively, the measures of two angles and the length of the corresponding side in another triangle, then the two triangles are congruent. AAS THEOREM

If the lengths of the hypotenuse and a leg of one triangle equal, respectively, the lengths of the hypotenuse and a leg of another triangle, then the triangles are congruent. HL THEOREM

If the length of the hypotenuse and the measure of an acute angle of one triangle equal, respectively, the length of the hypotenuse and the measure of an acute angle of another triangle, then the two triangles are congruent. HA THEOREM

Greater Side Theorem

Given two sides of a triangle and the angles opposite those sides, then the angle that has the greater measure is opposite the side that has the greater length.

Greater Angle Theorem

Given two angles of a triangle and the sides opposite those angles, the side with the greater length is opposite the angle with the greater measure.

In a right triangle, the hypotenuse is the longest side.

Triangular Inequality Theorem

The sum of the lengths of two sides of a triangle is always greater than the length of the remaining side of the triangle.

Addition Theorem of Inequality

If $a > b$ and $c > d$, then $a + c > b + d$

Corresponding Angles/Parallel Lines Theorem

If two lines are cut by a transversal so that corresponding angles have equal measures, then the two lines are parallel.

Alternate Interior Angle/Parallel Lines Theorem

If two lines are cut by a transversal so that alternate interior angles have equal measures, then the lines are parallel.

Same Side Interior Angles/Parallel Lines Theorem

If two lines are cut by a transversal so that same side interior angles are supplementary, then the lines are parallel.

In a plane, if two lines are perpendicular to the same line, then the two lines are parallel.

The shortest distance from a point not on a line to the line is the length of the perpendicular line segment joining the point to the line.

Chapter Eight

Parallelograms

Objectives

- List the properties of a quadrilateral
- List the properties of a parallelogram
- List the properties of a rhombus
- List the properties of a rectangle
- List the properties of a square
- List the ways to prove that a quadrilateral is a parallelogram
- List the ways to prove that a quadrilateral is a rhombus
- List the ways to prove that a quadrilateral is a rectangle
- List the ways to prove that a quadrilateral is a square
- Prove and apply theorems about quadrilaterals

Fibonacci Numbers

1	1	2	3	5	8	13	21	34	?	?	?	?

Can you find the next four numbers?

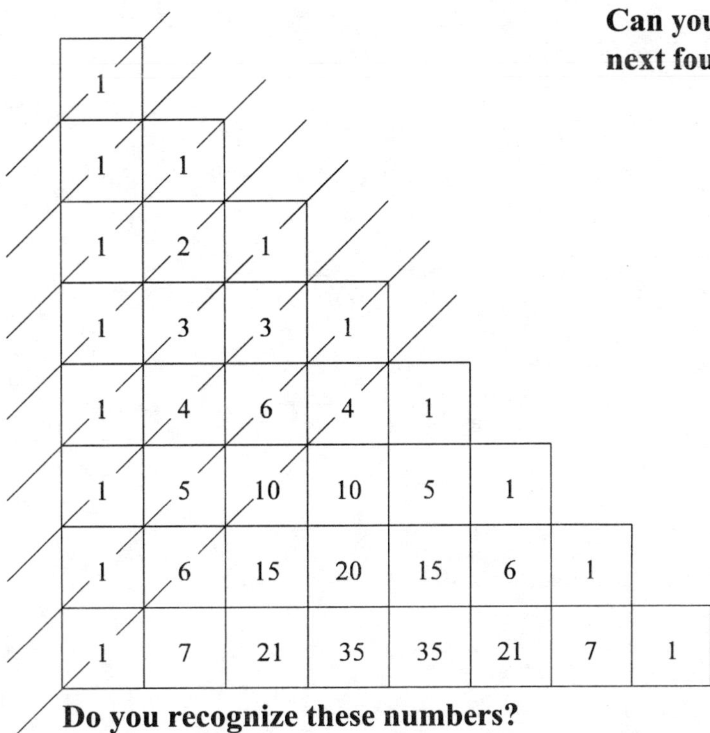

Do you recognize these numbers?
Find the sum of the numbers along each diagonal.

In this chapter we will study different kinds of quadrilaterals and parallelograms and investigate their properties. We will also examine the methods for proving that a given quadrilateral is a parallelogram.

Before we begin, we will review the definitions for the polygons that we will be using in this chapter.

Quadrilateral	A polygon with four sides
Parallelogram	A quadrilateral with two pairs of parallel sides
Rhombus	A quadrilateral with all sides of equal length (An equilateral quadrilateral)
Rectangle	A quadrilateral with all angles of equal measure (An equiangular quadrilateral)
Square	A quadrilateral with all angles of equal measure and all angles of equal length (An equiangular and equilateral quadrilateral)
Isosceles Trapezoid	A trapezoid whose legs have equal measures.

Consecutive angles of a quadrilateral are pairs of angles that are next to each other, that is, their vertices are the endpoints of a side of the quadrilateral.

Remember that with any definition we may interchange the hypothesis and conclusion. For example, given the definition of a parallelogram, we may write these two statements:

1. If a polygon is a quadrilateral with two pairs of parallel sides, then it is a parallelogram.
2. If a polygon is a parallelogram, then it is a quadrilateral with two pairs of parallel sides.

You should always be prepared to use either form of a definition when it is necessary to do so in your proofs.

Section 8.1 Characteristics of Parallelograms

The following theorems deal with the characteristics of quadrilaterals and parallelograms. You should have no difficulty filling in the missing reasons in these proofs.

THEOREM:
The sum of the measures of the angles of a quadrilateral is 360.

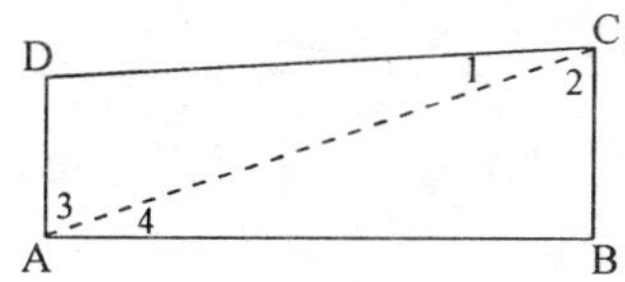

GIVEN:
Quadrilateral ABCD

PROVE:
$\angle A + \angle B + \angle C + \angle D = 360$

STATEMENTS	REASONS
1. Quadrilateral ABCD with angles $\angle$ A, $\angle$ B, $\angle$ C and $\angle$ D	1. ?
2. Construct line segment AC.	2. ?
3. $\angle 1 + \angle 3 + \angle D = 180$ in $\triangle$ ADC $\angle 2 + \angle 4 + \angle B = 180$ in $\triangle$ BAC	3. ?
4. $\angle 1 + \angle 3 + \angle D + \angle 2 + \angle 4 + \angle B = 360$	4. ?
5. $\angle 1 + \angle 2 = \angle DCB$ $\angle 3 + \angle 4 = \angle DAB$	5. ?
6. In quadrilateral ABCD, $\angle A + \angle B + \angle C + \angle D = 360$	6. Substitution Postulate of Equality ($\angle$ DCB for $\angle 1 + \angle 2$ in step 4) ($\angle$ DAB for $\angle 3 + \angle 4$ in step 4)

THEOREM:
A diagonal of a parallelogram separates it into two congruent triangles.

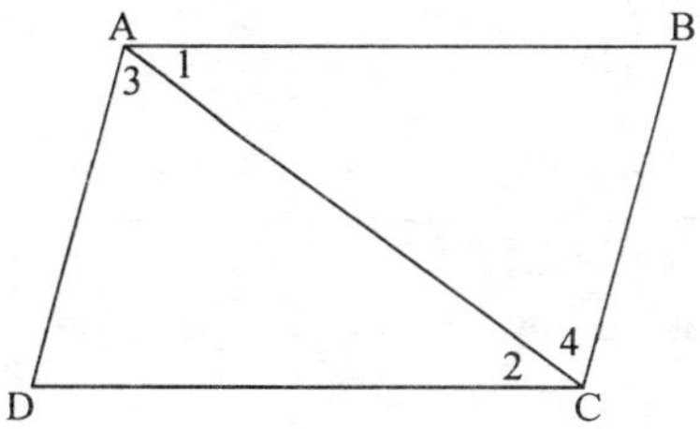

GIVEN:
Parallelogram ABCD with diagonals $\overline{AC}$ and $\overline{BD}$

PROVE:
$\triangle ABC \cong \triangle CDA$; $\triangle ADB \cong \triangle CBD$

STATEMENTS	REASONS
1. ABCD is a parallelogram.	1. ?
2. $\overline{AB} \parallel \overline{DC}$ and $\overline{AD} \parallel \overline{BC}$	2. Definition of Parallelogram
3. $\angle 1 = \angle 2$ and $\angle 3 = \angle 4$	3. Alternate Interior Angles Theorem
4. AC = AC	4. ?
5. $\triangle ABC \cong \triangle CDA$	5. ?

Now, can you write a similar proof for $\triangle ADB \cong \triangle CBD$

THEOREM:
The diagonals of a parallelogram bisect each other.

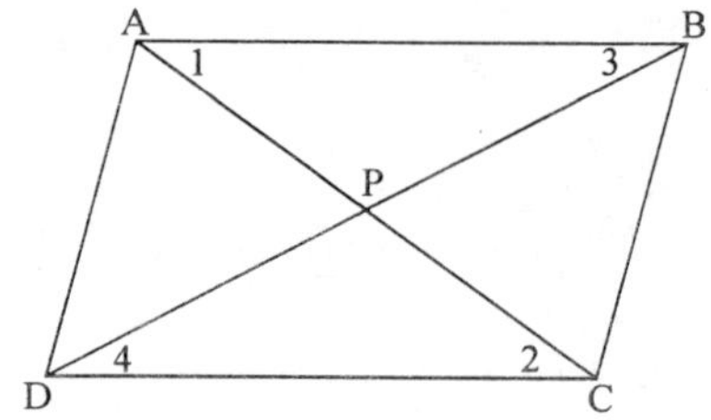

GIVEN:
Parallelogram ABCD with diagonals $\overline{AC}$ and $\overline{BD}$

PROVE:
$\overline{AC}$ and $\overline{BD}$ bisect each other.

STATEMENTS	REASONS
1. ABCD is a parallelogram with diagonals $\overline{AC}$ and $\overline{BD}$.	1. ?
2. △ ADB ≅ △ CBD	2. Theorem: A diagonal of a parallelogram separates it into congruent triangles.
3. AB = CD	3. ?
4. $\overline{AB} \parallel \overline{DC}$	4. Definition of Parallelogram
5. ∠ 3 = ∠ 4 and ∠ 1 = ∠ 2	5. ?
6. △ APB ≅ △ CPD	6. ?
7. AP = CP and BP = DP	7. ?
8. $\overline{AC}$ and $\overline{BD}$ bisect each other.	8. Definition of bisector of a line segment

THEOREM:
The opposite sides of a parallelogram have equal lengths.

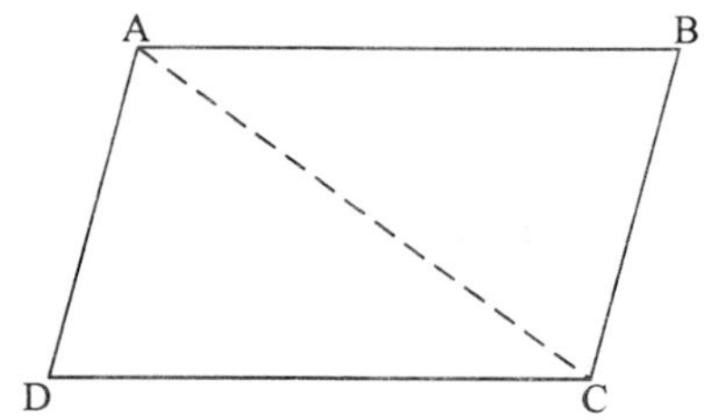

GIVEN:
Parallelogram ABCD

PROVE:
AB = CD and AD = BC

STATEMENTS	REASONS
1. ABCD is a parallelogram.	1. ?
2. Construct diagonal $\overline{AC}$.	2. ?
3. △ ABC ≅ △ CDA	3. ?
4. AB = CD and AD = BC	4. ?

THEOREM:
The opposite angles of a parallelogram have equal measures.

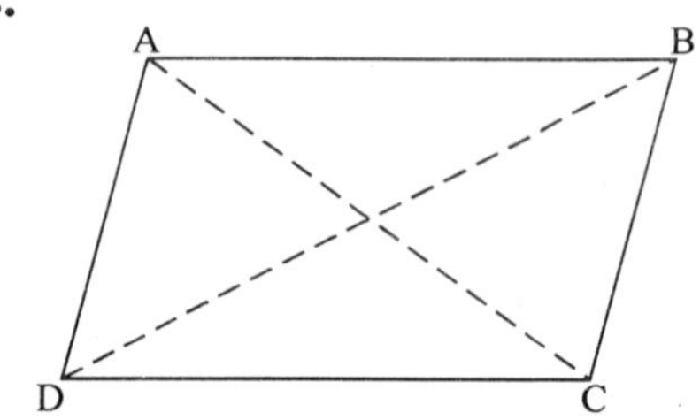

GIVEN:
ABCD is a parallelogram.

PROVE:
∠ A = ∠ C and ∠ B = ∠ D

STATEMENTS	REASONS
1. ABCD is a parallelogram.	1. ?
2. Construct diagonals $\overline{AC}$ and $\overline{BD}$.	2. ?
3. △ ABC ≅ △ CDA ; △ ADB ≅ △ CBD	3. ?
4. ∠ ABC = ∠ CDA; ∠ DAB = ∠ BCD	4. ?

THEOREM:
The consecutive angles of a parallelogram are supplementary.

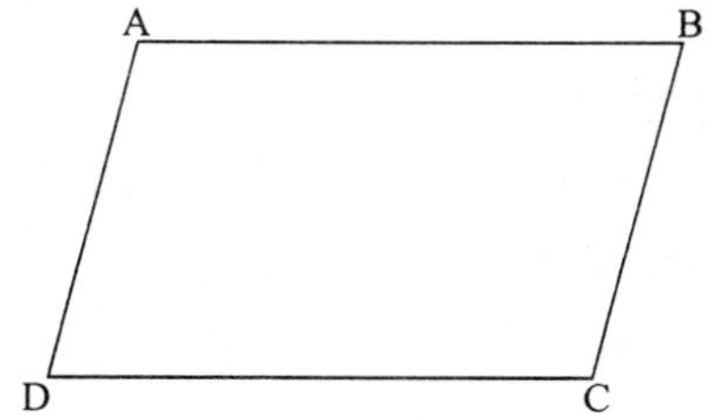

GIVEN:
Parallelogram ABCD

PROVE:
∠ A and ∠ D are supplementary.
∠ B and ∠ C are supplementary.
∠ A and ∠ B are supplementary.
∠ D and ∠ C are supplementary.

STATEMENTS	REASONS
1. ABCD is a parallelogram	1. ?
2. $\overline{AB} \parallel \overline{DC}$	2. Definition of parallelogram
3. ∠ A and ∠ D are supplementary. ∠ B and ∠ C are supplementary.	3. Same Side Interior Angles Theorem
4. $\overline{AD} \parallel \overline{BC}$	4. ?
5. ∠ A and ∠ B are supplementary. ∠ D and ∠ C are supplementary.	5. ?

Section 8.2 Proving That a Quadrilateral Is a Parallelogram

The next four theorems provide us with methods for proving that a quadrilateral is a parallelogram. As usual, you should supply the missing reasons

THEOREM:
A quadrilateral is a parallelogram if both pairs of opposite sides have equal lengths.

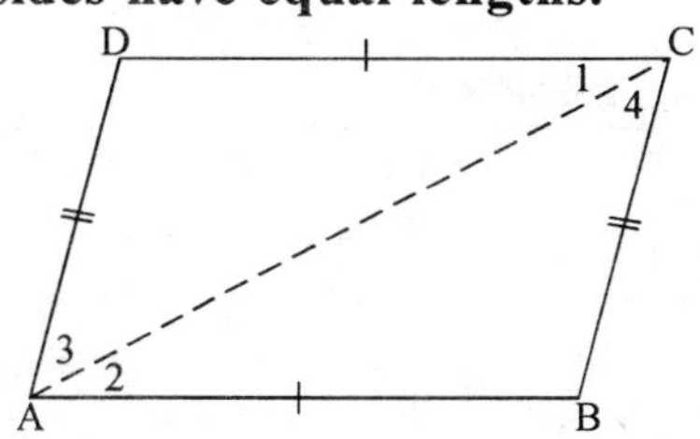

GIVEN:
Quadrilateral ABCD with AB = DC and AD = BC

PROVE:
Quadrilateral ABCD is a parallelogram.

STATEMENTS	REASONS
1. In quadrilateral ABCD, AB = DC and AD = BC	1. ?
2. Construct diagonal AC.	2. ?
3. AC = AC	3. ?
4. △ ADC ≅ △ CBA	4. ?
5. ∠ DCA = ∠ BAC (∠ 1 = ∠ 2)	5. ?
6. $\overline{DC} \parallel \overline{AB}$	6. ?
7. ∠ DAC = ∠ BCA (∠ 3 = ∠ 4)	7. ?
8. $\overline{DA} \parallel \overline{BC}$	8. ?
9. ABCD is a parallelogram.	9. Definition of Parallelogram

THEOREM:
A quadrilateral is a parallelogram if two sides have equal lengths and are parallel.

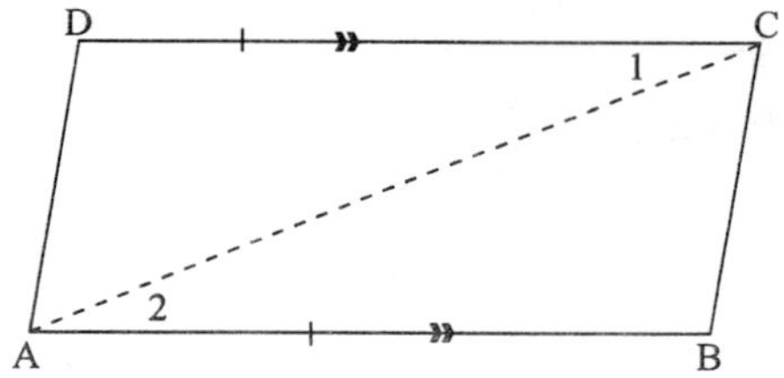

GIVEN:
Quadrilateral ABCD with DC = AB and $\overline{DC} \parallel \overline{AB}$

PROVE:
ABCD is a parallelogram.

STATEMENTS	REASONS
1. In quadrilateral ABCD, DC = AB and $\overline{DC} \parallel \overline{AB}$	1. ?
2. Construct diagonal $\overline{AC}$.	2. ?
3. ∠ 1 = ∠ 2	3. ?
4. AC = AC	4. ?
5. △ ADC ≅ △ CBA	5. SAS (Steps 1, 3, 4)
6. AD = BC	6. ?
7. ABCD is a parallelogram.	7. Theorem: A quadrilateral is a parallelogram if both pairs of opposite sides have equal lengths.

THEOREM:
A quadrilateral is a parallelogram if both pairs of opposite angles have equal measures.

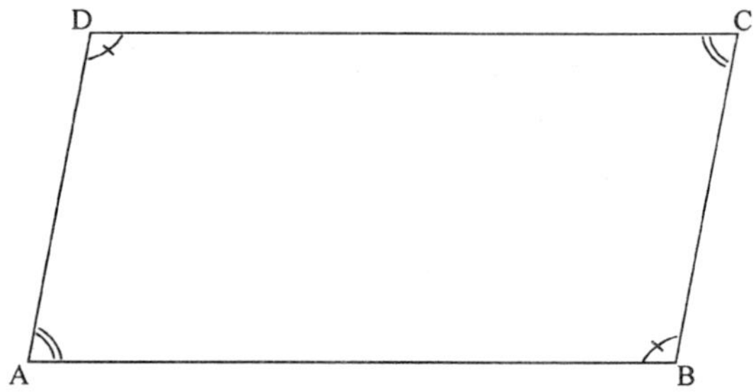

GIVEN:
Quadrilateral ABCD with ∠ A = ∠ C and ∠ D = ∠ B

PROVE:
Quadrilateral ABCD is a parallelogram.

STATEMENTS	REASONS
1. ABCD is a quadrilateral with ∠ A = ∠ C and ∠ D = ∠ B.	1. ?
2. ∠ A + ∠ B + ∠ C + ∠ D = 360	2. ?
3. ∠ A + ∠ B + ∠ A + ∠ B = 360	3. Substitution Postulate of Equality
4. 2 ∠ A + 2 ∠ B = 360	4. Substitution Postulate of Equality
5. ∠ A + ∠ B = 180	5. Division Postulate of Equality
6. ∠ A and ∠ B are supplementary.	6. ?
7. $\overline{AD} \parallel \overline{BC}$	7. Same Side Interior Angles/Parallel Lines Theorem
8. ∠ A + ∠ D = 180	8. Substitution Postulate of Equality
9. ∠ A and ∠ D are supplementary.	9. ?
10. $\overline{DC} \parallel \overline{AB}$	10. ?
11. ABCD is a parallelogram.	11. Definition of Parallelogram

THEOREM:
A quadrilateral is a parallelogram if its diagonals bisect each other.

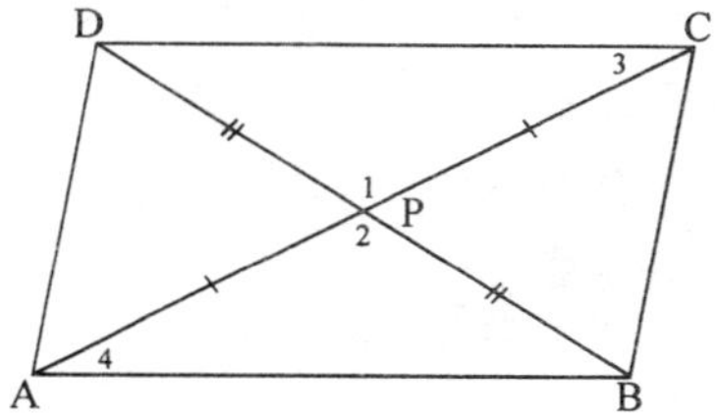

GIVEN:
Quadrilateral ABCD with diagonals $\overline{DB}$ and $\overline{AC}$
Diagonals $\overline{DB}$ and $\overline{AC}$ bisect each other.

PROVE:
Quadrilateral ABCD is a parallelogram.

STATEMENTS	REASONS
1. The diagonals of quadrilateral ABCD bisect each other.	1. ?
2. DP = BP and AP = CP	2. ?
3. $\angle 1 = \angle 2$	3. ?
4. $\triangle$ APB $\cong$ $\triangle$ CPD	4. ?
5. $\overline{AB} = \overline{CD}$	5. ?
6. $\angle 3 = \angle 4$	6. ?
7. $\overline{DC} \parallel \overline{AB}$	7. ?
8. Quadrilateral ABCD is a parallelogram.	8. Theorem: A quadrilateral is a parallelogram if two sides have equal lengths and are parallel.

Section 8.3 Rhombuses, Rectangles and Squares

Our next two theorems deal with the characteristics of a rhombus. The first theorem states that a rhombus must be a parallelogram. Because this theorem is true, we may conclude that all the theorems that we proved about a parallelogram are true also for a rhombus. The second theorem states that the diagonals of a rhombus are perpendicular to each other.

THEOREM:
A rhombus is a parallelogram.

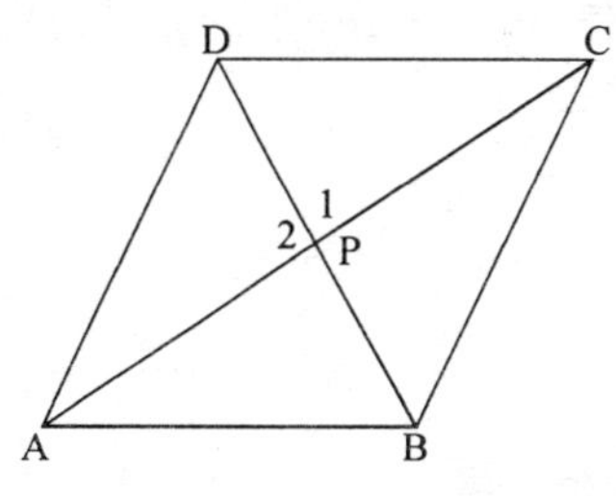

GIVEN:
Quadrilateral ABCD is a rhombus.

PROVE:
Quadrilateral ABCD is a parallelogram.

STATEMENTS	REASONS
1. Quadrilateral ABCD is a rhombus.	1. ?
2. AB = DC and AD = BC	2. Definition of rhombus
3. Quadrilateral ABCD is a parallelogram.	3. ?

THEOREM:
The diagonals of a rhombus are perpendicular.

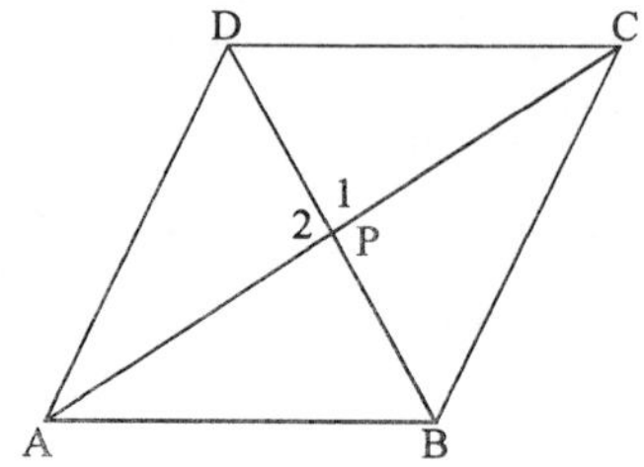

GIVEN:
Rhombus ABCD with diagonals $\overline{AC}$ and $\overline{DB}$

PROVE:
$\overline{AC} \perp \overline{BD}$

STATEMENTS	REASONS
1. ABCD is a rhombus with diagonals $\overline{AC}$ and $\overline{DB}$.	1. ?
2. AB = BC = DC = AD	2. Definition of a Rhombus
3. ABCD is a parallelogram.	3. Theorem: A rhombus is a parallelogram
4. $\overline{AC}$ and $\overline{BD}$ bisect each other.	4. ?
5. AP = CP	5. ?
6. DP = DP	6. ?
7. $\triangle$ DPA $\cong$ $\triangle$ DPC	7. ?
8. $\angle 1 = \angle 2$	8. ?
9. $\angle$ 1 and $\angle$ 2 form a linear pair.	9. Definition of Linear Pair
10. $\angle$ 1 and $\angle$ 2 are right angles.	10. ?
11. $\overline{DB} \perp \overline{AC}$	11. ?

The next theorem provides another way to prove that under certain conditions a parallelogram is a rhombus.

THEOREM:
If the diagonals of a parallelogram are perpendicular, then the parallelogram is a rhombus.

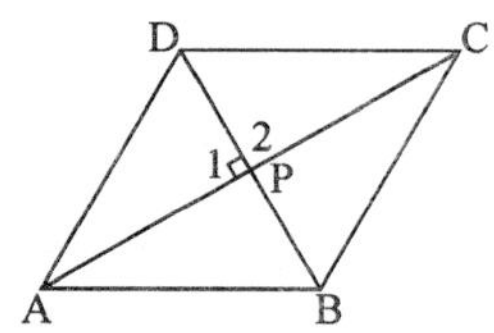

GIVEN:
ABCD is a parallelogram with diagonals $\overline{AC}$ and $\overline{DB}$; $\overline{AC} \perp \overline{DB}$

PROVE:
ABCD is a rhombus.

STATEMENTS	REASONS
1. ABCD is a parallelogram with diagonal $\overline{AC}$ perpendicular to diagonal $\overline{BD}$.	1. ?
2. $\angle$ 1 and $\angle$ 2 are right angles.	2. ?
3. $\angle 1 = \angle 2$	3. ?
4. DP = DP	4. ?
5. AP = CP	5. ?
6. ?	6. ?
7. AD = CD	7. cpcte
8. AD = BC and AB = DC	8. Theorem: The opposite sides of a parallelogram have equal measures.
9. AB = BC = DC = AD	9. Substitution Postulate of Equality
10. ABCD is a rhombus.	10. Definition of a Rhombus

The next three theorems deal with the characteristics of a rectangle. Notice that a rectangle is also a parallelogram, so we may conclude that it shares all the properties of a parallelogram. Notice also that the diagonals of a rectangle have equal lengths and each of its angles is a right angle.

THEOREM:
A rectangle is a parallelogram.

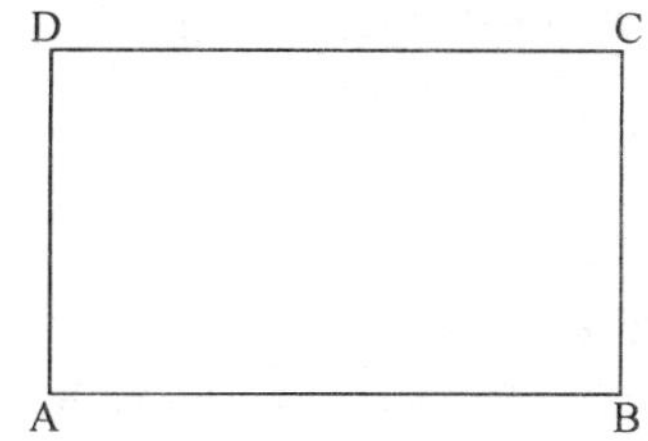

GIVEN:
Quadrilateral ABCD is a rectangle.

PROVE:
Quadrilateral ABCD is a parallelogram.

STATEMENTS	REASONS
1. Quadrilateral ABCD is a rectangle.	1. ?
2. $\angle A = \angle C$ and $\angle B = \angle D$	2. Definition of rectangle
3. Quadrilateral ABCD is a parallelogram.	3. ?

THEOREM:
The angles of a rectangle are right angles.

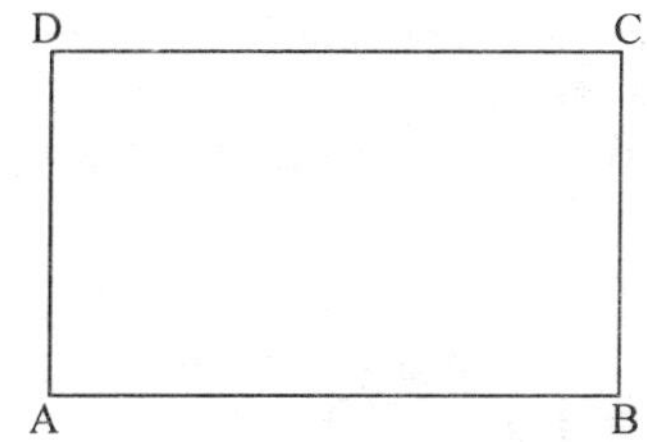

GIVEN:
Rectangle ABCD

PROVE:
∡ A, ∡ B, ∡ C and ∡ D are right angles.

STATEMENTS	REASONS
1. Quadrilateral ABCD is a rectangle.	1. ?
2. $\angle A = \angle B = \angle C = \angle D$	2. Definition of Rectangle
3. $\angle A + \angle B + \angle C + \angle D = 360$	3. ?
4. $\angle A + \angle A + \angle A + \angle A = 360$	4. Substitution Postulate of Equality
5. $4 \angle A = 360$	5. Substitution Postulate of Equality
6. $\angle A = 90$	6. ?
7. $\angle B = 90$, $\angle C = 90$, $\angle D = 90$	7. Substitution Postulate of Equality
8. ∡ A, ∡ B, ∡ C and ∡ D are right angles.	8. ?

THEOREM:
The diagonals of a rectangle have equal lengths.

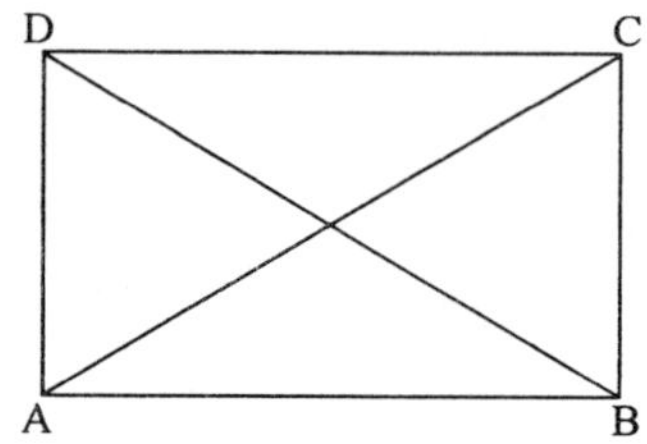

GIVEN:
Quadrilateral ABCD with diagonals $\overline{AC}$ and $\overline{BD}$

PROVE:
AC = BD

STATEMENTS	REASONS
1. Quadrilateral ABCD is a rectangle with diagonals $\overline{AC}$ and $\overline{BD}$.	1. ?
2. Quadrilateral ABCD is a parallelogram.	2. Theorem: A rectangle is a parallelogram
3. AD = BC	3. ?
4. DC = DC	4. ?
5. ∠ ADC = ∠ BCD	5. Definition of a rectangle
6. ?	6. ?
7. AC = BD	7. cpcte

The theorem that we will prove next gives us a way to prove that certain parallelograms must be rectangles.

THEOREM:
If the diagonals of a parallelogram have equal measures, then the parallelogram is a rectangle.

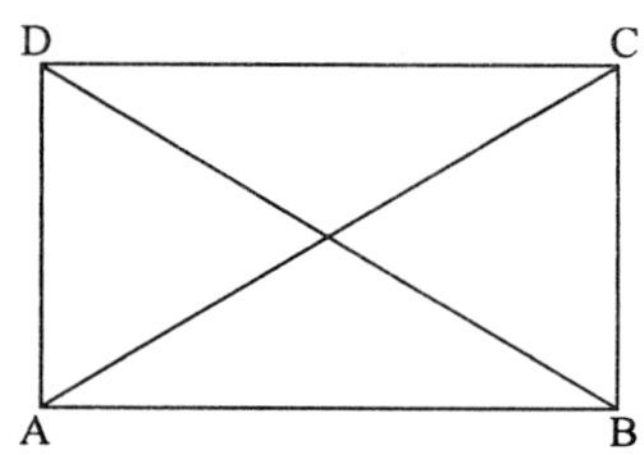

GIVEN:
ABCD is a parallelogram with diagonals $\overline{AC}$ and $\overline{BD}$;
AC = BD.

PROVE:
Parallelogram ABCD is a rectangle.

STATEMENTS	REASONS
1. ABCD is a parallelogram with diagonals $\overline{AC}$ and $\overline{BD}$.	1. ?
2. AD = BC	2. ?
3. AC = BD	3. ?
4. AB = AB	4. ?
5. ?	5. ?
6. ∠ DAB = ∠ CBA, that is, ∠ A = ∠ B in parallelogram ABCD	6. cpcte
7. In parallelogram ABCD, ∠ A = ∠ C and ∠ D = ∠ B	7. ?
8. In parallelogram ABCD, ∠ A = ∠ B = ∠ C = ∠ D	8. Substitution Postulate of Equality
9. Parallelogram ABCD is a rectangle.	9. Definition of Rectangle

Since a square is equiangular it is a rectangle; and since it is equilateral, it is a rhombus. We may conclude then that a square has all of the properties of a rhombus, a rectangle and a parallelogram. The proofs of the theorems about squares are left up to you.

THEOREM: A square is a parallelogram.

THEOREM: A square is a rhombus.

THEOREM: A square is a rectangle.

Now let us take time to summarize our work on quadrilaterals and parallelograms.

Properties of a Quadrilateral

The sum of its angles is 360.

Properties of a Parallelogram

1. The opposite sides are parallel.
2. The opposite sides have equal lengths.
3. The opposite angles have equal measures.
4. The consecutive angles are supplementary.
5. The diagonals bisect each other.
6. Each diagonal separates it into two congruent triangles.

Properties of a Rhombus

1. All of the properties of a parallelogram (It is a parallelogram.)
2. The diagonals are perpendicular.

Properties of a Rectangle

1. All of the properties of a parallelogram (It is a parallelogram.)
2. The diagonals have equal lengths.

Properties of a Square

1. All of the properties of a rhombus (It is a rhombus.)
2. All of the properties of a rectangle (It is a rectangle.)

Ways to Prove That a Quadrilateral Is a Parallelogram

1. Prove that both pairs of opposite sides are parallel.
2. Prove that both pairs of opposite sides have equal lengths.
3. Prove that both pairs of opposite angles have equal measures.
4. Prove that two sides are parallel and equal in length.
5. Prove that its diagonals bisect each other.

Ways to Prove That a Quadrilateral Is a Rhombus

1. Prove that it is equilateral.
2. Prove that it is a parallelogram with perpendicular diagonals.

Ways to Prove That a Quadrilateral Is a Rectangle

1. Prove that it is equiangular.
2. Prove that it is a parallelogram with diagonals of equal lengths.

Ways to Prove That a Quadrilateral Is a Square

1. Prove that it is both equiangular and equilateral.
   OR
   Prove that it is both a rhombus and a rectangle.

Exercises Chapter 8

(*Figures are not drawn to scale*)

1. List the properties of a parallelogram.

2. How may we prove that a quadrilateral is a parallelogram?

3. Place a check in the appropriate space if the indicated property holds for the given polygon.

	parallelogram	rectangle	rhombus	square
opposite sides parallel				
opposite sides equal lengths				
opposite angles equal measures				
consecutive angles supplementary				
diagonals form congruent triangles				
diagonals bisect each other				
diagonals have equal lengths				
diagonals are perpendicular				
all angles are right angles				
all sides have equal lengths				

4. Respond true or false. Support your response.
   a. Every square is a rhombus.
   b. Every rhombus is a square.
   c. Every square is a rectangle.
   d. Every rectangle is a square.
   e. Every rhombus is a rectangle.
   f. Every rectangle is a rhombus.
   g. Every rectangle is a parallelogram.
   h. Every rhombus is a parallelogram.
   i. Every parallelogram is a square.
   j. The diagonals of a quadrilateral bisect each other.
   k. The diagonals of a parallelogram have equal lengths.

5. Find the measures of all the angles of a parallelogram.
   a. If one of its angles has a measure of 30.

   b. If one of its angles has a measure of 45.

   c. If one of its angles has a measure of 72.

   d. If one of its angles has a measure of 90.

6. Find the measures of all the angles of a parallelogram.
   a. If one of its angles has a measure of 150.

   b. If one of its angles has a measure of 60.

   c. If one of its angles has a measure of 82.

   d. If one of its angles has a measure of 50.

7. Find the measure of each angle of parallelogram ABCD
   a. If $\angle D = x + 40$ and $\angle B = 2x$

   b. If $\angle D = 3x + 5$ and $\angle B = 7x - 13$

c. If $\angle A = 5x + 7$ and $\angle D = 3x + 13$

d. If $\angle A = 2x + 7$ and $\angle D = 3x + 13$

8. a. If $\angle D = x + 20$ and $\angle B = 2x$

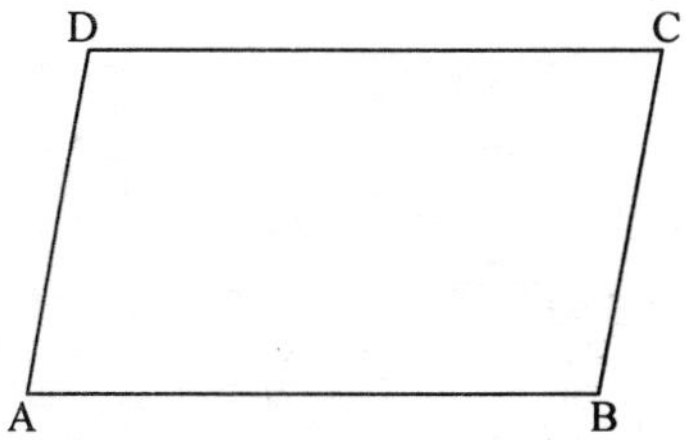

b. If $\angle D = 4x + 3$ and $\angle B = 6x - 11$

c. If $\angle A = 7x + 5$ and $\angle D = 3x + 25$

d. If $\angle A = 2x + 3$ and $\angle D = 3x + 2$

9. **GIVEN:** Quadrilateral ABCD with $\overline{AB} \parallel \overline{DC}$ and $\angle A = \angle C$

PROVE: Quadrilateral ABCD is a parallelogram.

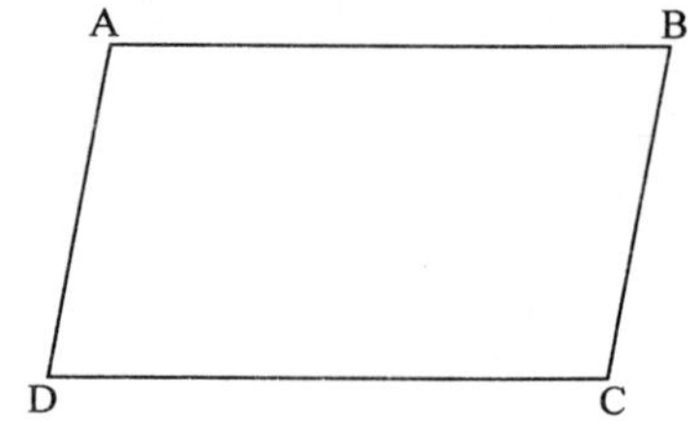

10. **GIVEN:** $\angle C = \angle M$, $\angle A = \angle E$ and EC = AC in Quadrilateral MACE.

PROVE: Quadrilateral MACE is a rhombus.

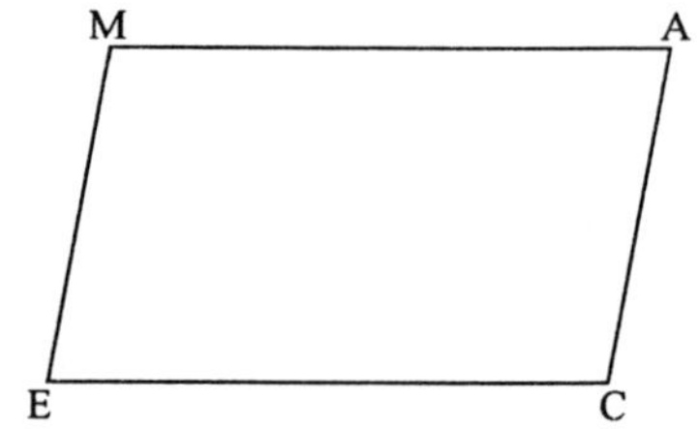

11. **GIVEN:** $\triangle$ AFB $\cong$ $\triangle$ DCB. D is the midpoint of $\overline{EC}$.

PROVE: Quadrilateral AFED is a parallelogram.

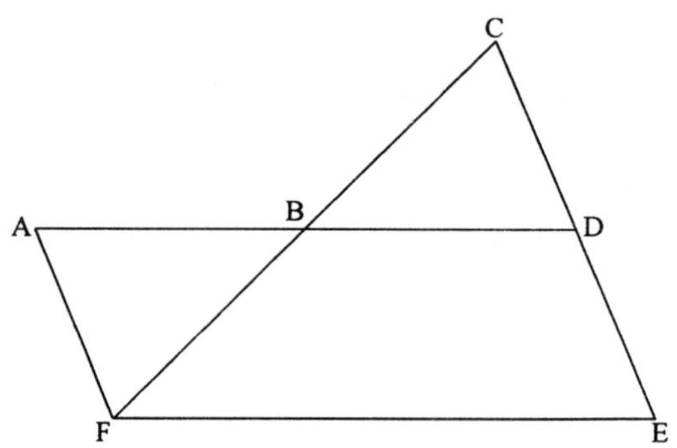

12. **GIVEN:** Square ABCD with diagonals $\overline{DB}$ and $\overline{AC}$ intersecting at point P

PROVE: $\triangle$ APB is an isosceles right triangle.

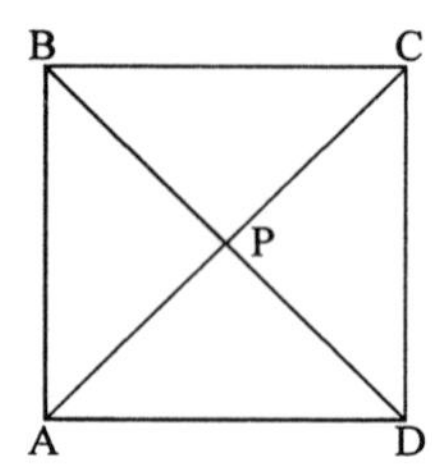

13. **GIVEN:** Parallelogram ABCD with $\overline{DE} \parallel \overline{BF}$

PROVE: DE = BF

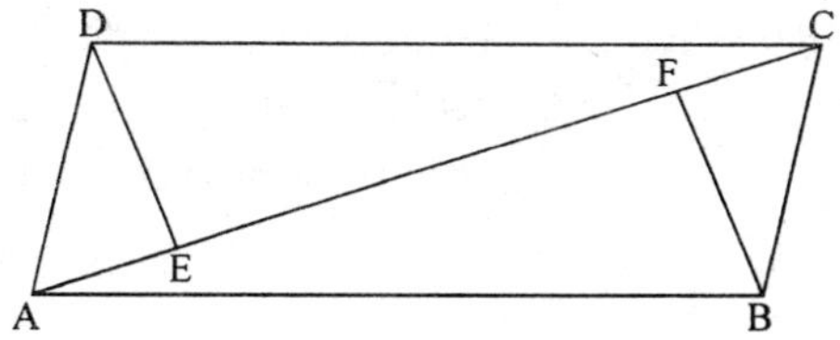

14. **GIVEN:** Quadrilateral AECB is a rectangle.
Quadrilateral EDCB is a parallelogram.

PROVE: △ ADC is isosceles.

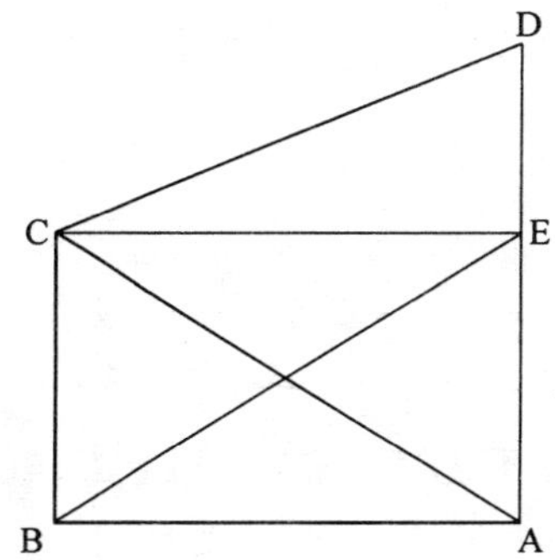

15. Given: Isosceles trapezoid ABCD
Prove: △ DEC ~ △ ABE

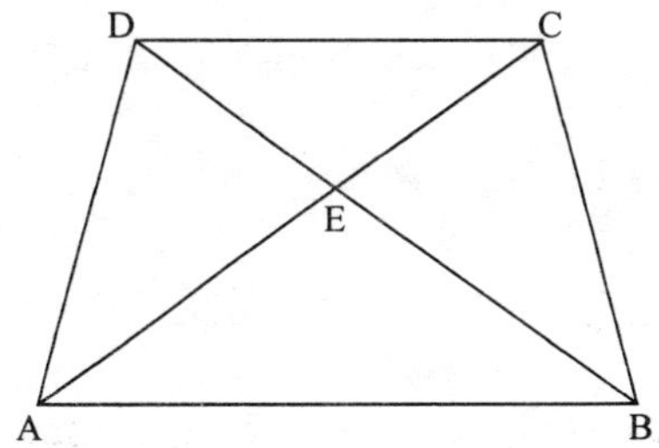

16.*Prove each theorem:

a. **THEOREM: Parallel lines are always the same distance apart.**
(Hint: Prove that AB = CD)

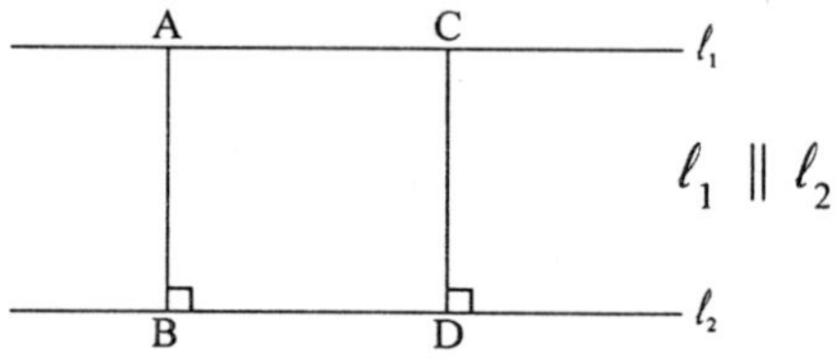

b. **THEOREM: The base angles of an isosceles trapezoid have equal measures.**
(You may want to use auxiliary lines)

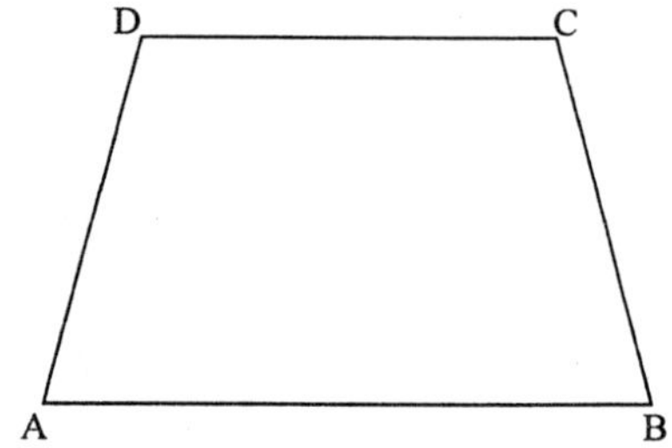

c. **THEOREM: The diagonals of an isosceles trapezoid are equal in length.**

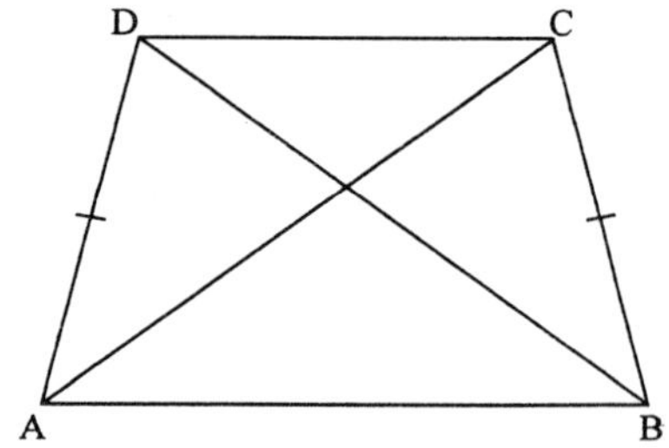

17. **GIVEN:** Parallelogram ABCD with E the midpoint of side $\overline{BC}$ and F the midpoint of side $\overline{AD}$

PROVE: $\angle 1 = \angle 2$

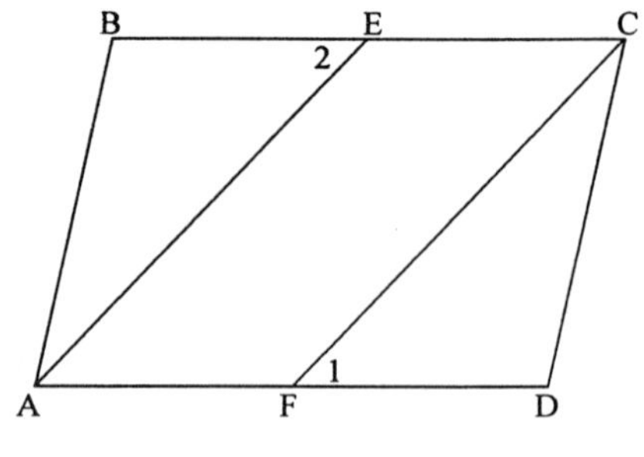

18. Show that the quadrilateral with sides determined by the following equations is a parallelogram.

$$3x - y + 2 = 0 \qquad x + y + 1 = 0 \qquad 6x - 2y = 1 \qquad 2x + 2y = 7$$

19. Show that the points A (2, 3), B (4, 1), C (8, 2) and D (6, 4) are the vertices of a parallelogram.

20. Show that the points A(0, 2), B(0, 0), C(2, 0) and D(2, 2) are the vertices of a parallelogram.

21. If we divide a line segment into two parts so that the ratio of the whole segment to its longer part is equal to the ratio of the longer part of the segment to its shorter part, then the segment is said to be divided into the golden ratio.

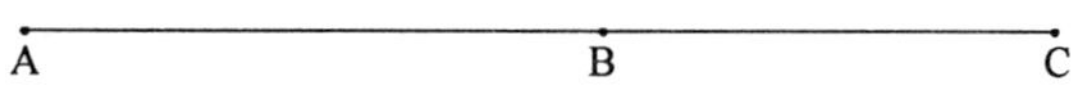

$$\frac{\text{whole segment}}{\text{longer part}} = \frac{\text{longer part}}{\text{shorter part}}$$

$$\frac{AC}{AB} = \frac{AB}{BC}$$

If we construct a rectangle using the lengths of the two parts of the original segment as the lengths of its adjacent sides, the resulting rectangle is called a Golden Rectangle.

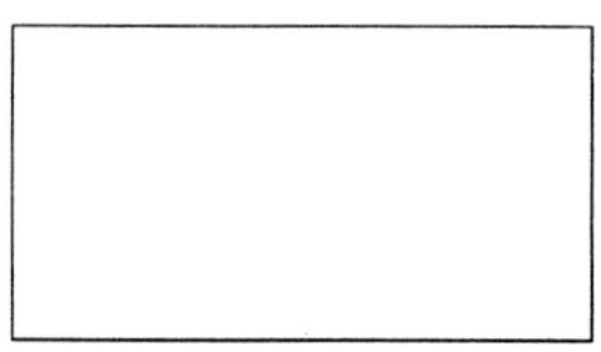

Such a rectangle is thought to be the most visually pleasing of all geometric figures. Golden Rectangles appear throughout history in the design of edifices and in the works of artists. They also appear quite regularly in nature.

Constructing a Golden Rectangle is not difficult:

a. Construct a square with sides of any length.
Label its vertices PADS as shown

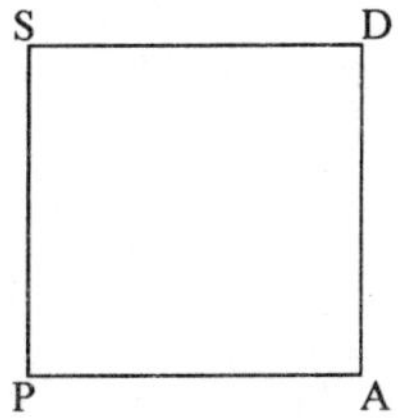

b. Find the midpoint of any side of the square.
We will call the selected side the base of the square, and we will call its midpoint O

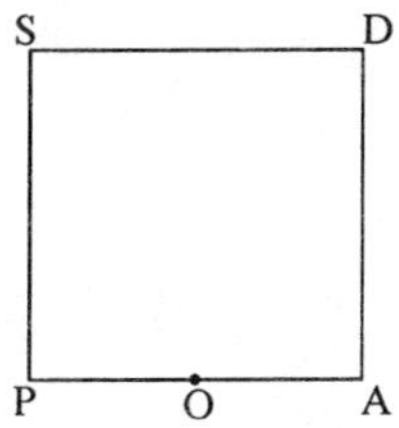

c. Using the midpoint O as the center and OD as a radius, construct an arc that will intersect the extended base of the square.

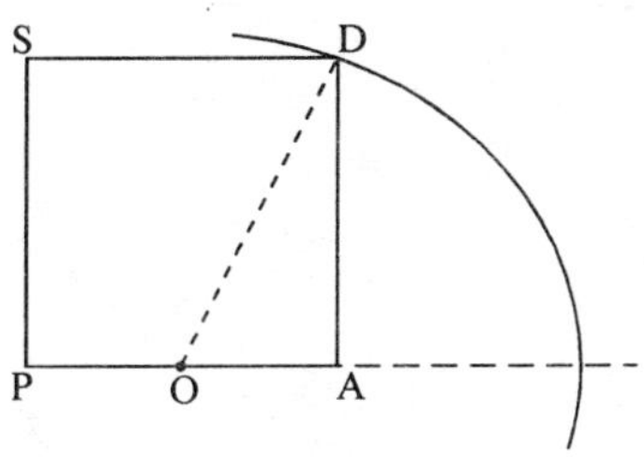

d. Extend the base of the square, so that it intersects the arc.
We will call the point of intersection B.
(This extension will be the base of the Golden Rectangle.)

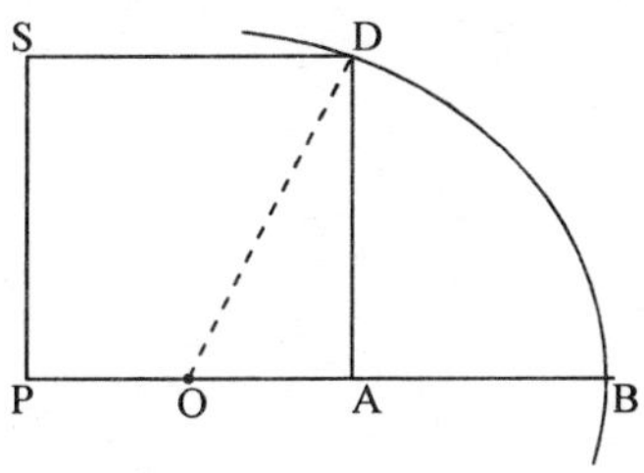

e. Construct a perpendicular segment from point B. Make its length equal to the length of a side of the square.

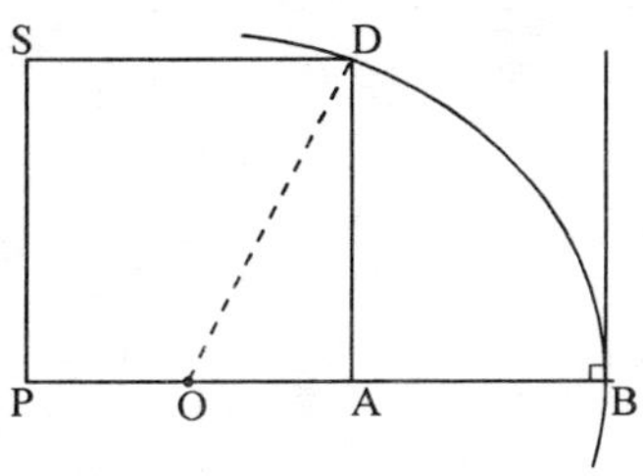

f. Extend the side of the square opposite the base so that it intersects the perpendicular line segment constructed from point B. We will call the point of intersection C.

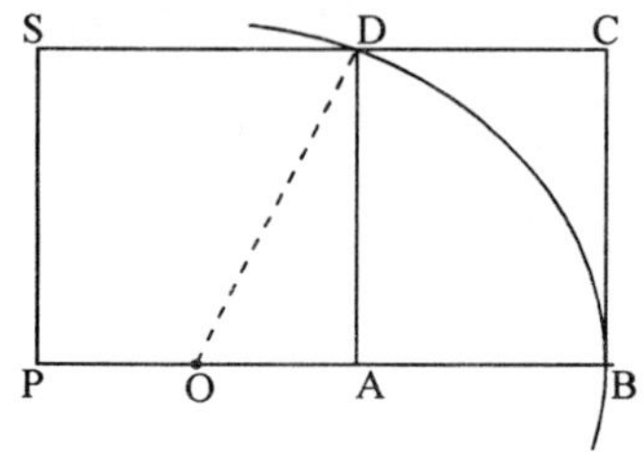

g. Erase three sides of the square as shown. The rectangle that remains is a Golden Rectangle.

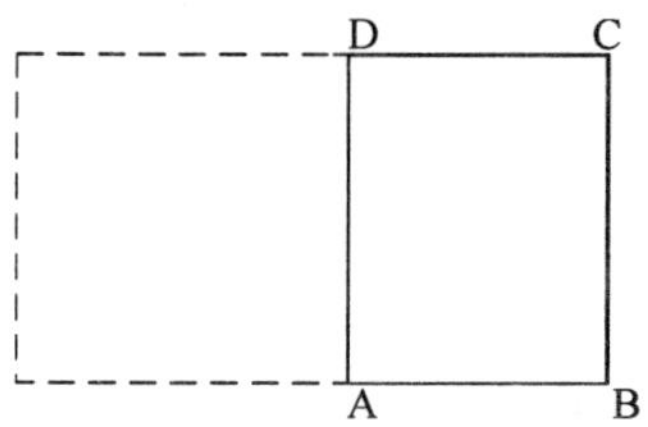

Now that we know how to construct a Golden Rectangle, we can use the construction to calculate the ratio of its adjacent sides. We want the ratio of PB to PA and the ratio of PA to AB. These two ratios should be equal.

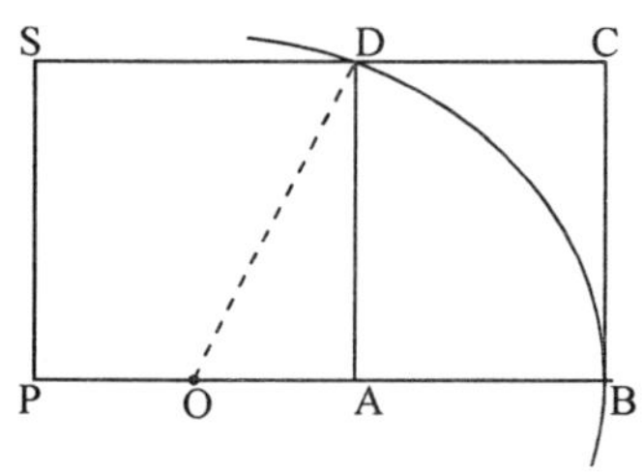

First, we let PA = 1. Then OA = OP = 1/2

Now we use the Pythagorean Theorem to find the length of $\overline{OD}$.

$(OD)^2 = (OA)^2 + (AD)^2$

$(OD)^2 = (1/2)^2 + 1^2$

$(OD)^2 = 1/4 + 1$

$(OD)^2 = 5/4$

$(OD)^2 = \frac{\sqrt{5}}{2}$

In the construction, $PB = PO + OB$

$PB = \frac{1}{2} + \frac{\sqrt{5}}{2}$ (OD = OB, radii of same circle)

$PB = \frac{1 + \sqrt{5}}{2}$

Since PA is 1, the desired ratio of PB to PA is $\frac{\frac{1 + \sqrt{5}}{2}}{1}$ or $\frac{1 + \sqrt{5}}{2}$

The approximate value of this ratio is 8/5.

Now why don't you try to calculate the ratio of PA to AB? Remember the two ratios should be equal.

Chapter Eight Highlights

Can You Define These Words?

quadrilateral
parallelogram
rhombus
rectangle
square
diagonal
consecutive angles

Do You Know and Understand This Postulate?

The Unique Line Postulate: Two Points Determine a Line

Do You Know and Understand These Theorems?

The sum of the angles of a quadrilateral is 360 degrees.

A diagonal of a parallelogram separates it into two congruent triangles.

The diagonals of a parallelogram bisect each other.

The opposite sides of a parallelogram are equal in length.

The opposite angles of a parallelogram are equal in measure.

The consecutive angles of a parallelogram are supplementary.

A quadrilateral is a parallelogram if both pairs of opposite sides have equal lengths.

A quadrilateral is a parallelogram if both pairs of opposite angles have equal measures.

A quadrilateral is a parallelogram if two sides have equal lengths and are parallel.

A quadrilateral is a parallelogram if the diagonals bisect each other.

A rhombus is a parallelogram.

The diagonals of a rhombus are perpendicular.

If the diagonals of a parallelogram are perpendicular then the parallelogram is a rhombus.

A rectangle is a parallelogram.

The angles of a rectangle are right angles.

The diagonals of a rectangle have equal lengths.

If the diagonals of a parallelogram have equal lengths, then the parallelogram is a rectangle.

A square is a parallelogram.

A square is a rhombus.

A square is a rectangle.

Parallel lines are always the same distance apart.

*The base angles of an isosceles triangle have equal measures.

*The diagonals of an isosceles trapezoid are equal in length.

Chapter Nine

Special Relationships in Triangles

Objectives

Prove and apply theorems related to the altitude to the hypotenuse of a right triangle.
Prove and apply theorems related to the 30-60 right triangle.
Prove and apply theorems related to the isosceles right triangle.
Prove and apply the Pythagorean Theorem and its converse.
Prove and appy theorems about the line segment joining the midpoints of two sides of a triangle.
Prove and apply theorems related to an angle bisector of a triangle.
Prove and appy theorems about a line parallel to a side of a triangle.

Section 9.1 Right Triangles

As you recall from our earlier work in Section 2.3, we may conclude that two triangles are similar if two angles of one triangle have the same measures, respectively, as two angles of the other triangle. In this chapter, we will use this statement as a postulate for our work. We will refer to it as the Angle Angle Postulate.

Angle Angle Postulate

If the measures of two angles of one triangle equal, respectively, the measures of two angles of another triangle, then the two triangles are similar. (AA POSTULATE)

For some of our exercises, we will need a postulate that we will call the Area Addition Postulate. This postulate allows us to obtain the total area of a polygonal region by dividing that region into smaller regions.

Area Addition Postulate

If a region is the union of two or more regions that have no interior points in common, then its area is the sum of their areas.

For example, in the illustration the area of polygon ACDE is the sum of the areas of the three triangles.

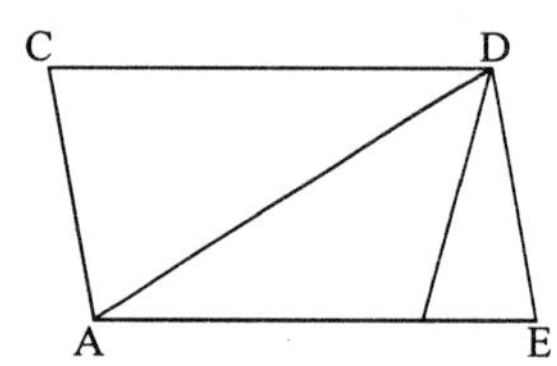

In this chapter and the next chapter, we will use some of the theorems on proportions that we have already proved deductively in Appendix A. For convenience, we will list them here.

THEOREM: In a proportion, the product of the extremes equals the product of the means.

THEOREM: We may interchange the extremes of a proportion.

THEOREM: We may interchange the means of a proportion.

THEOREM: We may interchange both the extremes and the means of a proportion.

THEOREM: If $\frac{a}{b} = \frac{c}{d}$ **then** $\frac{a+b}{b} = \frac{c+d}{d}$ **and** $\frac{a-b}{b} = \frac{c-d}{d}$

THEOREM: If $\frac{a}{b} = \frac{c}{d} = \frac{e}{f} = \frac{g}{h} = k$ **Then** $k = \frac{a+c+e+g}{b+d+f+h}$

We begin by proving some theorems about right triangles. You will notice that some details have been left out of many of the arguments. This is done so that we can avoid exceedingly long proofs.

We will call the first theorem the **Altitude to the Hypotenuse Theorem.**

THEOREM:
The altitude to the hypotenuse of a right triangle separates the triangle into two triangles that are similar to the original triangle and to each other.

GIVEN:
The figure with right triangle ACB
Right Angle ∠ ACB
Altitude to the Hypotenuse, $\overline{CD}$

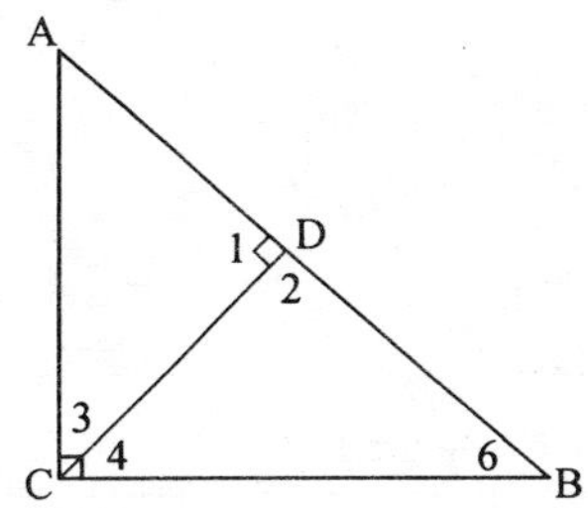

PROVE:
△ CDB ~ △ ADC
△ ACB ~ △ ADC
△ ACB ~ △ CDB

STATEMENTS	REASONS
1. The figure with right triangle ACB Right Angle ∠ ACB Altitude to the Hypotenuse $\overline{CD}$	1. ?
2. CD ⊥ AB	2. Definition of Altitude of a Triangle
3. 1 and 2 are right angles.	3. Definition of Perpendicular
4. ∠ ACB = ∠ 1 = ∠ 2	4. Theorem: All right angles have equal measures
5. ∠ A = ∠ A	5. ?
6. △ ACB ~ △ ADC	6. Angle Angle Postulate (Steps 4, 5)
7. ∠ B = ∠ B	7. ?
8. △ ACB ~ △ CDB	8. ?
9. ∠ 3 = ∠ 6	9. Definition of Similar Triangles (Step 6)
10. △ CDB ~ △ ADC	10. AA Postulate (Steps 9 and 4)

Now that we have proved this theorem, we can prove the following corollary.

COROLLARY:
The altitude to the hypotenuse of a right triangle separates the hypotenuse of the triangle into two segments (line segments) for which the following is true:

1. **Each leg of the triangle is a mean proportional between the hypotenuse and the segment of the hypotenuse next to that leg.**
2. **The altitude is the mean proportional between the two segments of the hypotenuse.**

GIVEN:

The figure with altitude $\overline{CD}$ to the hypotenuse $\overline{AB}$

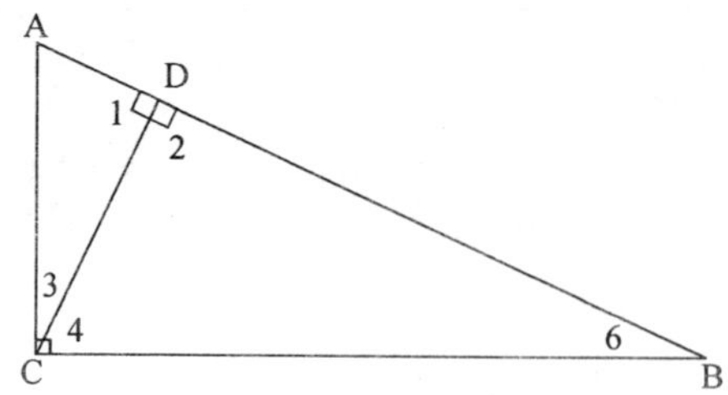

PROVE:

$\frac{AB}{AC} = \frac{AC}{AD}$

$\frac{AB}{BC} = \frac{BC}{BD}$

$\frac{AD}{CD} = \frac{CD}{DB}$

STATEMENTS	**REASONS**
1. The figure with altitude $\overline{CD}$ to the hypotenuse $\overline{AB}$	1. ?
2. $\triangle ACB \sim \triangle ADC$	2. Altitude to Hypotenuse Theorem
3. $\frac{AB}{AC} = \frac{AC}{AD}$	3. Definition of Similar Triangles
4. $\triangle ACB \sim \triangle CDB$	4. ?
5. $\frac{AB}{BC} = \frac{BC}{BD}$	5. ?
6. $\triangle ADC \sim \triangle CDB$	6. ?
7. $\frac{AD}{CD} = \frac{CD}{DB}$	7. ?

Now let us see how these results are used to find missing parts of triangles.

We will use this triangle as our model.

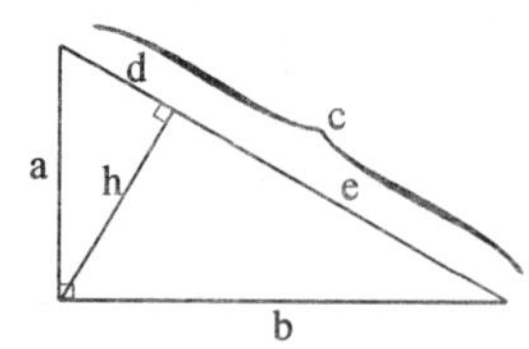

Each leg is a mean proportional. $\frac{c}{a} = \frac{a}{d}$

$\frac{c}{b} = \frac{b}{e}$

The altitude is a mean proportional. $\frac{d}{h} = \frac{h}{e}$

EXAMPLE 1:

Find the value of x.

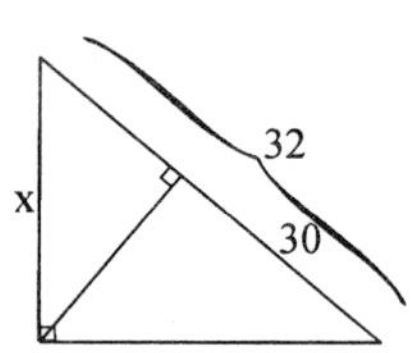

$$\frac{32}{x} = \frac{x}{2}$$
$$x \cdot x = 2 \cdot 32$$
$$x^2 = 64$$
$$x = \pm 8$$
$$x = 8$$

EXAMPLE 2:

Find the value of x.

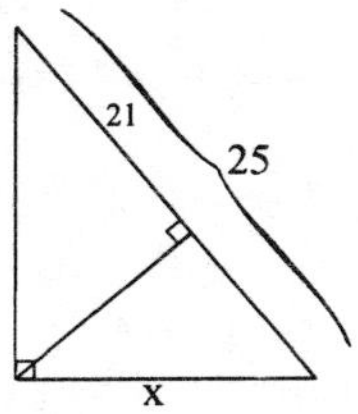

$$\frac{25}{x} = \frac{x}{4}$$
$$x^2 = 100$$
$$x = \pm 10$$
$$x = 10$$

EXAMPLE 3:

Find the value of x.

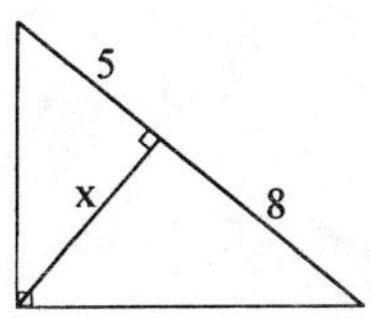

$$\frac{5}{x} = \frac{x}{8}$$
$$x^2 = 40$$
$$x = \pm\sqrt{40}$$
$$x = \pm 2\sqrt{10}$$
$$x = 2\sqrt{10}$$

EXAMPLE 4:

Find the value of x.

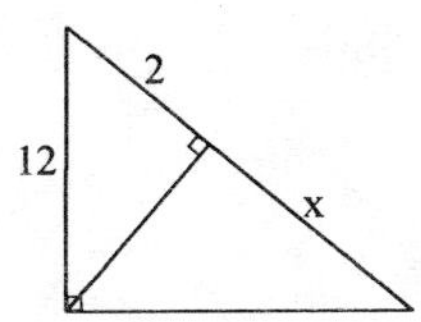

$$\frac{x+2}{12} = \frac{12}{2}$$
$$\frac{x+2}{12} = \frac{6}{1}$$
$$x + 2 = 72$$
$$x = 70$$

We can use the result of the first corollary to prove the well-known Pythagorean Theorem.

THE PYTHAGOREAN THEOREM

In a right triangle, the square of the hypotenuse equals the sum of the squares of its legs.

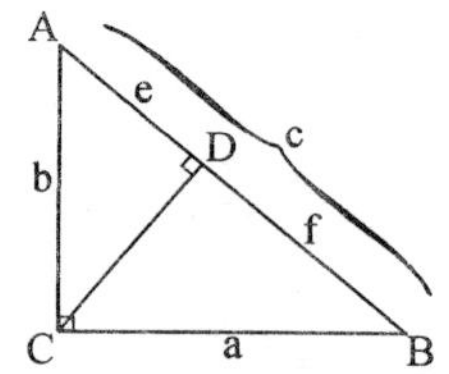

GIVEN:
Right triangle ACB with legs of lengths a and b and hypotenuse of length c

PROVE:
$c^2 = a^2 + b^2$

STATEMENTS	REASONS
1. Construct $\overline{CD}$ perpendicular to $\overline{AB}$	1. Perpendicular Postulate Given a line and a point not on the line, there is exactly one line through the point that is perpendicular to the given line.
2. $\frac{c}{b} = \frac{b}{e}$ and $\frac{c}{a} = \frac{a}{f}$	2. Corollary: Each leg is a mean proportional between the hypotenuse and the segment of the hypotenuse next to that leg.
3. $b^2 = ce$; $a^2 = cf$	3. Theorem: Product of Extremes Equals Product of Means
4. $a^2 + b^2 = ce + cf$	4. Addition Postulate of Equality
5. $a^2 + b^2 = c(e + f)$	5. Substitution Postulate of Equality
6. $e + f = c$	6. Whole Equals Sum of Parts Postulate
7. $a^2 + b^2 = c \cdot c$	7. Substitution Postulate of Equality
8. $a^2 + b^2 = c^2$	8. Substitution Postulate of Equality

This is only one of many ways that the Pythagorean Theorem may be proved. In your exercises, you will be asked to examine some of these proofs.

The converse of the Pythagorean Theorem is also true. The proof will involve the construction of a triangle that is congruent to the given triangle.

CONVERSE OF THE PYTHAGOREAN THEOREM.

THEOREM:

If the sum of the squares of two sides of a triangle equals the square of the remaining side of the triangle, then the triangle is a right triangle.

GIVEN:

△ ACB with sides of lengths a, b and c
c is the greatest length.
$a^2 + b^2 = c^2$

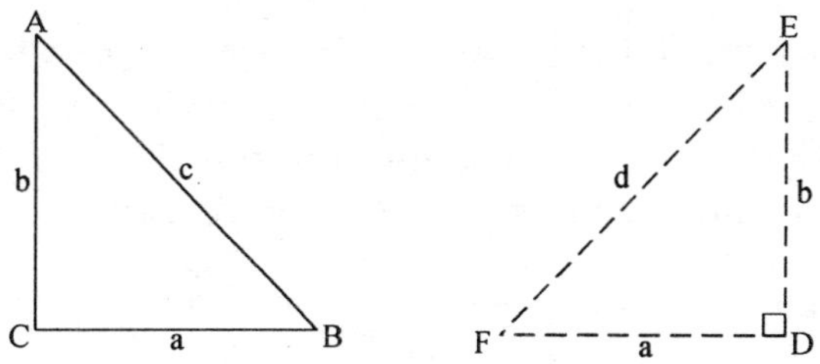

PROVE:

△ ACB is a right triangle.

STATEMENTS	REASONS
1. △ ACB with sides of lengths a, b and c; c is the greatest length. $a^2 + b^2 = c^2$	1. Given
2. Construct a right triangle, △ EDF with legs of lengths a and b. Let d represent the length of the hypotenuse and ∠ EDF = 90.	2. ?
3. $a^2 + b^2 = d^2$	3. Pythagorean Theorem
4. But $a^2 + b^2 = c^2$	4. Given
5. $c^2 = d^2$	5. Transitive Postulate of Equality
6. $c = \pm d$ or $c = d$ since c and d are positive	6. Roots Postulate of Equality
7. △ ACB ≅ △ EDF	7. SSS (steps 1,2,6)
8. ∠ ACB = ∠ EDF	8. CPCTE
9. ∠ ACB = 90	9. Transitive Postulate of Equality
10. ∠ ACB is a right angle.	10. Definition of Right Angle
11. △ ACB is a right triangle.	11. Definition of Right Triangle

Now let us turn our attention to some properties of special right triangles. We want to prove that in a right triangle with acute angles of measures 30° and 60° (a 30-60 right triangle), the length of the hypotenuse is twice the length of the side opposite the 30° angle. In addition, we want to prove that in an isosceles right triangle (a 45-45 right triangle), the length of the hypotenuse equals the product of $\sqrt{2}$ and the length of one of the legs of the triangle.

We will use a discussion format for this proof. Be sure to provide explanations in the form of definitions, postulates, theorems and given information, when you are asked to do so.

We will call this theorem the **30-60 RIGHT TRIANGLE THEOREM**

THEOREM:

In a right triangle with an angle of 30 degrees, the length of the hypotenuse is twice the length of the side opposite the 30 degree angle. The length of the other side of the triangle, that is, the side opposite the 60 degree angle, is the product of $\sqrt{3}$ and the length of the side opposite the 30 degree angle.

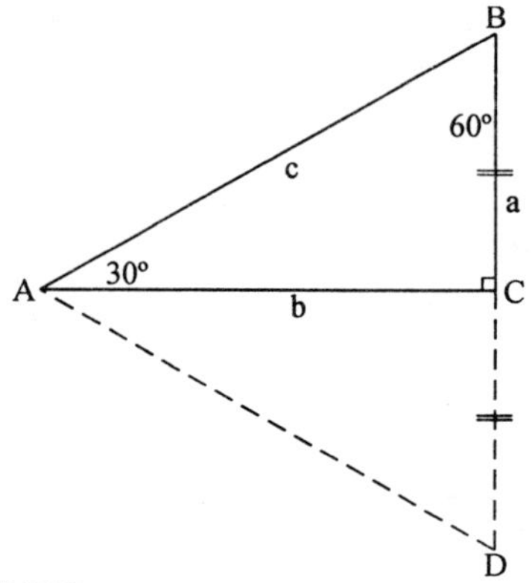

GIVEN:

△ ABC with right angle ACB
a is the length of the side opposite the 30 degree angle.
b is the length of the side opposite the 60 degree angle.
c is the length of the hypotenuse.

PROVE:

$c = 2a$ and $b = a\sqrt{3}$

STATEMENTS	REASONS

First we will use the Unique Point Postulate to locate a point, say D, on the ray opposite ray $\overrightarrow{CB}$ so that BC = CD. Construct $\overline{AD}$.

1. Can you explain why △ ACB is congruent to △ ACD?
2. Can you explain why △ DAB is equiangular and therefore equilateral?
3. Now can you explain why $c = 2a$? Remember that △ DAB is equilateral?
4. Can you use the Pythagorean Theorem to prove that $b = a\sqrt{3}$?

You will be asked to prove the next theorem as one of your exercises.

We will call this theorem the **45-45 RIGHT TRIANGLE THEOREM.**

THEOREM:

In a right triangle with an angle of 45 degrees, the length of the hypotenuse is the product of $\sqrt{2}$ and the length of a leg of the triangle.

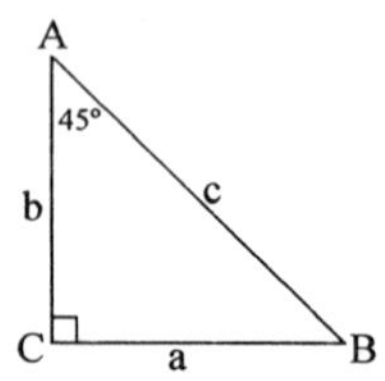

GIVEN:

△ ACB with right angle ACB
a is the length of the leg opposite the 45 degree angle.
b is the length of the other leg of the triangle.
c is the length of the hypotenuse.

PROVE:

$c = a\sqrt{2}$

Let us see how we can apply these theorems to find missing parts of triangles. We can use these triangles as models.

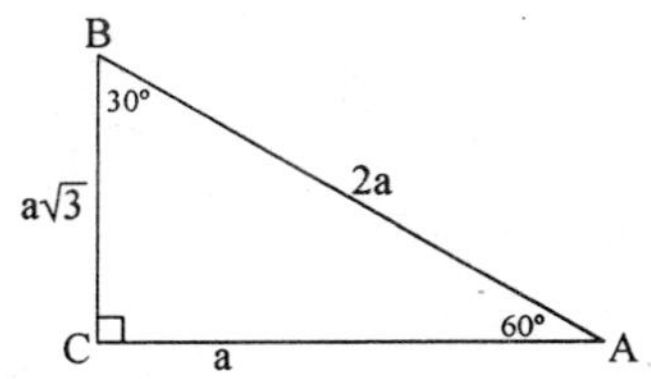

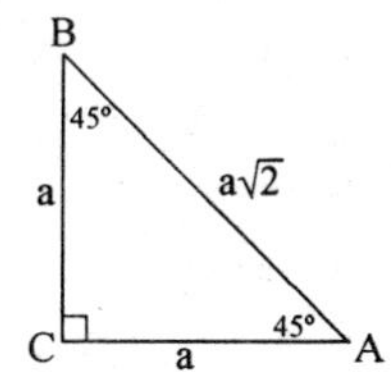

EXAMPLE 1

Find the values of b and c if the side opposite the 30 degree angle in right triangle ABC has a length of 10.

The length of the hypotenuse is twice the length of the side opposite the 30 degree angle.

$$c = 2a$$
$$c = 2 \cdot 10 = 20$$

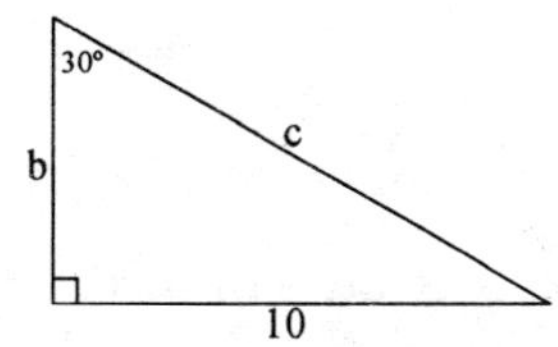

The length of the side opposite the 60 degree angle is the product of the length of the side opposite the 30 degree angle and $\sqrt{3}$.

$$b = a\sqrt{3}$$
$$b = 10 \cdot \sqrt{3} = 10\sqrt{3}$$

When we are given the side opposite the 30 degree angle, we multiply to find the other sides of the triangle.

EXAMPLE 2

Find the values of a and b if c is 12 in a 30-60 right triangle.

Since c is twice the value of a, we divide c by 2 to obtain the value of a.

$$a = \frac{12}{2} = 6$$

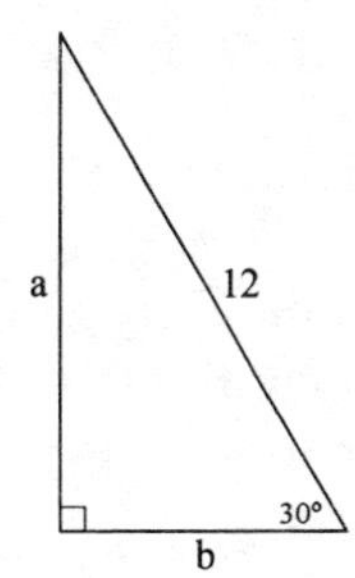

Once we have the value of a, we find b by multiplying by $\sqrt{3}$, as we did in example 1.

$$b = 6 \cdot \sqrt{3} = 6\sqrt{3}$$

When we are given the hypotenuse, we divide to find the length of the side opposite the 30 degree angle. Once we have that length, we can use it to find the other length.

EXAMPLE 3

Find the values of a and c if b is 9 in a 30-60 right triangle.
We divide 9 by $\sqrt{3}$ to obtain the value of a.

$$a = \frac{9}{\sqrt{3}}$$

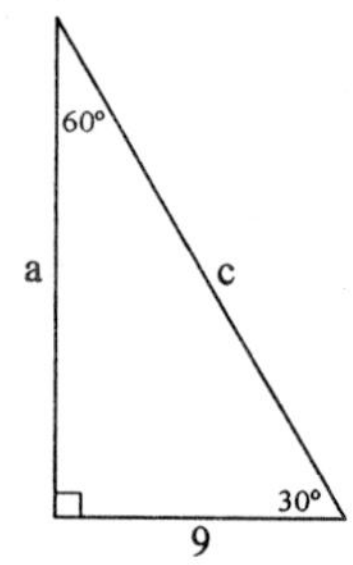

Rationalizing this value we get $\dfrac{9\sqrt{3}}{3\sqrt{3}} = 3\sqrt{3}$

Therefore $a = 3\sqrt{3}$

Once we have the value of a, we find the length of the hypotenuse by multiplying a by 2.

$$c = 2a$$
$$c = 2(3\sqrt{3}) = 6\sqrt{3}$$

When we are given the side opposite the 60 degree angle, first we divide to find a. After we find a, we can multiply to find the other length.

EXAMPLE 4

Find the length of the hypotenuse of an isosceles right triangle with a leg of length 15.

An isosceles right triangle is a 45-45 right triangle, therefore we may use our theorem for the 45-45 right triangle.

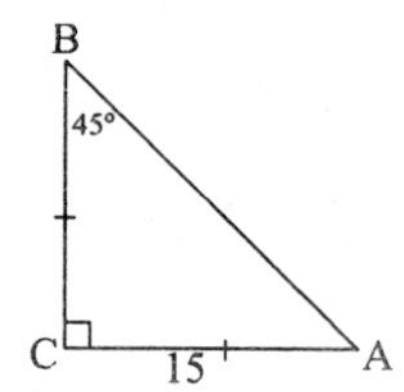

a represents the length of a leg of the triangle so $a = 15$.

The hypotenuse in this triangle has a length equal to the product of $\sqrt{2}$ and the length of a leg.

$$c = a\sqrt{2}$$
$$c = 15\sqrt{2}$$

For both theorems, the side opposite the 30 degree angle in the 30-60 right triangle and the side opposite a 45 degree angle in the 45-45 right triangle are used as references. When we know the lengths of these sides, we only need to multiply to find the other sides of the triangle. When we are not given the lengths of these sides, we can divide to find them, and afterwards find the remaining side by multiplying.

Exercises 9.1

For some of these exercises, you may want to use auxiliary lines, rays or line segments.(Figures are not drawn to scale.)

1. Prove the following theorem:

THEOREM:

In a right triangle with an angle of 45 degrees, the length of the hypotenuse is the product of $\sqrt{2}$ and the length of a leg of the triangle.

GIVEN:

$\triangle$ ACB with right angle ACB

a is the length of the leg opposite the 45 degree angle.

b is the length of the other leg of the triangle.

c is the length of the hypotenuse.

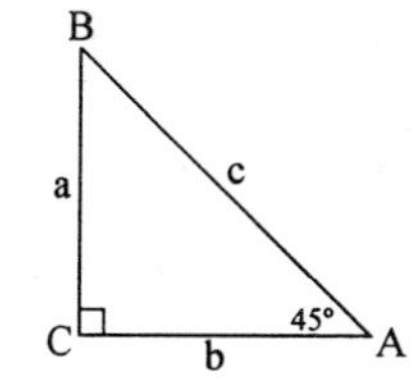

PROVE:

$c = a\sqrt{2}$

STATEMENTS	REASONS

2. Use the areas of polygons in the figure to prove that

$c^2 = a^2 + b^2$

We begin by constructing square ABCD with each side of length (a + b) as shown.

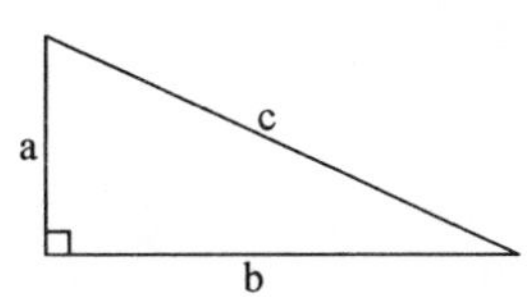

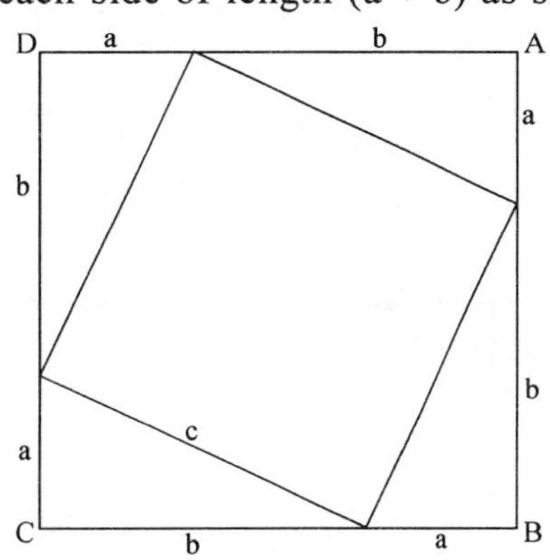

a. Why are the four triangles congruent to the given triangle and congruent to each other?

____

b. What is the area of each triangle? ______________________

c. What is the sum of the areas of the four triangles?______________________

d. Can you prove that the quadrilateral formed by the hypotenuses of the four triangles is a square?

____

e. What is the area of this square?______________________

f. What is the area of the square with sides of length $(a + b)$?______________________

Use your algebra to simplify the expression that represents this area. ______________________

g. What is the combined sum of the areas of the four triangles and the area of the smaller square?

____

h. Note that the area of the larger square is equal to the combined sum of the areas of the four triangles and the area of the smaller square.

Can you write an algebraic equality for this relationship?

____

i. What result is obtained when the equation is simplified?

____

3. Use the areas of polygons in the figure to prove that

$c^2 \ = \ a^2 + b^2.$

This proof was discovered by President James Garfield.
First we construct trapezoid EDCA with bases of lengths a and b and altitude of length $(a + b)$ as shown in the figure.

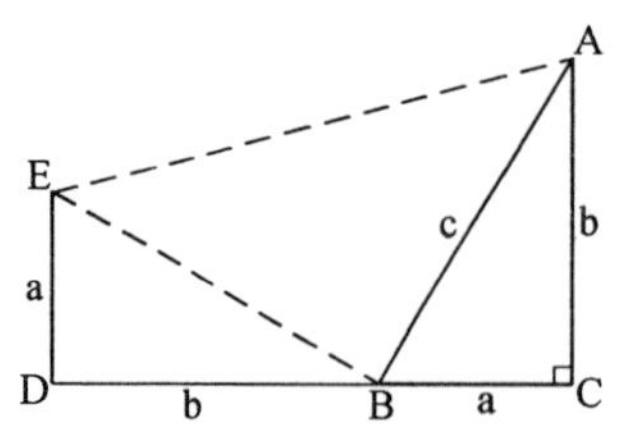

(Hint: Prove that the area of the trapezoid equals the sum of the area of the three triangles)

4. If m and n represent positive integers and $m > n$, can you use your algebra to prove that a triangle with sides of lengths $(m^2 + n^2)$, $(m^2 - n^2)$ and $(2mn)$ must be a right triangle?

5. Use exercise 4 to generate lengths for the sides of ten right triangles. You may replace m and n with any positive integers you choose as long as m is greater than n. The lengths for one triangle have been generated for you where m = 3 and n = 1.

		leg	leg	hypotenuse
m	n	$m^2 - n^2$	2mn	$m^2 + n^2$
3	1	8	6	10

6. Find AB and AC

a. If BC = 6
b. If BC = 13
c. If BC = 15
d. If BC = 1
e. If BC = $2\sqrt{6}$

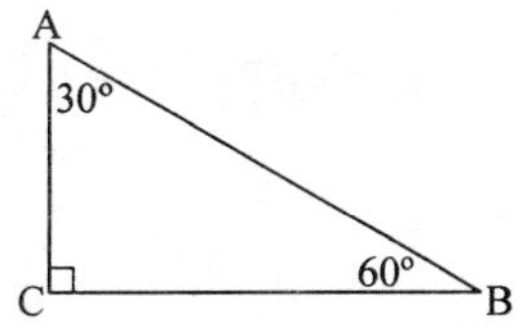

7. Find BC and AC

a. If AB = 10
b. If AB = 8
c. If AB = $4\sqrt{2}$
d. If AB = $\sqrt{15}$

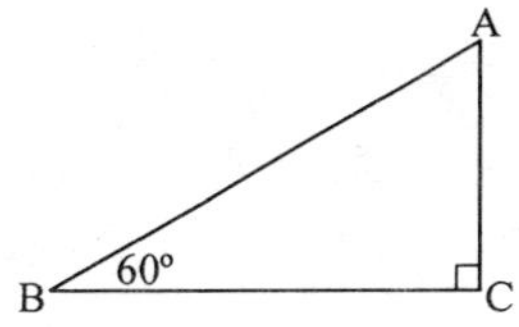

8. Find BC and AB

a. If AC = $\sqrt{15}$
b. If AC = $\sqrt{3}$
c. If AC = 10
d. If AC = 12

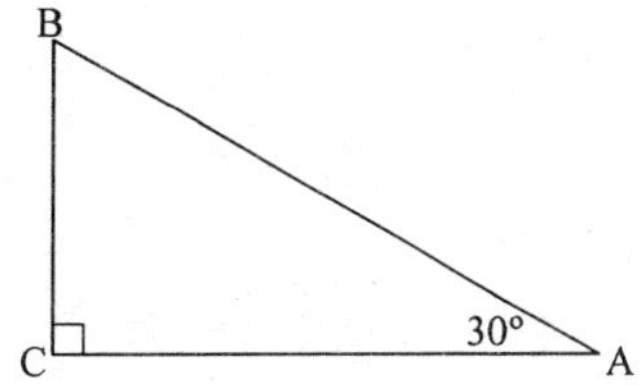

9. Find AB and BC

a. If AC = 1
b. If AC = 12
c. If AC = $\sqrt{3}$
d. If AC = $2\sqrt{6}$
e. If AC = $\sqrt{10}$

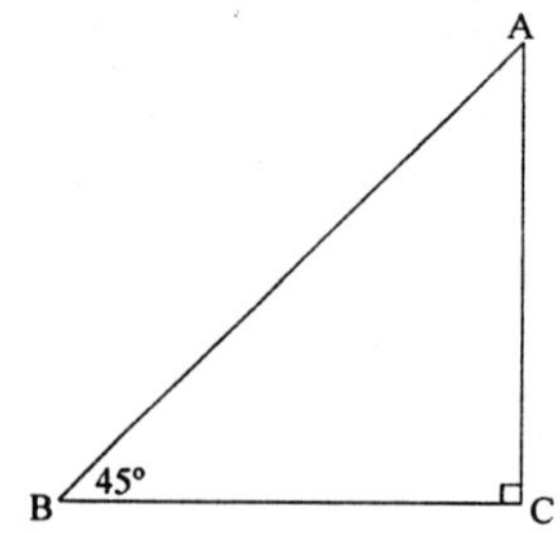

10. Find AC and BC

a. If AB = $\sqrt{10}$
b. If AB = $\sqrt{6}$
c. If AB = 12
d. If AB = 4

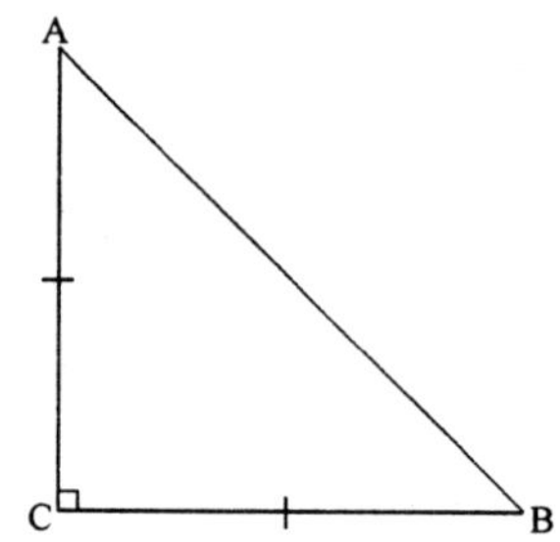

11. Find the area of an isosceles triangle that has a leg of length 10 and a base angle with each of the following measures:

a. 45 b. 30 c. 60

12. Find the area of an isosceles triangle that has a leg of length 30 and a base angle with each of the following measure:

a. 45 b. 30 c. 60

13. Find the area of an isosceles triangle that has a base of 6 if each of its base angles has the following measure:

a. 45 b. 30 c. 60

14. Find the area of an isosceles triangle that has a base of 20 if each of its base angles has the following measure:

a. 45 b. 30 c. 60

15. Each of the following numbers represents the length of the diagonal of a square. In each case, find the area of the square.

a. $5\sqrt{2}$ b. $6\sqrt{2}$ c. $8\sqrt{6}$ d. 12 e. 10

16. Each of the following numbers represents the length of the diagonal of a square. In each case, find the area of the square.

a. $7\sqrt{2}$ b. $4\sqrt{14}$ c. $3\sqrt{5}$ d. 14 e. 4

17. a. Find the area of $\triangle$ BAC if AD = 8.

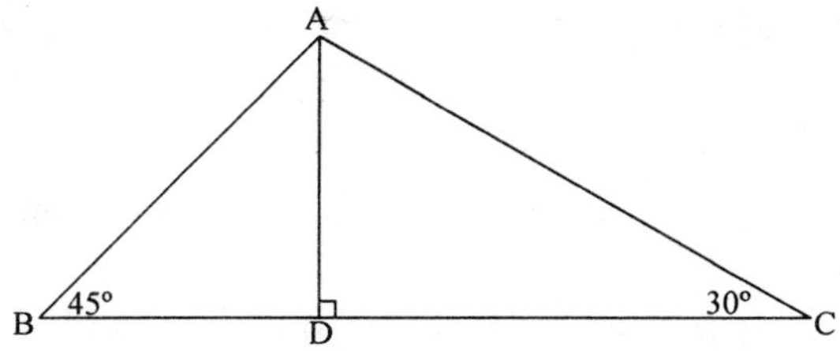

b. Find the perimeter of $\triangle$ BAC if BD = 4.

c. Find the area of $\triangle$ BAC if AC = 12.

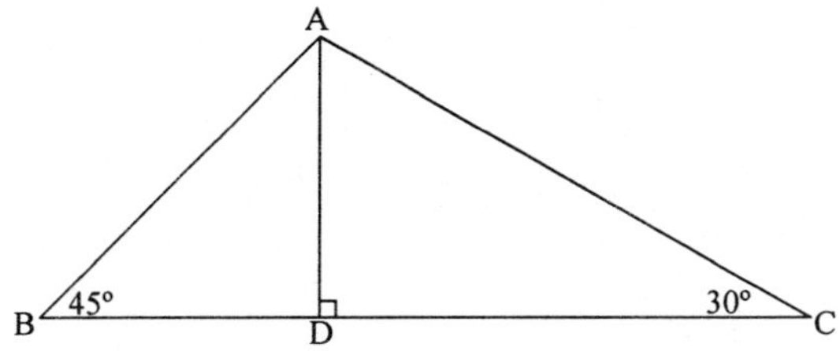

d. Find the area of $\triangle$ BAC if AB = $4\sqrt{2}$.

18. Find the area and the length of each diagonal of a rhombus:

a. If one of its angles has a measure of 60 and one of its sides has a length of 10.

b. If one of its angles has a measure of 60 and one of its sides has a length of 12.

c. If one of its angles has a measure of 120 and one of its sides has a length of 8.

d. If one of its angles has a measure of 60 and one of its sides has a length of 6.

e. If one of its angles has a measure of 60 and one of its sides has a length of x.

19. a. Find the area of the rhombus ABCD if one of its diagonals has a length of x and its other diagonal has a length of y. (Hint: Recall the relationship between the diagonals of a parallelogram and the relationship between the diagonals of a rhombus)

b. Use the result of exercise 19a to fill in the blank:

The area of a rhombus is equal to ½ the product of the lengths of its____________________

20. Find the area of a parallelogram that has sides of lengths 12 and 4 if the measure of one of its angles is as follows:

a. 150 b. 45 c. 60 d. 120

21. Construct each of the following:

a. A parallelogram with sides of lengths 9 and 4 and an angle of measure 30.

b. A rhombus with a side of length 6 and an angle of measure 30.

22. Find the area of an equilateral triangle if each side has the following length:

a. 12   b. 8   c. 4   d. a

23. a. A man travels 8 miles due north, then 6 miles due east. How far is he from his starting point?

b. A woman travels 14 miles due north, then travels 6 miles due east, and then travels 6 miles due south. How far is she from her starting point?

24. Each side of a baseball diamond is 90 ft long. How far is it from homeplate to second base?

25. Use the Mean Proportion Corollaries to find the values of x and y.

a.

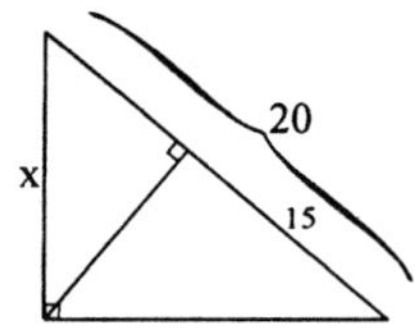

b.

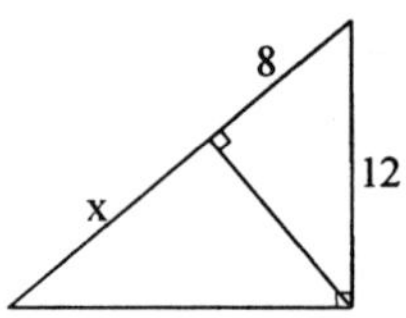

c.

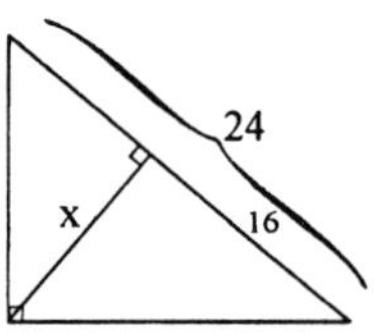

d.

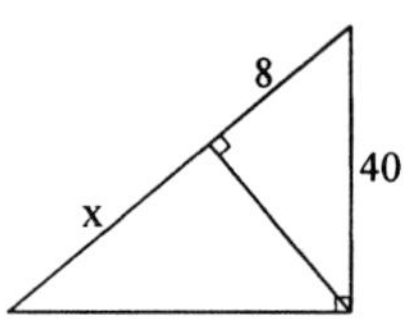

e.

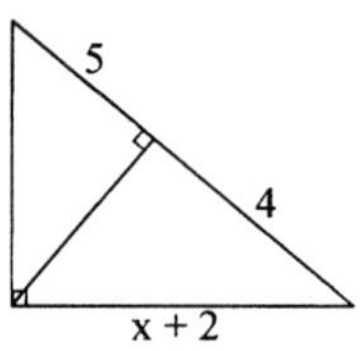

f.

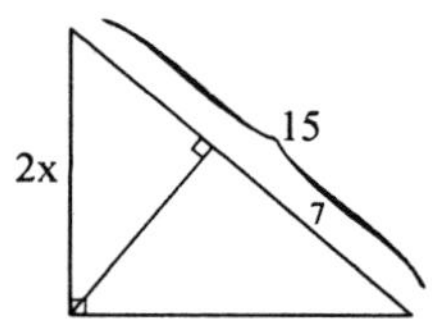

g.

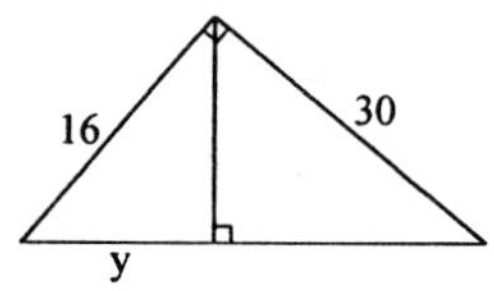

h.

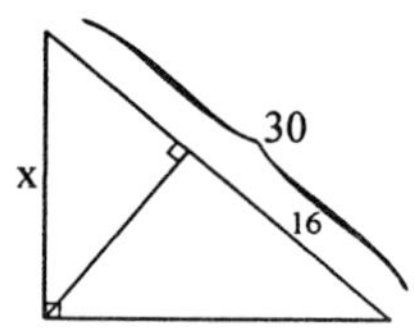

i.

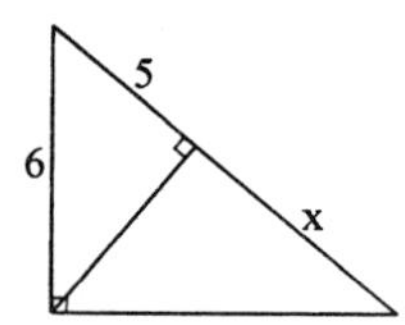

j.

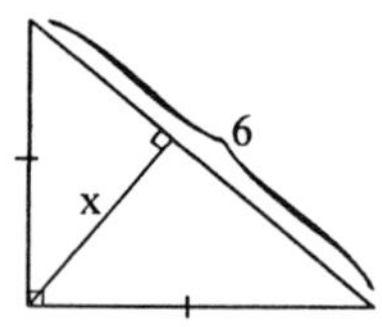

k.

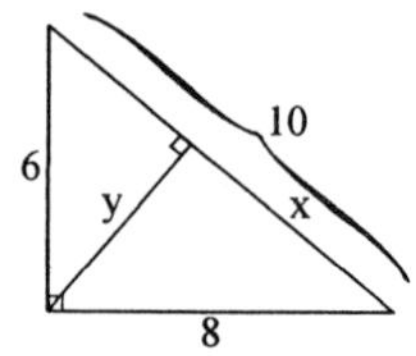

l.

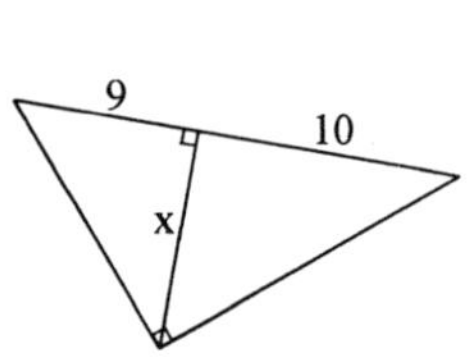

Section 9.2 General Properties Of Triangles

If we join the midpoints of two sides of a triangle, we get a rather surprising result.

We will call this theorem the **TRIANGLE MIDPOINT THEOREM**

THEOREM:

The line segment joining the midpoints of two sides of a triangle is parallel to the remaining side and one-half as long.

GIVEN:

The figure
D is the midpoint of $\overline{AB}$ in △ ABC.
E is the midpoint of $\overline{AC}$ in △ ABC.

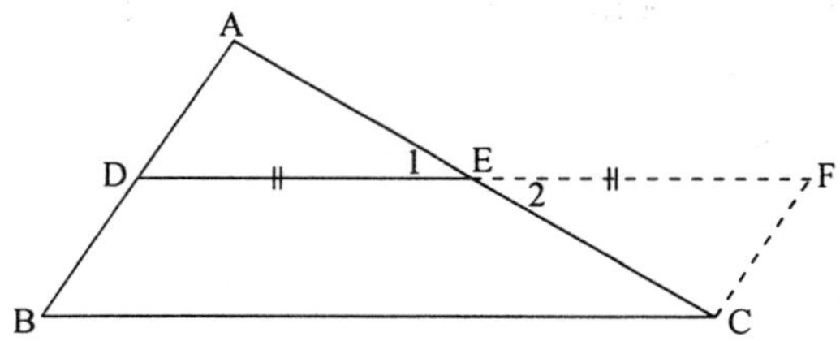

PROVE:

$\begin{cases} DE = \frac{1}{2} BC \\ \overline{DE} \parallel \overline{BC} \end{cases}$

STATEMENTS	REASONS
1. The figure D is the midpoint of $\overline{AB}$ in △ ABC. E is the midpoint of $\overline{AC}$ in △ ABC.	1. ?
2. Locate F on the ray opposite $\overrightarrow{ED}$ so that EF = DE.	2. Unique Point Postulate
3. Construct $\overline{CF}$.	3. Postulate: Two points determine a line
4. ∠ 1 = ∠ 2	4. ?
5. E is the midpoint of $\overline{AC}$.	5. Given
6. AE = CE	6. ?
7. △ AED ≅ △ CEF	7. ?
8. ∠ DAE = ∠ FCE	8. ?
9. $\overleftrightarrow{DA} \parallel \overleftrightarrow{CF}$	9. ?
10. AD = CF	10. CPCTE
11. D is the midpoint of $\overline{AB}$.	11. ?
12. AD = BD	12. ?
13. BD = CF	13. ?
14. Quadrilateral BDFC is a parallelogram.	14. Theorem: If a quadrilateral has two sides that are parallel and of equal lengths then it is a parallelogram.
15. $\overline{DE} \parallel \overline{BC}$	15. Definition of a parallelogram
16. DE + EF = DF	16. ?
17. DF = BC	17. Theorem: Opposite sides of a parallelogram have equal lengths.
18. DE + EF = BC	18. Substitution Postulate of Equality
19. DE + DE = BC	19. Substitution Postulate of Equality
20. 2 DE = BC	20. Substitution Postulate of Equality
21. DE = $\frac{1}{2}$ BC	21. Multiplication Postulate of Equality

The theorem that we are going to prove next will lead to some very useful proportion theorems.

THEOREM:

A line parallel to one side of a triangle cuts off a triangle that is similar to the given triangle.

GIVEN:

The figure with $\overleftrightarrow{ST} \parallel \overleftrightarrow{QR}$ in $\triangle PQR$

PROVE:

$\triangle PST \sim \triangle PQR$

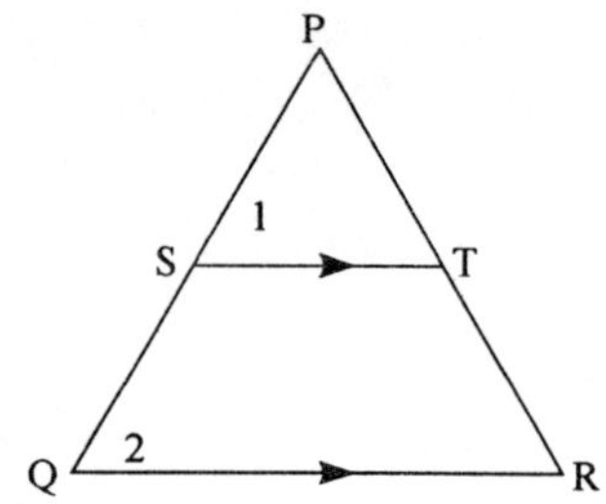

STATEMENTS	REASONS
1. The figure with $\overleftrightarrow{ST} \parallel \overleftrightarrow{QR}$ in $\triangle PQR$	1. Given
2. $\angle P = \angle P$	2. Reflexive Postulate of Equality
3. $\overline{ST} \parallel \overline{QR}$	3. Given
4. $\angle 1 = \angle 2$	4. ?
5. $\triangle PST \sim \triangle PQR$	5. AA Postulate for Similar Triangles

Now see if you can use this theorem along with some of the theorems on proportions to finish proving the next theorem.

THEOREM:

A line parallel to one side of a triangle cuts the other two sides of the triangle proportionally.

GIVEN:

$\overleftrightarrow{ST} \parallel \overleftrightarrow{QR}$ in $\triangle PQR$

PROVE:

$\frac{PS}{SQ} = \frac{PT}{TR}$

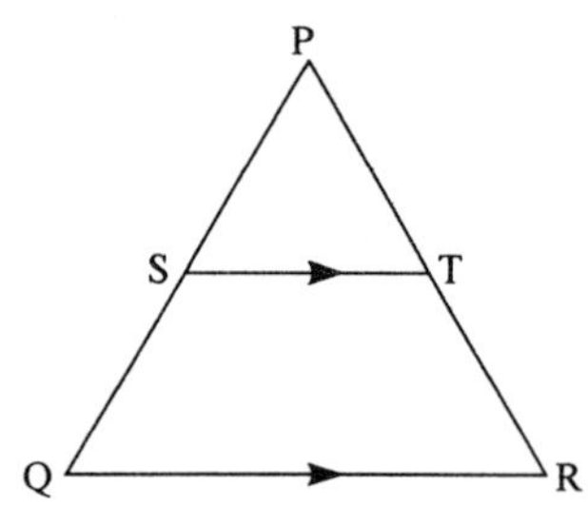

STATEMENTS	REASONS
1. In $\triangle PQR$, $\overleftrightarrow{ST} \parallel \overleftrightarrow{QR}$	1. Given
2. $\triangle PQR \sim \triangle PST$	2. The last theorem that we proved
3. $\frac{PQ}{PS} = \frac{PR}{PT}$	3. Definition of Similar Triangles
4. $\frac{PQ - PS}{PS} = \frac{PR - PT}{PT}$	4. A Proportion Theorem?

Can you finish this proof?

THEOREM:

The bisector of one angle of a triangle divides the opposite side of the triangle in the same ratio as the other two sides.

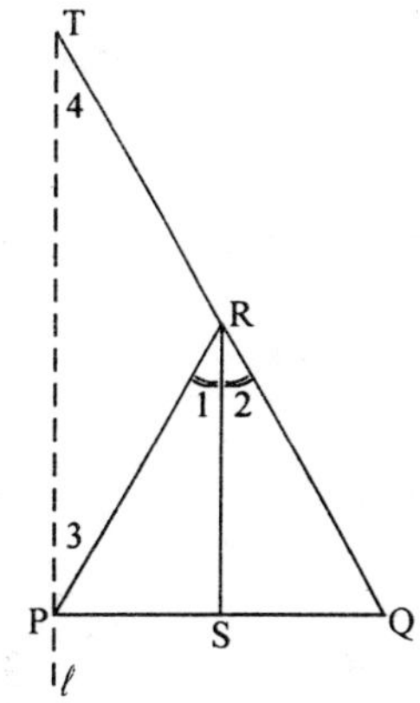

GIVEN:

$\overrightarrow{RS}$ bisects angle PRQ

PROVE:

$$\frac{QR}{RP} = \frac{QS}{PS}$$

STATEMENTS	REASONS
1. $\overrightarrow{RS}$ bisects angle PRQ	1. Given
2. Through P construct ℓ parallel to $\overline{RS}$	2. The Parallel Postulate
3. Let T be the point of intersection of $\overline{RQ}$ with ℓ.	3. Theorem: Two lines intersect in exactly one point.
4. $\angle 2 = \angle 4$	4. ?
5. $\angle 3 = \angle 1$	5. ?
6. $\angle 1 = \angle 2$	6. Definition of Angle Bisector
7. $\angle 3 = \angle 4$	7. ?
8. RT = RP	8. ?
9. $\frac{QR}{RT} = \frac{QS}{PS}$ for $\triangle$ QTP	9. Theorem: A line parallel to one side of a triangle divides the other two sides proportionally.
10. $\frac{QR}{RP} = \frac{QS}{PS}$	10. Substitution Postulate of Equality

EXAMPLE 1

In $\triangle$ MNO, S is the midpoint of side $\overline{MO}$, and T is the midpoint of side $\overline{ON}$. Find the length of side $\overline{ST}$ if the length of side $\overline{MN}$ is 35.

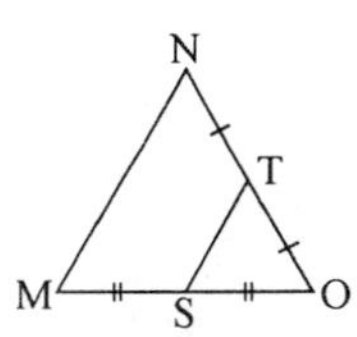

The length of line segment $\overline{ST}$ is 1/2 the length of line segment $\overline{MN}$ in $\triangle$ MNO.

ST = 1/2 $\overline{MN}$

ST = 1/2 (35)

ST = 17.5

EXAMPLE 2

$\overline{RP}$ is parallel to side $\overline{SA}$ of $\triangle$ TSA. Find the length of $\overline{TR}$ if RS = 15, TP = 5 and PA = 7.

$\overline{RP}$ cuts sides of triangle TSA proportionally, therefore we have the following proportion.

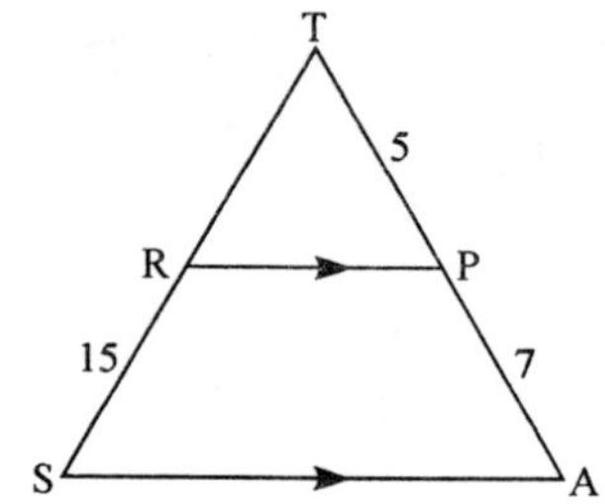

$$\frac{TR}{RS} = \frac{TP}{PA}$$

Substituting the given values we obtain $\frac{TR}{15} = \frac{5}{7}$

Solving for TR we get

$$7\ TR = 5 \cdot 15$$
$$7\ TR = 75$$
$$TR = \frac{75}{7}$$
$$TR = 10\frac{5}{7}$$

EXAMPLE 3

In the figure, Ray $\overrightarrow{AD}$ is the bisector of angle PAR.
If AP = 12, AR = 10 and PD = 5, find the length of $\overline{RD}$.

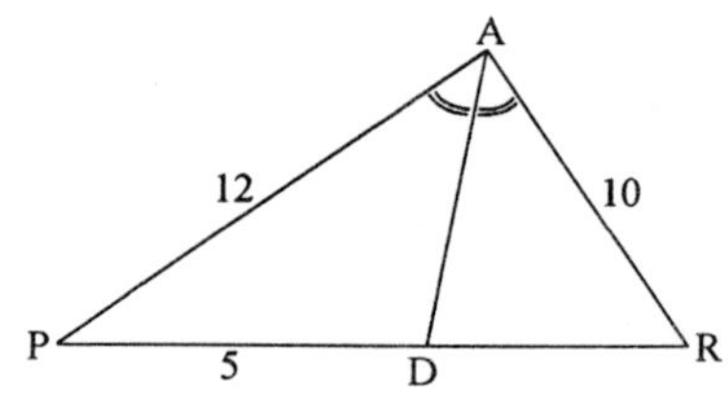

The bisector divides side $\overline{PR}$ of $\triangle$ PAR in the same ratio as sides $\overline{PA}$ and $\overline{RA}$. Therefore we have the proportion $\frac{PA}{RA} = \frac{PD}{RD}$. Substituting the given values we get $\frac{12}{10} = \frac{5}{RD}$.

Solving for RD, we obtain $RD = 4\frac{1}{6}$.

Exercises 9.2

(*Figures are not drawn to scale*)

1. In $\triangle$ PQR, line segment $\overline{ST}$ is parallel to line segment $\overline{RQ}$.

   a. Find TQ if PR = 10, PS = 4 and PT = 8.

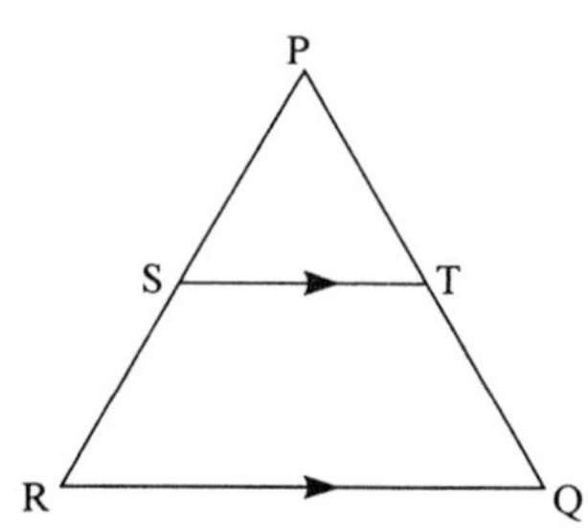

b. Find SR if PR = 9, PT = 3 and PQ = 18.

c. Find PQ if SR = PT, PS = 4 and TQ = 9.

d. Find PQ if PS = 5, PR = 17 and TQ is 2 more than PT.

e. Find PQ if PS = 5, TQ = 10 and SR = PT.

2. In △ ABC, line segment $\overline{DE}$ is parallel to line segment $\overline{BC}$.

a. Find AE if AD = 16, AB = 40 and EC = 12.

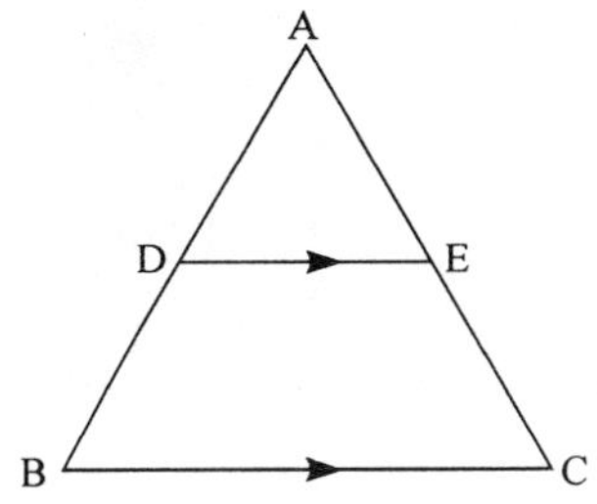

b. Find EC if AB = 12, AD = 8 and AE = 16.

c. Find AC if AD = 3, AB = 11 and EC is 4 more than AE.

d. Find AC if AE = BD, AD = 2 and EC = 8.

e. Find AC if AD = EC, BD = 6 and AE = 8.

f. Find AD, if AD = $x + 3$, AB = $3x + 3$, AE = $3x - 6$ and EC = $2x + 7$.

3. Given the following conditions, prove that the perimeter of $\triangle$ DEF is 1/2 the perimeter of $\triangle$ ABC:

D is the midpoint of line segment $\overline{AB}$
E is the midpoint of line segment $\overline{BC}$
F is the midpoint of line segment $\overline{AC}$

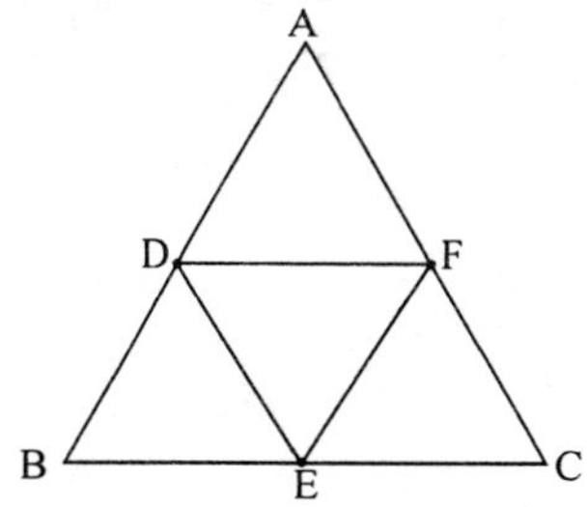

4. If points P, Q, S and R are the midpoints of the sides of a rhombus (as shown in the figure), prove that Quadrilateral PQRS is a rectangle.

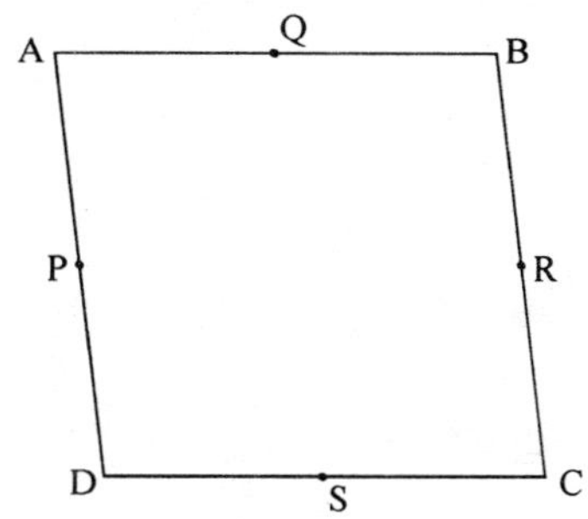

5. If points D, E and F are midpoints of the sides of $\triangle$ ABC, prove that quadrilateral DEFB is a rhombus.

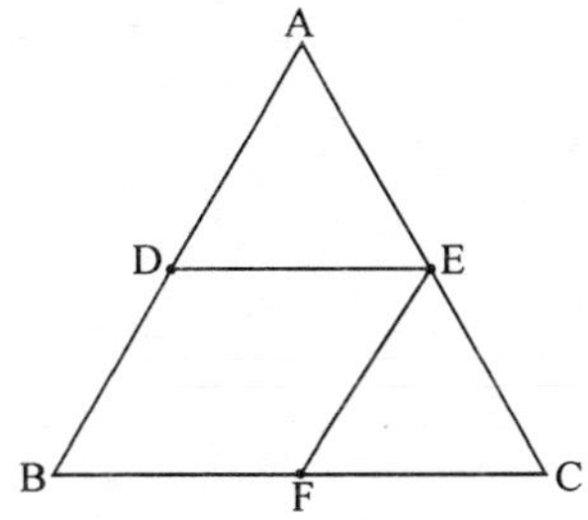

6. If points D, E and F are the midpoints of the sides of equilateral triangle ABC, prove that $\triangle$ DEF is equilateral.

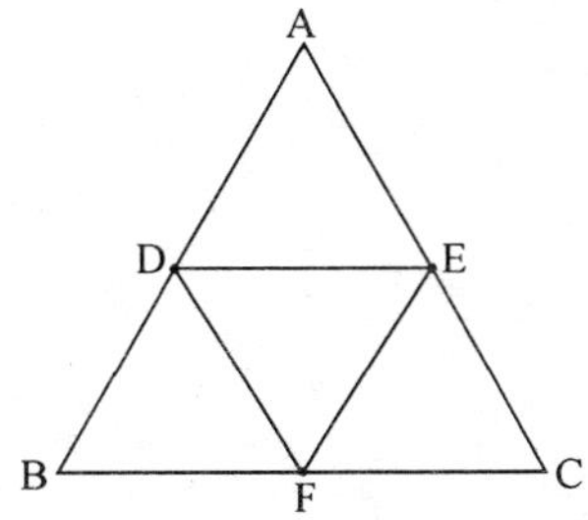

7. In the figure, ray $\overrightarrow{CD}$ is the bisector of angle ACB in $\triangle$ ABC.

a. If AC = 8, BC = 16 and AD = 6, find DB.

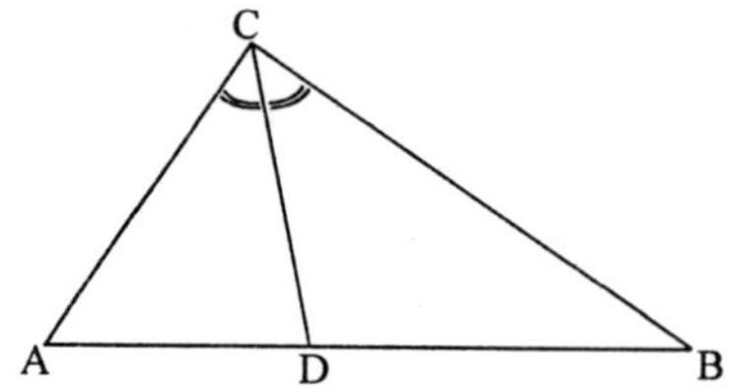

b. If AC = 15, BC = 24 and AD = 4, find DB.

c. If AC = $\frac{2}{3}$BC and BD = 48, find AD.

d. If BC = 4AC and AD = 15, find DB.

8. Use the figure in exercise 7.
In the figure, ray $\overrightarrow{CD}$ is the bisector of angle ACB in $\triangle$ ABC.

a. If AC = 12, BC = 8 and AD = 2, find DB.

b. If AC = 12, BC = 18 and AD = 2, find DB.

c. If AC = 1/4 BC and BD = 60, find AD.

d. If BC = 5AC and BD = 60, find AD.

e. If AC = 12, BC = 18 and AD = 2, find AB.

9. Construct a triangle that is similar to △ SRT.

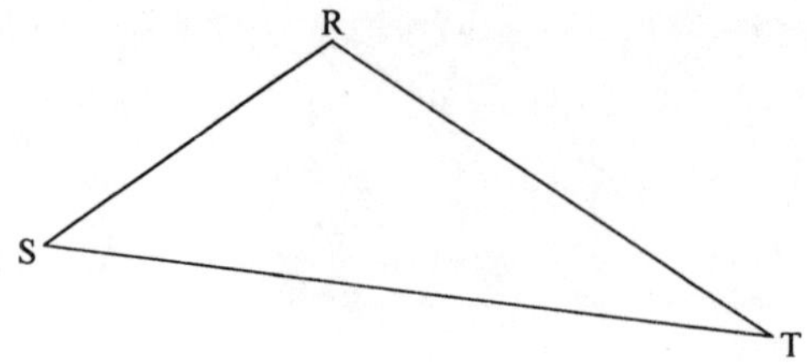

10. Construct a triangle that is similar to △ PQR.

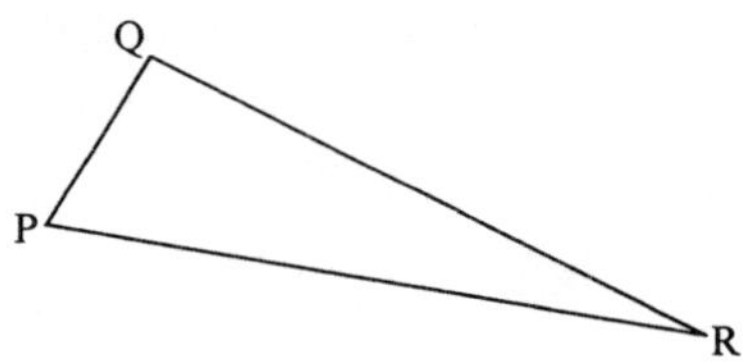

Chapter Nine Highlights

Do You Know and Understand These Postulates?

Angle Angle Postulate

If the measures of two angles of one triangle equal, respectively, the measures of two angles of another triangle, then the two triangles are similar. (AA POSTULATE)

Area Addition Postulate

If a region is the union of two or more regions that have no interior points in common, then its area is the sum of their areas.

Do You Know and Understand These Theorems?

In a proportion, the product of the extremes equals the product of the means.

We may interchange the extremes of a proportion.

We may interchange the means of a proportion.

We may interchange both the extremes and the means of a proportion.

If $\frac{a}{b} = \frac{c}{d}$ **then** $\frac{a+b}{b} = \frac{c+d}{d}$ **and** $\frac{a-b}{b} = \frac{c-d}{d}$

If $\frac{a}{b} = \frac{c}{d} = \frac{e}{f} = \frac{g}{h} = k$ **Then** $k = \frac{a+c+e+g}{b+d+f+h}$

The altitude to the hypotenuse of a right triangle separates the triangle into two triangles that are similar to the original triangle and to each other.

Altitude to the Hypotenuse Corollary.

The altitude to the hypotenuse of a right triangle separates the hypotenuse of the triangle into two segments (line segments) for which the following is true:

1. Each leg of the triangle is a mean proportional between the hypotenuse and the segment of the hypotenuse next to that leg.
2. The altitude is the mean proportional between the two segments of the hypotenuse.

The Pythagorean Theorem

In a right triangle, the square of the hypotenuse equals the sum of the squares of its legs.

Converse of the Pythagorean Theorem.

If the sum of the squares of two sides of a triangle equals the square of the remaining side of the triangle, then the triangle is a right triangle.

30-60 Right Triangle Theorem

In a right triangle with an angle of 30 degrees, the length of the hypotenuse is twice the length of the side opposite the 30 degree angle. The length of the other side of the triangle, that is, the side opposite the 60 degree angle, is the product of $\sqrt{3}$ and the length of the side opposite the 30 degree angle.

45-45 Right Triangle Theorem

In a right triangle with an angle of 45 degrees, the length of the hypotenuse is the product of $\sqrt{2}$ and the length of a leg of the triangle.

Triangle Midpoint Theorem

The line segment joining the midpoints of two sides of a triangle is parallel to the remaining side and one-half as long.

A line parallel to one side of a triangle cuts off a triangle that is similar to the given triangle.

A line parallel to one side of a triangle cuts the other two sides of the triangle proportionally.

The bisector of one angle of a triangle divides the opposite side of the triangle in the same ratio as the other two sides.

Chapter Ten

Proofs Involving Circles

Objectives

Prove and apply theorems about angles related to circles
Prove and apply theorems about lines and line segments related to circles
Prove and apply theorems about tangents

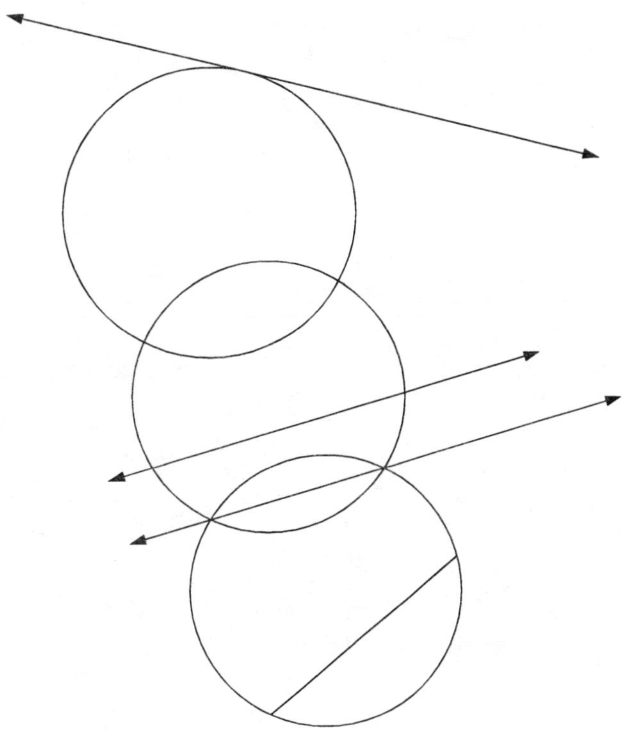

Section 10.1 Postulates and Definitions

Before beginning this chapter, we define some terms that we will need in the development of our proofs. Other definitions may be reviewed in Section 3.1 or in Appendix D.

A Secant of a Circle is a line that intersects the circle in two points.

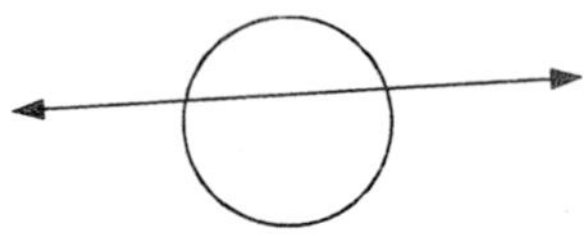

A Secant Ray: A ray is a secant ray if it intersects the circle in two points.

A Secant Segment: A segment is a secant segment if it contains two points of the circle.

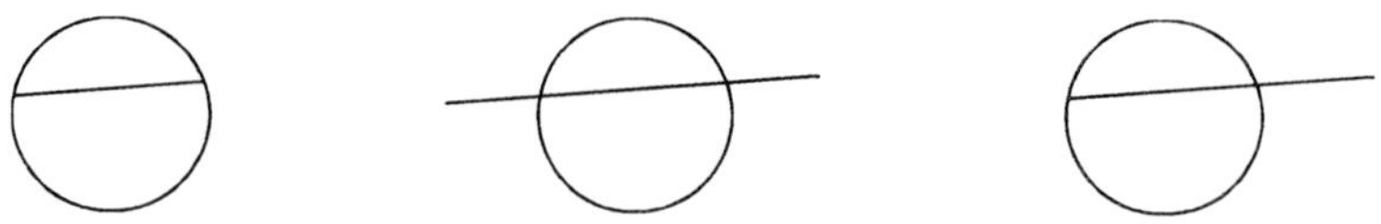

A Tangent to the Circle: is a line that intersects the circle in exactly one point.

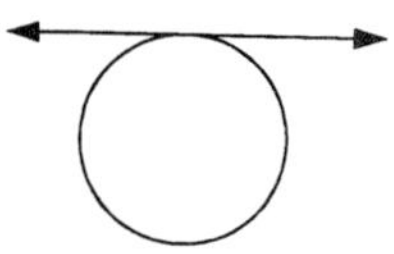

A Tangent Ray: A ray is a tangent ray to a circle if it intersects the circle in exactly one point and the line that contains it is a tangent of the circle.

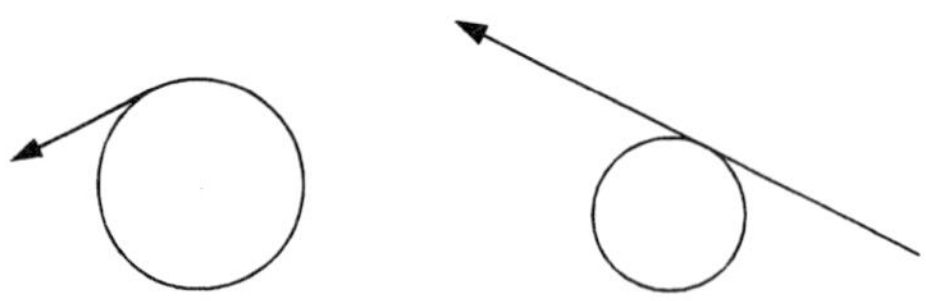

A Tangent Segment: A segment is a tangent segment to a circle if it intersects the circle in exactly one point and the line that contains it is a tangent of the circle.

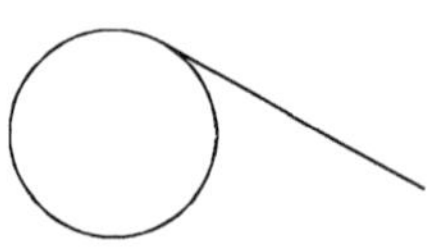

A Central Angle of a Circle: An angle is a central angle of a circle if its vertex is the center of the circle.

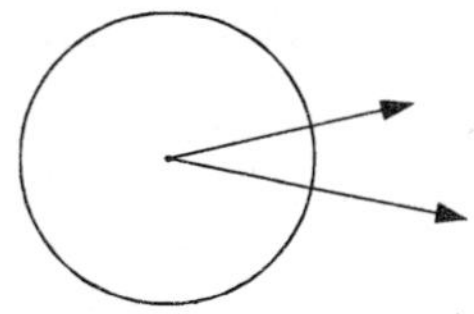

An Inscribed Angle of a Circle: is an angle whose vertex is on the circle and whose sides are two secant rays.

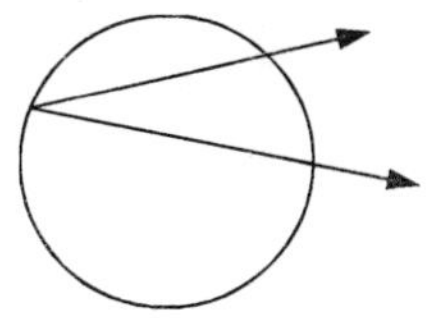

The Measure of a Central Angle of a Circle: is the measure of the minor arc that the central angle intercepts.

A Semicircle is an arc of the circle whose endpoints are the endpoints of a diameter of the circle.

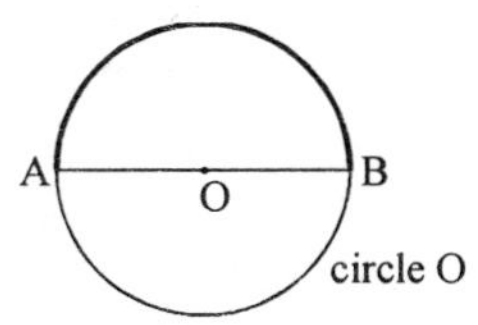

The Measure of a Semicircle is 180.

Next we have a postulate that we will need for our work.

Arc Addition Postulate or
Whole Equals Sum of Parts Postulate for Arcs
If A and C are endpoints of an arc and B is another point on the arc, then
$m\overset{\frown}{AB} + m\overset{\frown}{BC} = m\overset{\frown}{AC}$.

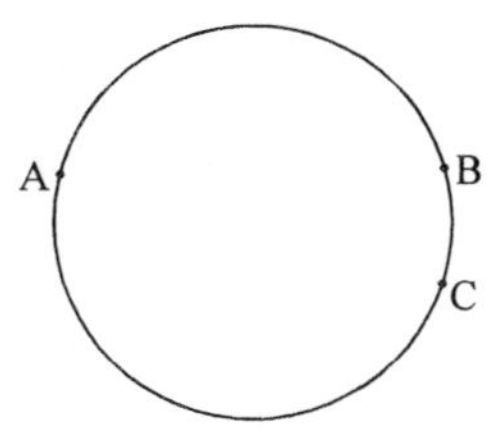

In some of the proofs in this chapter, we choose to leave out a few details. Again, the reason for this is we do not want our proofs to become bulky. We begin by proving some basic theorems.

This first theorem hardly deserves to be called a theorem, since it follows so directly from the definitions of radius and circle.

THEOREM:
Radii of the same circle are equal.

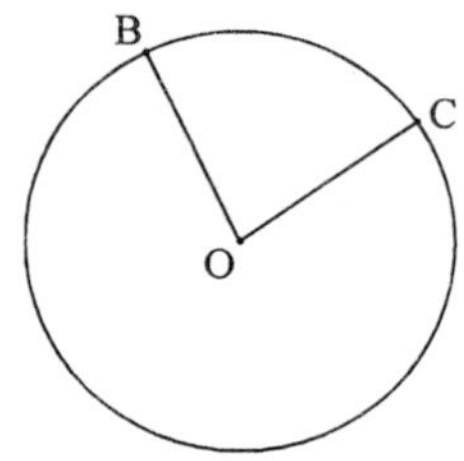

GIVEN:
Circle O with radii $\overline{OB}$ and $\overline{OC}$

PROVE:
OB = OC

STATEMENTS	REASONS
1. Circle O with radii $\overline{OB}$ and $\overline{OC}$	1. Given
2. $\overline{OB}$ and $\overline{OC}$ are line segments joining the center of the circle to points B and C of the circle	2. Definition of Radius
3. OB = OC	3. Definition of Circle

THEOREM:
The perpendicular from the center of a circle to a chord of the circle, bisects the chord.

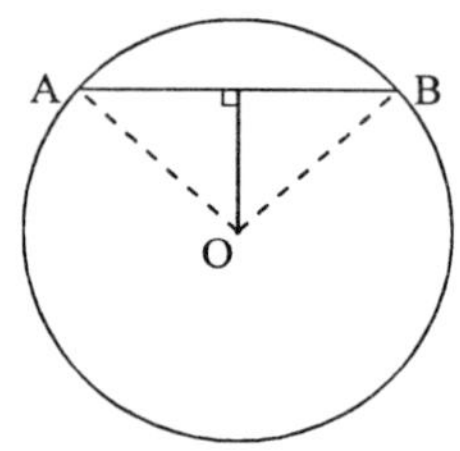

GIVEN:
Circle O with $\overline{OP} \perp \overline{AB}$

PROVE:
AP = BP

STATEMENTS	REASONS
1. Circle O with $\overline{OP} \perp \overline{AB}$	1. Given
2. Construct $\overline{OA}$ and $\overline{OB}$.	2. ?
3. OA = OB	3. Radii of a circle are equal.
4. OP = OP	4. Reflexive Postulate of Equality
5. $\angle$ OPA and $\angle$ OPB are right angles.	5. Perpendicular lines form right angles (Definition)
6. $\triangle$ AOP and $\triangle$ BOP are right triangles.	6. Definition of right triangle
7. $\triangle$ AOP is congruent to $\triangle$ BOP.	7. HL Theorem
8. AP = BP	8. CPCTE

We will call this next theorem the **TANGENT THEOREM.**

THEOREM:
A tangent of a circle is perpendicular to the radius at the point of tangency.

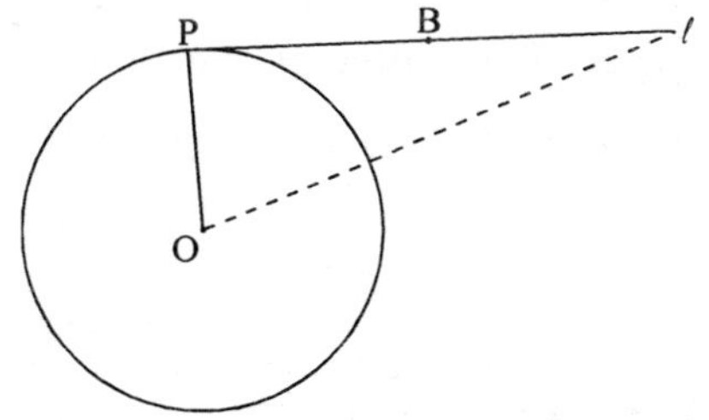

GIVEN:
Line l is a tangent to Circle O with point of contact P. $\overline{OP}$ is a radius drawn to the point of contact.

PROVE:
$\overline{OP}$ is perpendicular to line ℓ.

We will do an indirect proof, that is, let us suppose that $\overline{OP}$ is not perpendicular to line ℓ.

Locate point B on line ℓ so that $\overline{OB}$ is perpendicular to line ℓ. By one of our postulates, we know that such a line exists. Since $\overline{OB}$ is the perpendicular line segment joining point O to line ℓ, OB $<$ OP. This is true because the perpendicular line segment joining a point to a line is the shortest segment joining the point to the line. But, point B is an exterior point of Circle O.

Therefore OB $>$ OP by the definition of an exterior point of a circle. This is a contradiction of the Trichotomy Postulate so our supposition must not be true and $\overline{OP}$ is perpendicular to line ℓ.

Now we will use this theorem to prove another theorem about tangents.

THEOREM:
The two segments formed by two tangents to a circle from an external point of the circle have equal lengths.

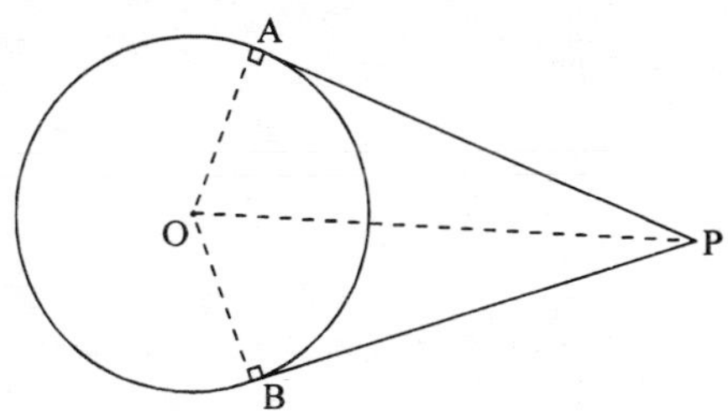

GIVEN:
$\overline{PA}$ tangent to Circle O at A
$\overline{PB}$ tangent to Circle O at B

PROVE:
PA = PB

STATEMENTS	REASONS
1. $\overline{PA}$ tangent to Circle O at A $\overline{PB}$ tangent to Circle O at B	1. Given
2. Construct OA and OB and OP.	2. ?
3. OA = OB	3. ?
4. OP = OP	4. ?
5. $\overline{OA} \perp \overline{PA}$ and $\overline{OB} \perp \overline{PB}$	5. Theorem: A tangent to a circle is perpendicular to the radius at the point of contact
6. Angle OAP and Angle OBP are right angles.	6. Definition of Perpendicular Lines
7. △ PAO and △ PBO are right triangles.	7. Definition of right triangle
8. △ PAO and △ PBO are congruent.	8. HL Theorem
9. PA = PB	9. CPCTE

In the group of proofs in the next two sections, we will simplify our notation by using variables to represent measures of arcs and angles. This should make the algebraic expressions involved in these proofs much easier to handle. Proofs in this group may be divided into three categories:

Section 10.2 Angles that have vertices on the circle.

Section 10.3 Angles that have vertices in the exterior of the circle

Section 10.3 Angles that have vertices in the interior of the circle

Section 10.2 Angles that Have Vertices on the Circle

As you recall, an inscribed angle has its vertex on the circle. Since inscribed angles will be used in many of our proofs, we begin by showing that their measures must be ½ the measure of their intercepted arcs.

Our first theorem about inscribed angles will require three proofs.

> **THEOREM:**
> **The measure of an inscribed angle is ½ the measure of the arc that it intercepts.**

We consider the following situations:

The center of the circle is located on the angle.

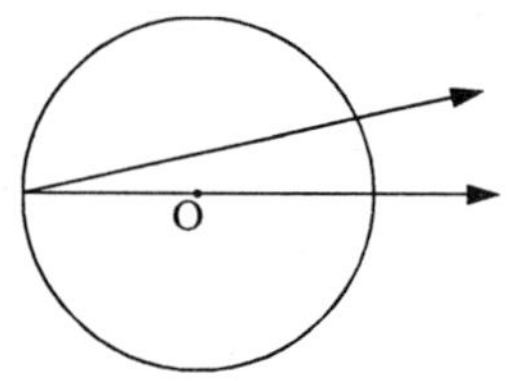

The center of the circle is located in the exterior of the angle.

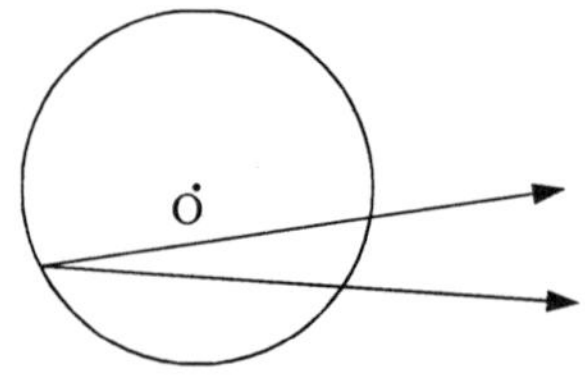

The center of the circle is located in the interior of the angle.

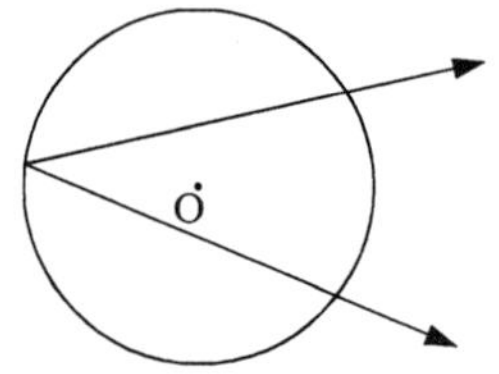

We will set up proofs for the first two situations, and you will be asked to provide a proof for the third situation as one of your exercises.

THEOREM:
The measure of an inscribed angle is ½ the measure of the arc that it intercepts.

CASE 1
The center of the circle is located on the angle.

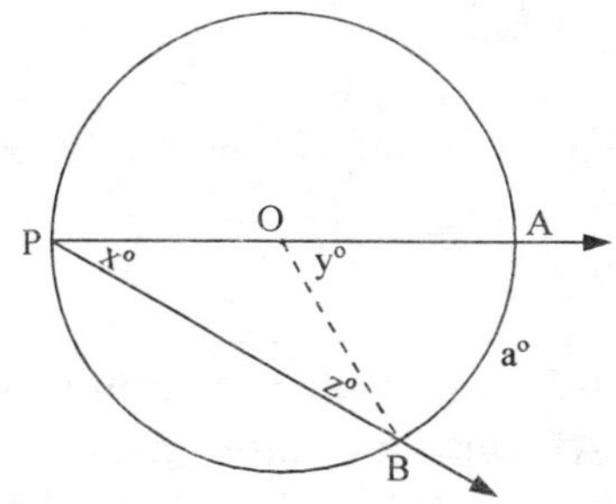

GIVEN:
Circle O with inscribed angle APB
arc AB with measure a

PROVE:

$\angle P = \frac{1}{2} a$

STATEMENTS	REASONS
1. Circle O with inscribed angle APB Arc AB with measure a	1. Given
2. Construct $\overline{OB}$ in Circle O	2. ?
3. OP = OB	3. ?
4. x = z	4. ?
5. y = x + z	5. ?
6. y = x + x	6. ?
7. y = 2x	7. ?
8. $x = \frac{1}{2} y$	8. ?
9. y = a	9. The measure of a central angle is equal to the measure of the arc that it intercepts.
10. $x = \frac{1}{2} a$	10. ?

CASE 2
The center of the circle is located in the exterior of the angle.

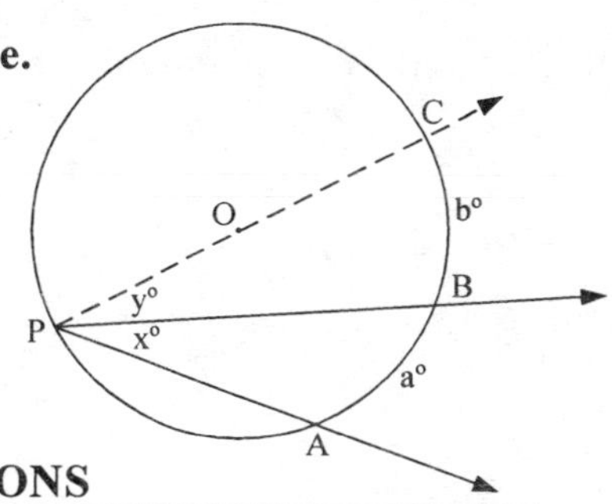

GIVEN:
Circle O with inscribed angle APB
Arc AB with measure a

PROVE:

$\angle APB = \frac{1}{2} a$

STATEMENTS	REASONS
1. Circle O with inscribed angle APB arc AB with measure a	1. Given
2. Construct ray PO. Call the point where it intersects the circle, Point C.	2. ?
3. $y = \frac{1}{2} b$	3. The last theorem we proved
4. $\angle APC = \frac{1}{2} (a + b)$	4. ?
5. $\angle APC = x + y$	5. ?
6. $x + y = \frac{1}{2} (a + b)$	6. ?
7. $x + y = \frac{1}{2} a + \frac{1}{2} b$	7. ?
8. $x + \frac{1}{2} b = \frac{1}{2} a + \frac{1}{2} b$	8. ?
9. $x = \frac{1}{2} a$	9. ?

Now we will prove another theorem about an angle that has its vertex on the circle. One side of the angle is a secant ray; while the other side of the angle is a tangent ray.

THEOREM:
The measure of an angle with its vertex on the circle and whose sides are secant and tangent rays of the circle is ½ the measure of the arc that it intercepts.

GIVEN:
Circle O with angle ABC
line BC is tangent to Circle O at point B.
Arc AB with measure a

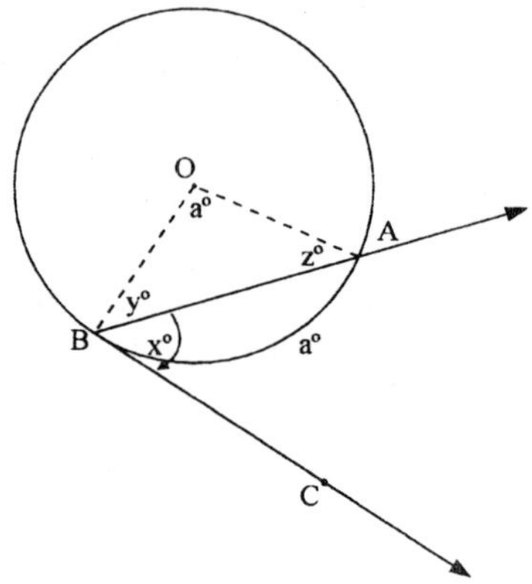

PROVE:

$\angle ABC = \frac{1}{2} a$

STATEMENTS	REASONS
1. Circle O with angle ABC line BC is tangent to Circle O at point B. Arc AB with measure a	1. ?
2. Construct line segments OB and OA	2. ?
3. In △ AOB, OB = OA	3. ?
4. In △ AOB, y = z	4. ?
5. Radius OB is perpendicular to tangent BC.	5. Tangent Theorem: A tangent to a circle is perpendicular to the radius drawn to the point of contact.
6. $\angle$ OBC = 90	6. ?
7. $\angle$ OBC = x + y	7. ?
8. x + y = 90	8. ?
9. x = 90 – y	9. ?
10. a + y + z = 180	10. ?
11. a + y + y = 180	11. ?
12. a + 2y = 180	12. ?
13. a = 180 – 2y	13. ?
14. a = 2(90 – y)	14. ?
15. a = 2x	15. ?
16. $x = \frac{1}{2} a$	16. ?

Section 10.3 Angles That Have Vertices in the Exterior and Interior of the Circle

Now we can use the theorems we have just proved to argue the next three theorems. These theorems involve angles that have their vertices in the exterior of the circle.The sides of these angles are secant or tangent rays.

THEOREM:

The measure of an angle whose vertex is an exterior point of the circle and whose sides are secant or tangent rays is ½ the absolute value of the difference in the measures of the arcs that it intercepts.

CASE 1

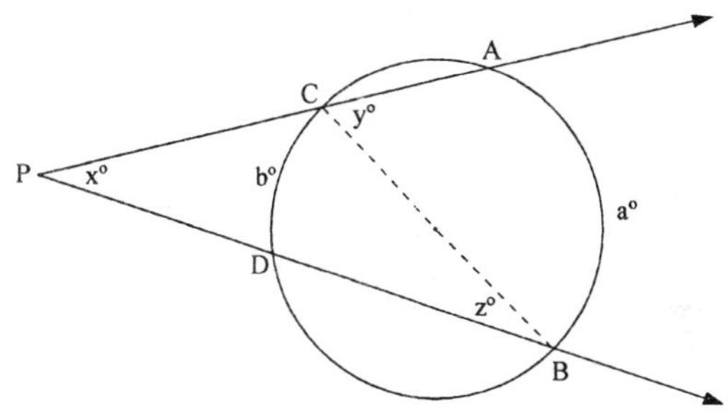

GIVEN:

Angle APB with measure x intercepting arc CD with measure b and arc AB with measure a, a > b

PROVE:

$x = \frac{1}{2}(a - b)$ *absolute value if we do not know which arc has greater measure*

STATEMENTS	REASONS
1. Angle APB with measure x intercepting arc CD with measure b and arc AB with measure a, a > b	1. ?
2. Construct $\overline{BC}$.	2. ?
3. $y = \frac{1}{2}a$	3. ?
4. $z = \frac{1}{2}b$	4. ?
5. $y = x + z$	5. ?
6. $x = y - z$	6. ?
7. $x = \frac{1}{2}a - \frac{1}{2}b$	7. ?
8. $x = \frac{1}{2}(a - b)$	8. ?

CASE 2

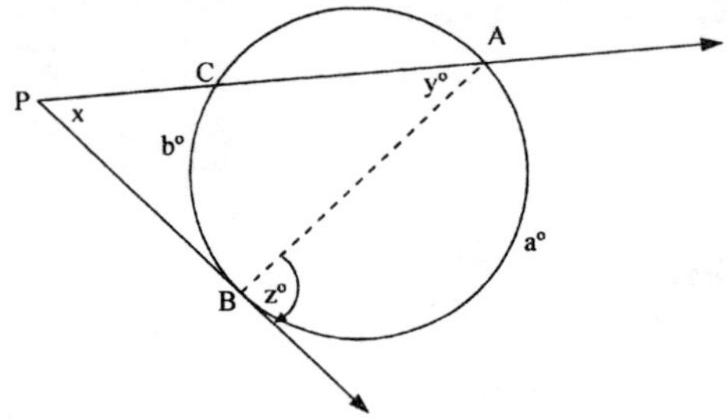

GIVEN:

Angle APB with measure x intercepting arc CB with measure b and arc AB with measure a, a > b

PROVE:

$x = \frac{1}{2}(a - b)$ *absolute value if we do not know which arc has greater measure*

STATEMENTS	REASONS
1. Angle APB with measure x intercepting arc CB with measure b and arc AB with measure a, a > b	You can provide the reasons.
2. Construct $\overline{AB}$.	
3. $y = \frac{1}{2} b$	
4. $z = \frac{1}{2} a$	
5. $z = x + y$	
6. $x = z - y$	
7. $x = \frac{1}{2} a - \frac{1}{2} b$	
8. $x = \frac{1}{2}(a - b)$	

CASE 3

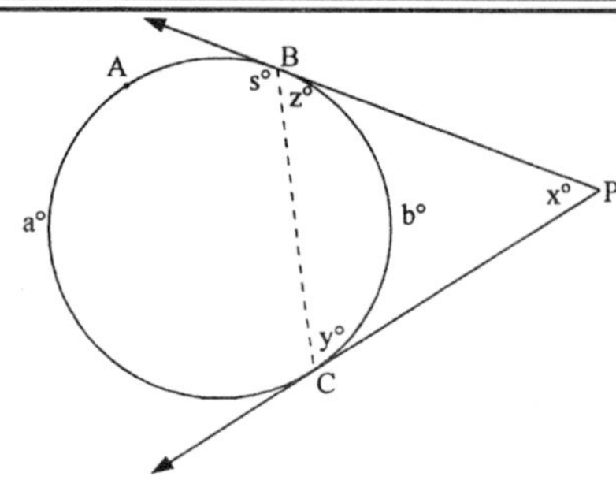

GIVEN:

Angle BPC with measure x intercepting arc CB with measure b and arc BAC with measure a, a > b

PROVE:

$x = \frac{1}{2}(a - b)$ *absolute value if we do not know which arc has greater measure*

STATEMENTS	REASONS
1. Angle BPC with measure x intercepting arc CB with measure b and arc BAC with measure a, a > b	You can provide the reasons.
2. Construct $\overline{BC}$.	
3. $z = \frac{1}{2} b$	
4. $s = \frac{1}{2} a$	
5. $s = x + y$	
6. $x = s - y$	
7. But $y = z$ since $PB = PC$. Why?	
8. $x = s - z$	
9. $x = \frac{1}{2} a - \frac{1}{2} b$	
10. $x = \frac{1}{2}(a - b)$	

THEOREM:

The measure of an angle whose vertex is in the interior of the circle and whose sides are formed by two intersecting secants is ½ the sum of the measures of the intercepted arcs.

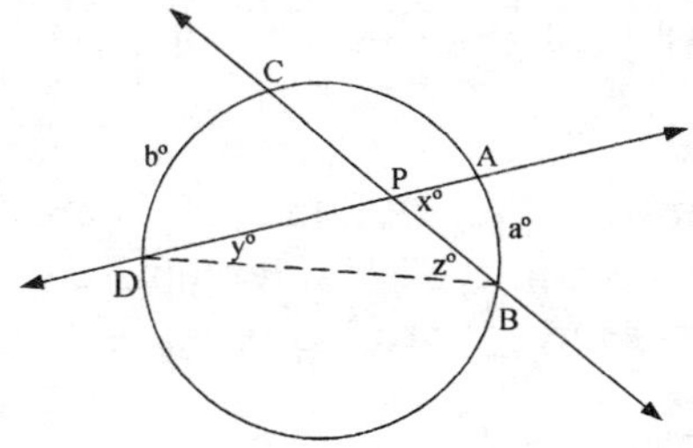

GIVEN:

Angle APB with measure x
Intercepted arcs are: Arc AB with measure a and Arc CD with measure b.

PROVE:

$$x = \frac{1}{2}(a + b)$$

STATEMENTS	REASONS
1. Angle APB with measure x Intercepted arcs are Arc AB with measure a and Arc CD with measure b.	Again, you may supply the reasons.
2. Construct line segment BD.	
3. $y = \frac{1}{2} a$	
4. $z = \frac{1}{2} b$	
5. $y + z = \frac{1}{2} a + \frac{1}{2} b$	
6. $x = y + z$	
7. $x = \frac{1}{2} a + \frac{1}{2} b$	
8. $x = \frac{1}{2}(a + b)$	

Summarizing our results, we have the following:

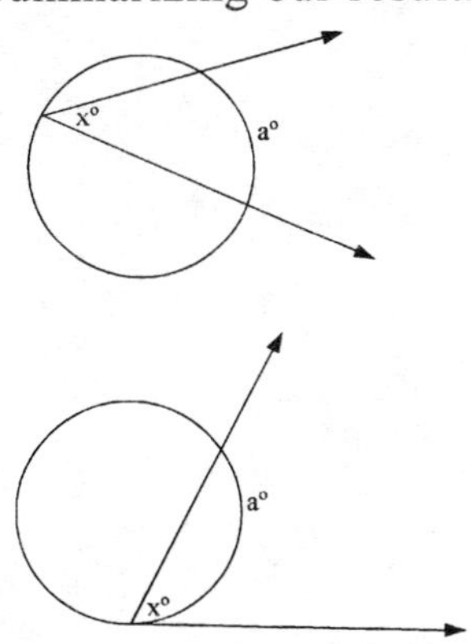

$x = \frac{1}{2} a$

vertex on the circle

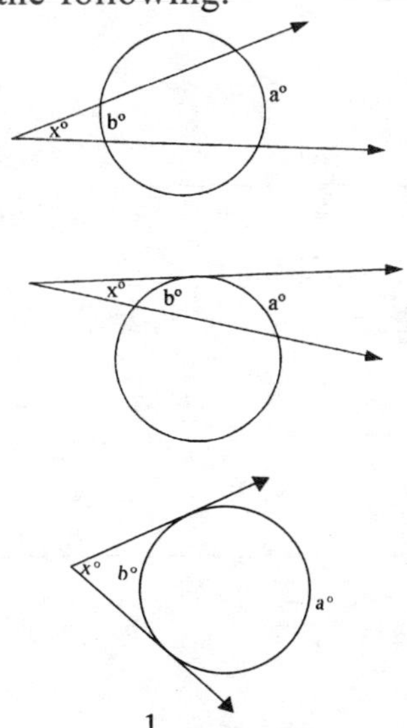

$x = \frac{1}{2} |a - b|$

vertex in the exterior of the circle

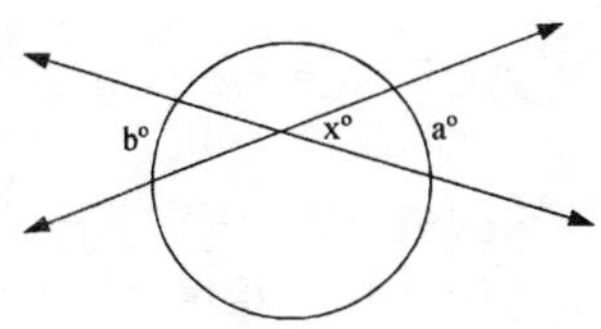

$x = \frac{1}{2}(a + b)$

vertex in the interior of the circle

The theorem that we will prove now is very easy to prove, and it will be quite useful in the proofs of other theorems.

THEOREM:
Angles inscribed in the same arc have equal measures.

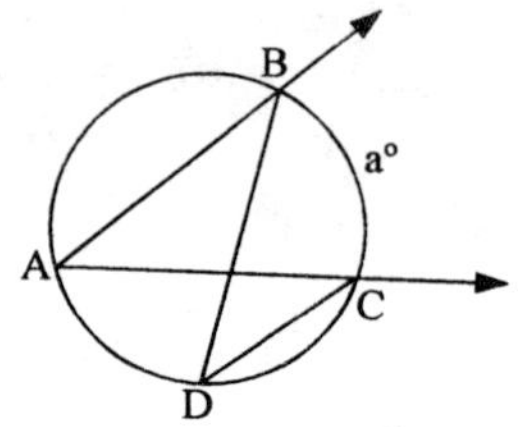

GIVEN:
Inscribed angle BAC intercepts arc BC.
Inscribed angle BDC intercepts arc BC.
The measure of arc BC is a.

PROVE:
$\angle BAC = \angle BDC$

STATEMENTS	REASONS
1. ?	1. ?
2. $\angle BAC = \frac{1}{2} a$	2. ?
3. $\angle BDC = \frac{1}{2} a$	3. ?
4. $\angle BAC = \angle BDC$	4. ?

We finish up this section by proving that an angle inscribed in a semicircle is a right angle.

THEOREM:
An angle inscribed in a semicircle is a right angle.

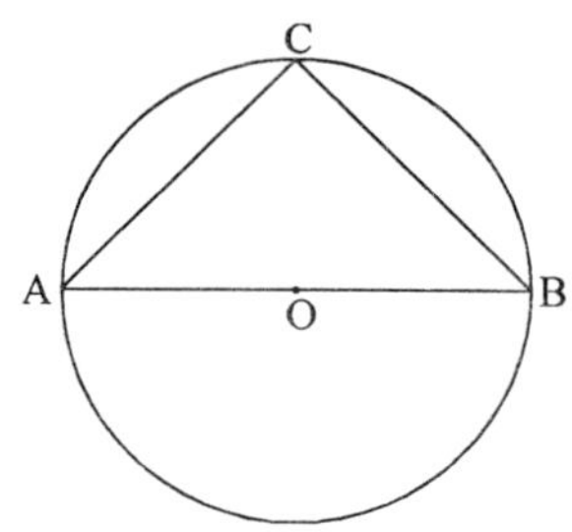

GIVEN:
Circle O with Angle ACB inscribed in semicircle AB

PROVE:
Angle ACB is a right angle.

STATEMENTS	REASONS
1. Circle O with Angle ACB inscribed in semicircle AB	1. Given
2. The measure of semicircle AB is 180.	2. Definition of measure of a semicircle
3. $\angle ACB = \frac{1}{2}(180) = 90$	3. ?
4. Angle ACB is a right angle.	4. ?

Section 10.4 Tangent and Secant Segments

In addition to the theorems involving angle measures, we also have theorems involving lengths when chords, secants and tangents intersect.

THEOREM:

If two chords intersect in the interior of a circle, the product of the lengths of the segments of one chord equals the product of the lengths of the segments of the other chord.

GIVEN:

$\overline{AB}$ and $\overline{CD}$ are chords of the circle.
Chords $\overline{AB}$ and $\overline{CD}$ intersect at point P

PROVE:

$AP \cdot PB = CP \cdot DP$

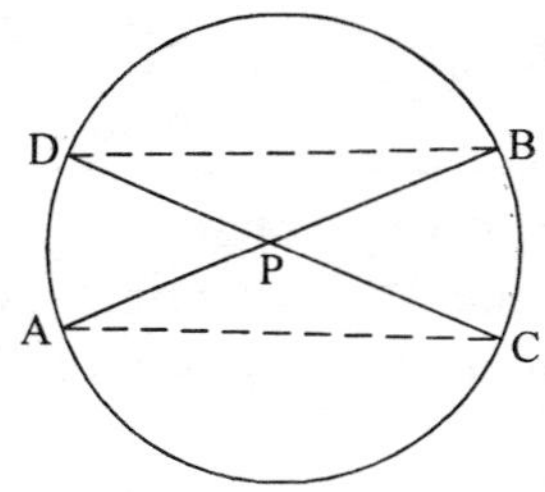

We will discuss this one informally.

We first construct $\overline{BD}$ and $\overline{AC}$.

Angle A and Angle D have equal measures because they intercept the same arc.

Likewise, Angle C and Angle B have equal measures because they intercept the same arc.

$\triangle$ APC is similar to $\triangle$ DPB because they have two pairs of angles with equal measures.

$\frac{AP}{DP} = \frac{PC}{PB}$ because corresponding sides of similar triangles are proportional.

$\frac{AP}{PC} = \frac{DP}{PB}$ because the means in a proportion may be interchanged.

$AP \cdot PB = DP \cdot PC$ because in a proportion the product of the extremes equals the product of the means.

THEOREM:

If a tangent segment and a secant segment intersect in an exterior point of the circle, the tangent segment is the mean proportional between the secant segment and the external segment of the secant segment.

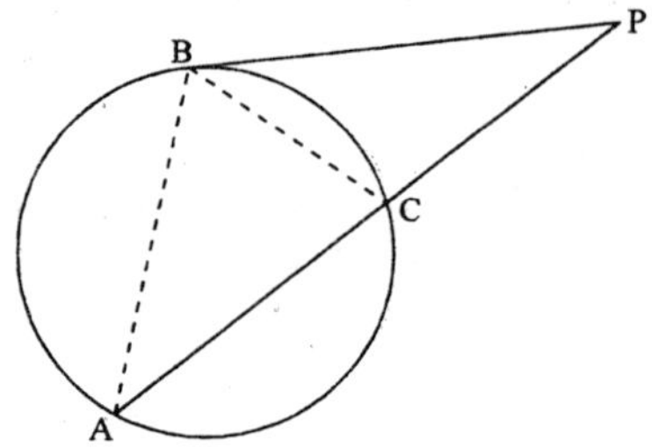

GIVEN:

The figure with tangent segment PB and secant segment PA intersecting at point P in the exterior of the circle

PROVE:

$\frac{PA}{PB} = \frac{PB}{PC}$

1. We begin by constructing line segments BA and BC.	Why?
2. The measure of angle A is 1/2 the measure of arc BC.	Why?
3. The measure of angle PBC is 1/2 the measure of arc BC.	Why?
4. $\angle A = \angle PBC$	Why?
5. $\angle P = \angle P$	Why?
6. Therefore, $\triangle$ PAB is similar to $\triangle$ PBC.	Why?
7. $\frac{PA}{PB} = \frac{PB}{PC}$	Why?

Our next theorem involves two secant segments intersecting outside a circle.

THEOREM:

If two secant segments intersect in an exterior point of the circle, the product of one of the segments and its external segment equals the product of the other secant segment and its external segment.

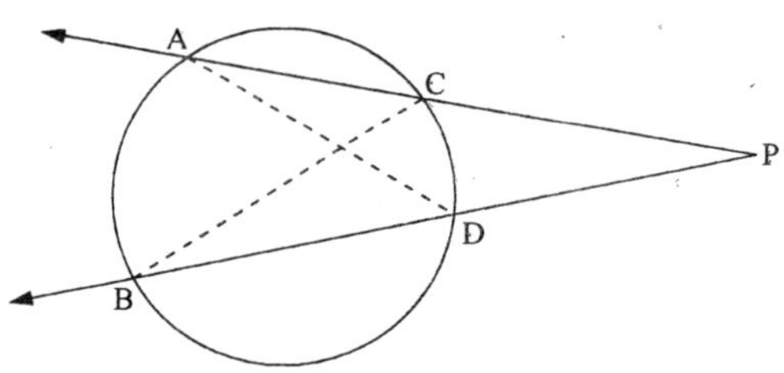

GIVEN:

The figure with secant segments $\overline{PA}$ and $\overline{PB}$ intersecting in exterior point P. $\overline{PA}$ intersects the circle at C; $\overline{PB}$ intersects the circle at D.

PROVE:

$PA \cdot PC = PB \cdot PD$

STATEMENTS	REASONS
1. The figure with secant segments $\overline{PA}$ and $\overline{PB}$ intersecting in exterior point P. $\overline{PA}$ intersects the circle at C; $\overline{PB}$ intersects the circle at D.	
2. $\angle A = \angle B$	Since they are inscribed angles that intercept the same arc.
3. $\angle P = \angle P$	Why?
4. $\triangle APD \sim \triangle BPC$	Why?
5. $\frac{PA}{PB} = \frac{PD}{PC}$	Why?
6. $PA \cdot PC = PB \cdot PD$	Why?

Section 10.5 Tangent Relationships for Two Circles

Before beginning this section, we will take time to define some terms that we will need in our discussion.

A Common Tangent of Two or More Circles: A line that is tangent to each of the circles. Line ℓ is a common tangent of Circle O and Circle P.

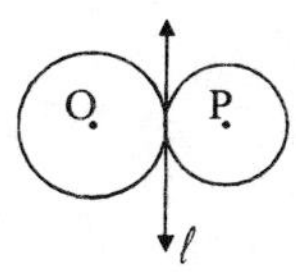

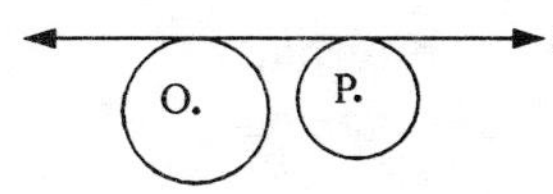

Tangent Circles: Circles that have the same tangent line at the same point of contact. Circle O and Circle P are tangent circles.

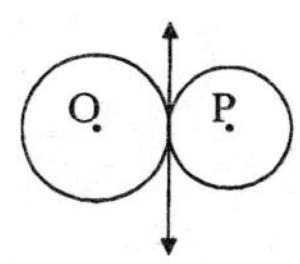

Line of Centers: The line containing the centers of two circles. $\overleftrightarrow{OP}$ is the line of centers for Circle O and Circle P.

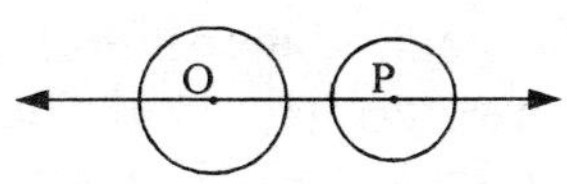

We may classify two coplanar circles by the number of points they have in common.

CASE 1

The circles may not intersect.
(No common points)

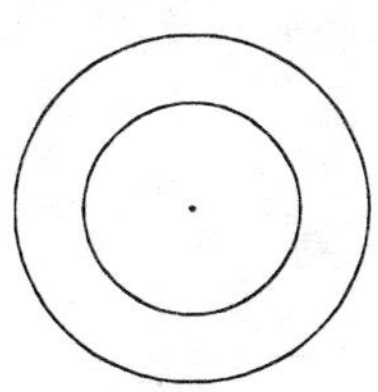

Concurrent
Circles
(same center)

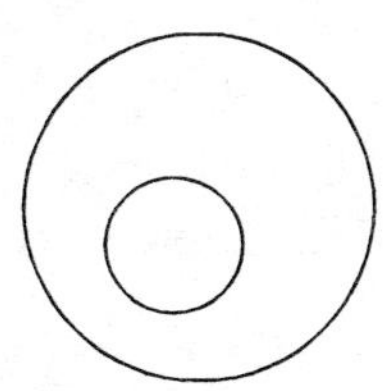

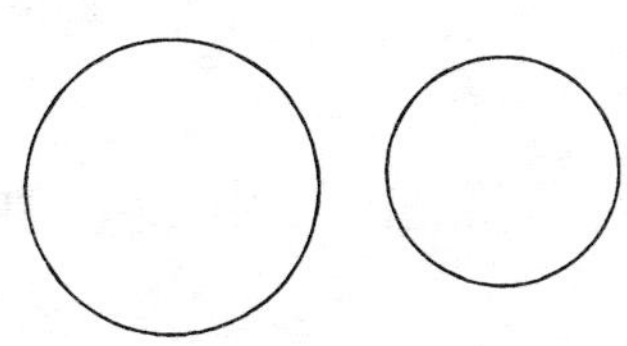

CASE 2

The circles may intersect in exactly one point

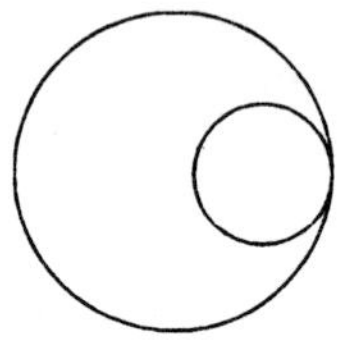

internally tangent

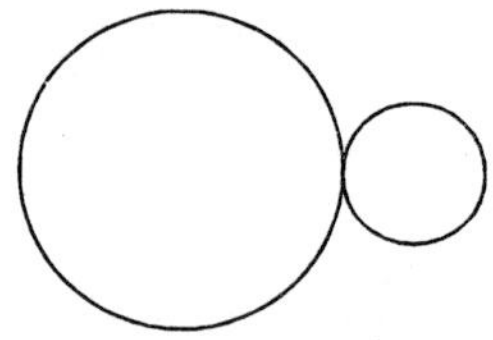

externally tangent

CASE 3

The circles may intersect in exactly two points

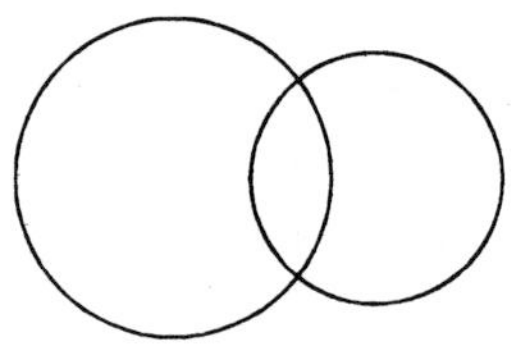

Now let us consider situations where two circles have a common tangent.

When we examine the pairs of circles in CASE 1, we note that these are the only possibilities for common tangents:

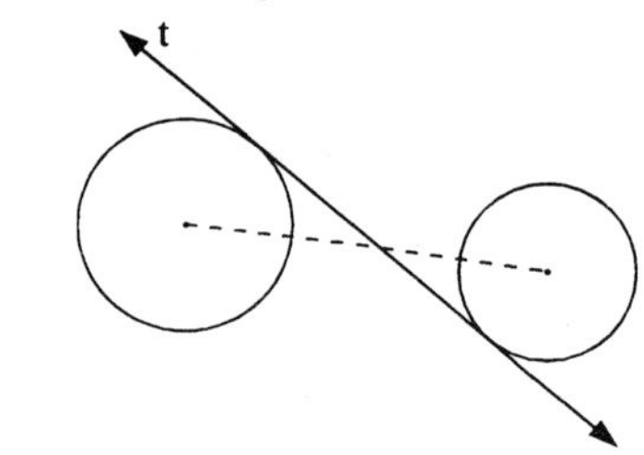

Line t is called a **common internal tangent** of the two circles. With the exception of the points of contact, the two circles lie on different sides of tangent t. Also, notice that line t intersects the line segment joining the two centers.

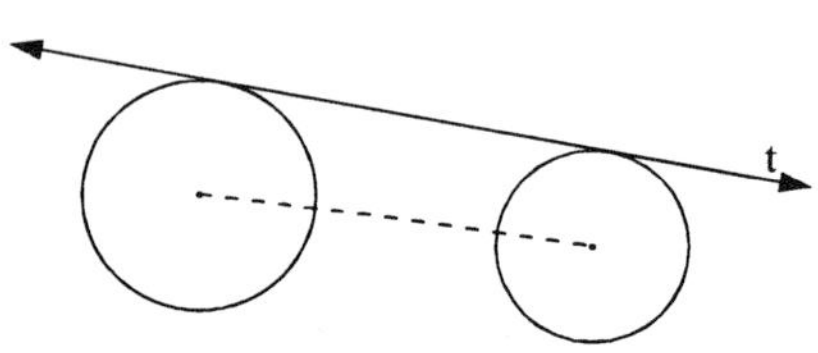

Line t is called a **common external tangent** of the two circles. With the exception of the points of contact, the two circles lie on the same side of tangent t. Also, notice that line t does not intersect the line segment joining the two centers.

When we examine the pairs of circles in **CASE 2,** we note these possibilities:

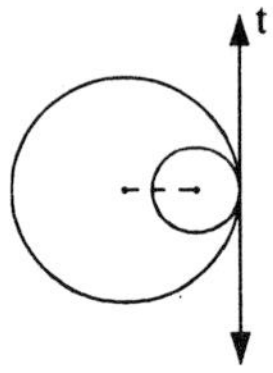

Line t is a common external tangent of the two circles.

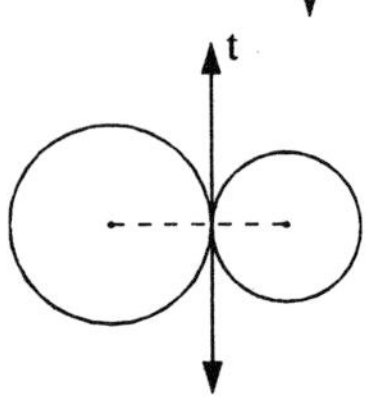

Line t is a common internal tangent of the two circles.

Finally, examining the pairs of circles in **CASE 3,** we see this possibility:

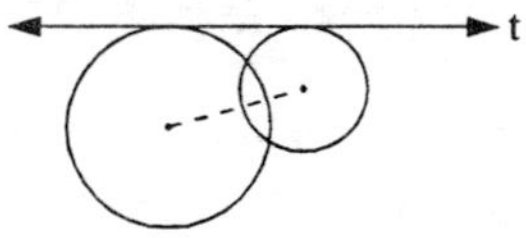

Line t is a common external tangent of the two circles.

Now we can look at some rather interesting theorems about common tangents of circles. Sketches of proofs of these theorems are given. As always, be sure to study each one and provide reasons when necessary.

THEOREM:

If m and n are two external tangents to two circles, then the line segment joining the points of tangency on line m has the same length as the line segment joining the points of tangency on line n.

GIVEN:

Line m and line n , common tangents to Circle O and Circle R

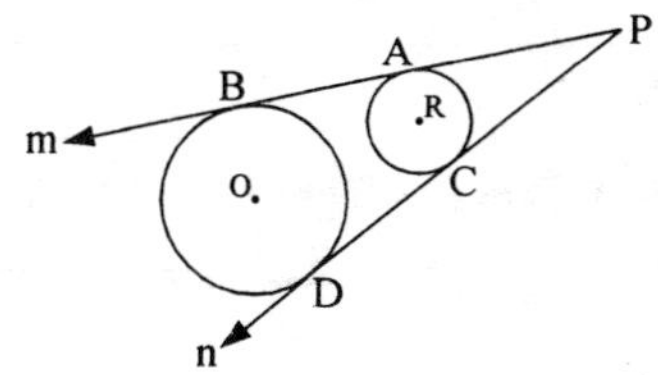

PROVE:

AB = CD

STATEMENTS	REASONS
1. Line m and line n , common tangents to Circle O and Circle R	1. ?
2. PB = PD and PA = PC	2. Why?
3. PB – PA = PD – PC	3. Subtraction Postulate Of Equality
4. PA + AB = PB PC + CD = PD	4. Whole Equals Sum Of Parts Postulate
5. PB – PA = AB PD – PC = CD	5. Subtraction Postulate Of Equality (step 4)
6. ∴ AB = CD	6. Substitution Postulate Of Equality (steps 3 and 5)

THEOREM:

If two circles are tangent, then their centers and their common point of tangency are collinear.

GIVEN:

Circle A and Circle B tangent to line ℓ at point P

PROVE:

Points A, B and P are collinear.

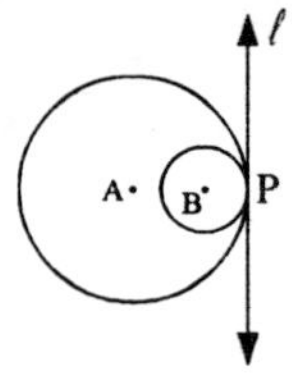

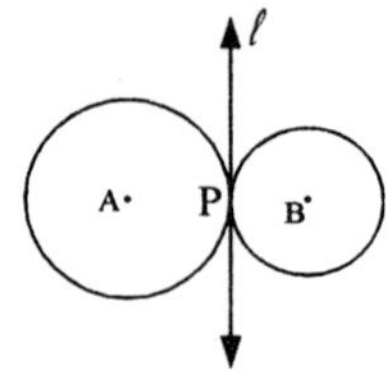

---

1. Circle A and Circle B tangent to line ℓ at point P
2. Through Point P, Line $\overleftrightarrow{AP}$ is perpendicular to line ℓ. Why?
3. Through Point P, Line $\overleftrightarrow{BP}$ is perpendicular to line ℓ. Why?
4. Line $\overleftrightarrow{AP}$ must be the same as Line $\overleftrightarrow{BP}$ because in a plane, from a point on a line, there is exactly one line that is perpendicular to the given line (Perpendicular Postulate).

THEOREM:

If two circles intersect in two points then their line of centers is the perpendicular bisector of the chord joining the points of intersection.

GIVEN:

Circle A and Circle B intersecting at Point C and Point D
Point E is the intersection of line AB and chord CD.

PROVE:

$\overleftrightarrow{CD} \perp \overleftrightarrow{AB}$ and DE = CE

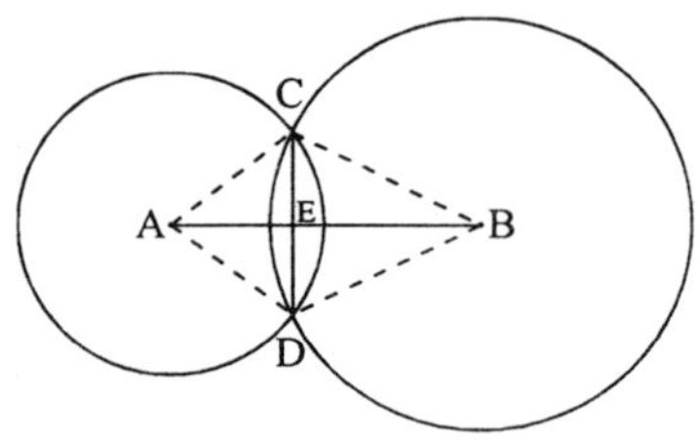

---

1. Circle A and Circle B intersecting at Point C and Point D
   Point E is the intersection of line $\overleftrightarrow{AB}$ and chord $\overline{CD}$.
2. Construct $\overline{AC}$, $\overline{AD}$, $\overline{BC}$ and $\overline{BD}$.
   Prove that $\triangle$ ACB $\cong$ $\triangle$ ADB.
3. AC = AD; BC = BD Why?
4. AB = AB Why?
5. $\triangle$ ACB $\cong$ $\triangle$ ADB Why?
6. $\angle$ CAB = $\angle$ DAB Why?
   Next, we prove that $\triangle$ AEC $\cong$ $\triangle$ AED.
7. AE = AE
8. $\triangle$ AEC $\cong$ $\triangle$ AED Why?

CAN YOU FINISH THIS PROOF?

Exercises Chapter 10

(*Figures are not drawn to scale*)

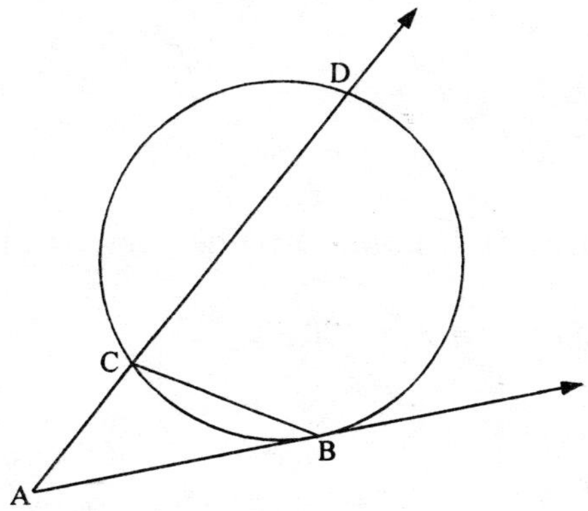

Use the above figure for Exercises 1–4. $\overleftrightarrow{AB}$ is tangent to the Circle at B

1. Find $m\overset{\frown}{BC}$ if $\angle A = 36$ and $m\overset{\frown}{BCD} = 230$.

2. Find $\angle A$ if $m\overset{\frown}{BD} = 160$ and $m\overset{\frown}{BC} = 12$.

3. Find $\angle DCB$ if $m\overset{\frown}{BD} = 142$.

4. Find $\angle A$ if $m\overset{\frown}{CD} = 210$ and $m\overset{\frown}{CB} = 70$.

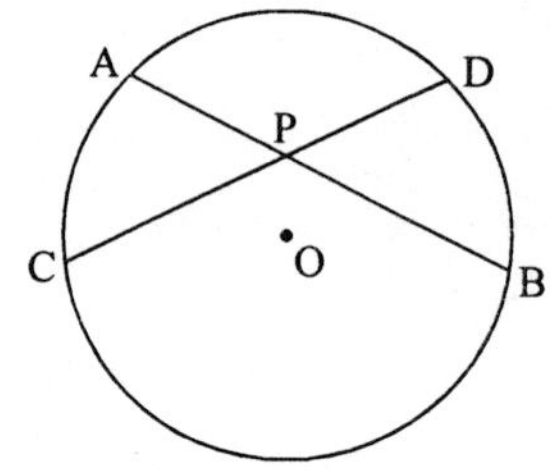

Use the above figure for Exercises 5–7. Point O is the center of the circle.

5. Find $\angle$ APC if $m\overset{\frown}{AC} = 20$ and $m\overset{\frown}{BD} = 80$.

6. Find $m\overset{\frown}{BC}$ if $\angle APC = 35$ and $m\overset{\frown}{AD} = 40$.

7. Find $m\overset{\frown}{BC}$ if $\angle APC = 48$ and $m\overset{\frown}{AD} = 60$.

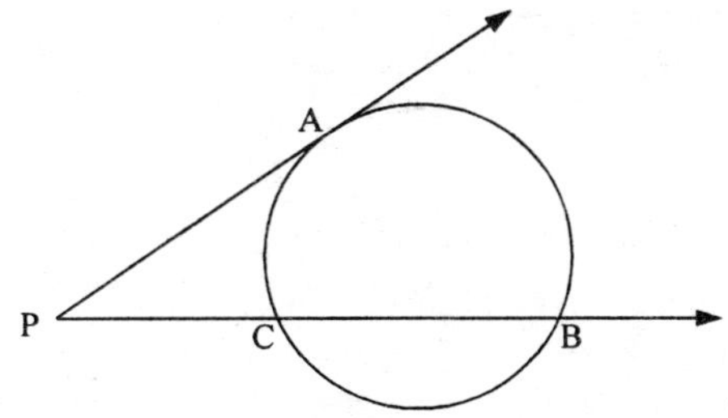

Use the above figure for Exercises 8–10. $\overleftrightarrow{AP}$ is tangent to the Circle at A.

8. If $m\widehat{AC} = 50$ and $m\widehat{AB} = 105$, find $\angle P$.

9. $\angle P = 30$ and $m\widehat{AC} = 60$, find $m\widehat{AB}$.

10. $\angle P = 42$ and $m\widehat{AC} = 130$, find $m\widehat{AB}$.

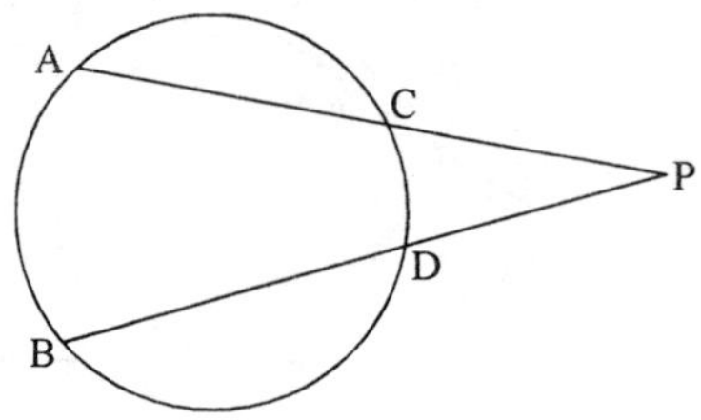

Use the above figure for Exercises 11 and 12.

11. a. Find $\angle P$ if $m\overset{\frown}{AB} = 80$ and $m\overset{\frown}{CD} = 20$.

b. Find $\angle P$ if $m\overset{\frown}{AB} = 72$ and $m\overset{\frown}{CD} = 40$.

c. Find $m\overset{\frown}{CD}$ if $\angle P = 28$ and $m\overset{\frown}{AB} = 90$.

d. Find $\angle P$ if $m\overset{\frown}{AC} = 20$, $m\overset{\frown}{BD} = 30$ and $m\overset{\frown}{CD} = 25$.

12. a. Find $\angle P$ if $m\overset{\frown}{AB} = 60$ and $m\overset{\frown}{CD} = 15$.

b. Find $m\overset{\frown}{CD}$ if $\angle P = 42$ and $m\overset{\frown}{AB} = 86$.

c. Find $m\overset{\frown}{AB}$ if $\angle P = 32$ and $m\overset{\frown}{CD} = 20$.

d. Find $\angle P$ if $m\overset{\frown}{AC} = 80$, $m\overset{\frown}{BD} = 120$ and $m\overset{\frown}{AB} = 100$.

13. Lines AP and PB are tangents to the circle. A and B are the points of tangency.

a. If $AP = 3x + 5$ and $BP = 8x + 2$, find the length of $\overline{AP}$.

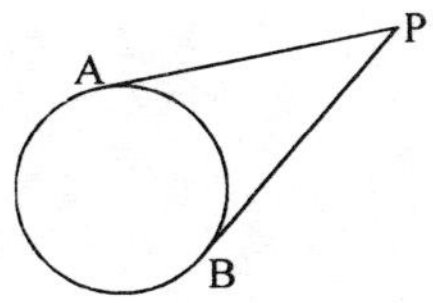

b. If $AP = 10$ and $BP = 2x + 1$, find the value of x.

14. If $\angle$ APB is an inscribed angle of Circle O, find the measure of $\angle$ APB.

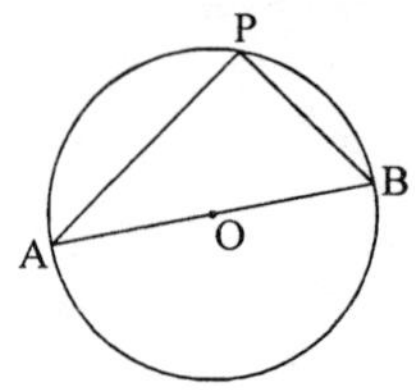

15. Find the measure of $\overset{\frown}{PB}$ in Circle O if PA = PB.

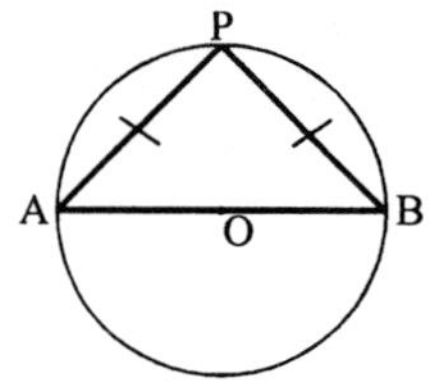

16. If $\angle$ A = 30 and AB = 12 in Circle P

a. Find the measure of $\overset{\frown}{AC}$.

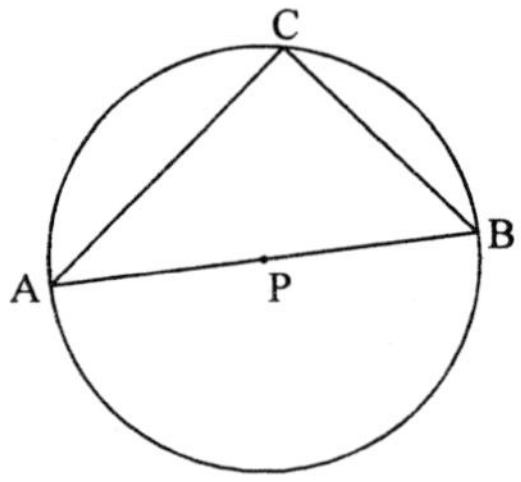

b. Find the length of $\overline{BC}$.

c. Find the length of $\overline{AC}$.

d. Find the measure of $\overset{\frown}{BC}$.

17. In Circle O, $\angle$ BCA = 45. Find the measure of angle BOA.

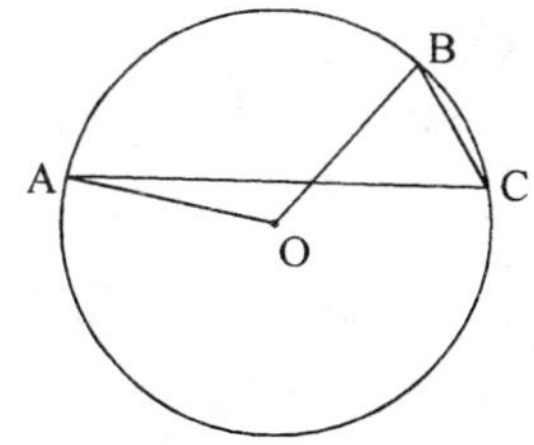

18. In the figure $\angle$ A = 120, $\angle$ BDC = 20 and $m\widehat{AD}$ = 80. Find each of the following.

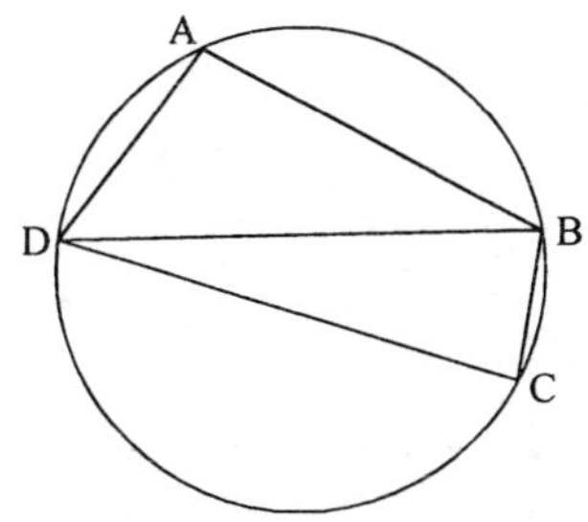

a. $m\widehat{DC}$ e. $m\widehat{BA}$

b. $\angle$ ADB f. $\angle$ C

c. $\angle$ ABC g. $m\widehat{DAB}$

d. $m\widehat{BC}$ h. $\angle$ DBC

19. Use the figure in exercise 18 to do these exercises:
$\angle$ C = 50 , $m\widehat{BC}$ = 60 and $m\widehat{AD}$ = 80

a. $m\widehat{DAB}$ e. $m\widehat{DC}$

b. $\angle$ A f. $\angle$ ADB

c. $\angle$ DBC g. $m\widehat{AB}$

d. $\angle$ DBC

20* Prove the following theorem:

THEOREM:
The measure of an inscribed angle is ½ the measure of the arc that it intercepts.

GIVEN:
Circle O with inscribed angle APB
Arc AB has measure a.

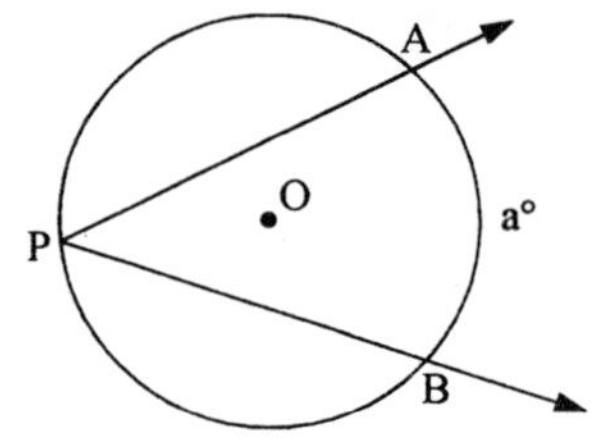

PROVE:

$\angle APB = \frac{1}{2} a$

21.* Prove the following theorems:

a. **THEOREM:**
Two chords of a circle have equal lengths iff their intercepted arcs have equal measures.

(*Iff* means you must prove two theorems)

We define the **Intercepted Arc of a Chord** to be the minor arc that has as its endpoints the endpoints of the chord.

b. **THEOREM:**
The opposite angles of an inscribed quadrilateral are supplementary.

22. **GIVEN:**
Line $\overleftrightarrow{PC}$ is tangent to Circle A.
Line $\overleftrightarrow{PD}$ is tangent to Circle B.
Line $\overleftrightarrow{PE}$ is tangent to Circle A and Circle B at Point E.

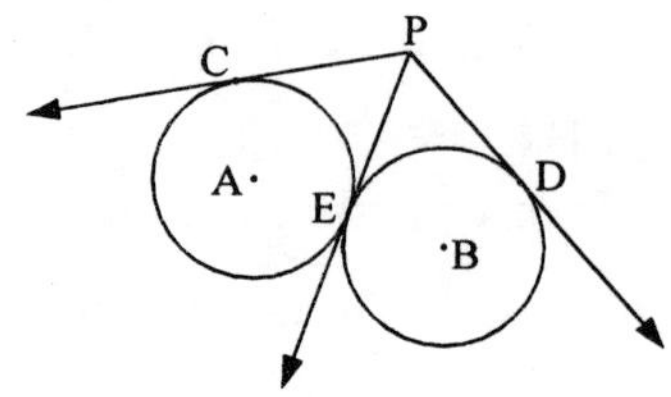

PROVE:
PC = PD

23. Draw two circles in the same plane so that:

a. They have exactly one common tangent.

b. They have exactly two common tangents.

c. They have exactly three common tangents.

d. They have exactly four common tangents.

24. Can you draw two coplanar circles that do not have a common tangent?

25. In the drawing, line $\overleftrightarrow{AD}$ is a common tangent of the two circles and line $\overleftrightarrow{BC}$ is a common tangent of the two circles.

Prove that AD = BC

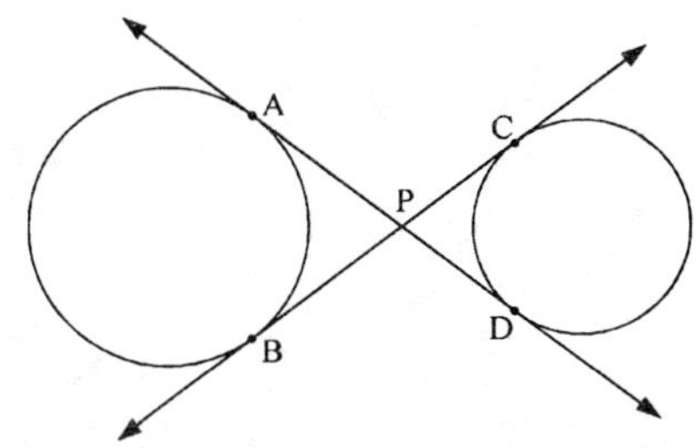

26. If one circle has a radius of 8, and another circle has a radius of 15, find the length of the line segment joining their centers:

a. If the circles are concurrent.

b. If the circles are externally tangent.

c. If the circles are internally tangent.

27. If one circle has a radius of 6, and another circle has a radius of 20, find the length of the line segment joining their centers:

a. If the circles are concurrent.

b. If the circles are externally tangent.

c. If the circles are internally tangent.

Chapter Ten Highlights

Do You Know and Understand These Terms?

A Secant of a Circle
A Secant Ray
A Secant Segment
A Tangent to the Circle
A Tangent Segment
A Tangent Ray
A Central Angle of a Circle
An Inscribed Angle of a Circle
The Measure of a Central Angle of a Circle
A Semicircle
The Measure of a Semicircle
Intercepted Arc of a Chord
A Common Tangent of Two or more Circles
Tangent Circles
Line of Centers
Common Internal Tangent
Common External Tangent

Do You Know and Understand This Postulate?

Arc Addition Postulate or
Whole Equals Sum of Parts Postulate for Arcs
If A and C are endpoints of an arc and B is another point on the arc, then $m\overset{\frown}{AB} + m\overset{\frown}{BC} = m\overset{\frown}{AC}$.

Do You Know and Understand These Theorems?

Tangent Theorem

Radii of the same circle are equal.

A tangent of a circle is perpendicular to the radius at the point of tangency.

The measure of an inscribed angle is ½ the measure of the arc that it intercepts.

The measure of an angle with its vertex on the circle and whose sides are secant and tangent rays of the circle is ½ the measure of the arc that it intercepts.

The measure of an angle whose vertex is an exterior point of the circle and whose sides are secant or tangent rays is ½ the absolute value of the difference in the measures of the arcs that it intercepts.

The measure of an angle whose vertex is in the interior of the circle and whose sides are formed by two intersecting secants is ½ the sum of the measures of the intercepted arcs.

Angles inscribed in the same arc have equal measures.

If two chords intersect in the interior of a circle, the product of the lengths of the segments of one chord equals the product of the lengths of the segments of the other chord.

If a tangent segment and a secant segment intersect in an exterior point of the circle, the tangent segment is the mean proportional between the secant segment and the external segment of the secant segment.

If two secant segments intersect in an exterior point of the circle, the product of one of the segments and its external segment equals the product of the other secant segment and its external segment.

An angle inscribed in a semicircle is a right angle.

The perpendicular from the center of a circle to a chord of the circle bisects the chord.

The two segments formed by two tangents to a circle from an external point of the circle have equal lengths.

Two chords of a circle have equal lengths iff their intercepted arcs have equal measures.

The opposite angles of an inscribed quadrilateral are supplementary.

If m and n are two external tangents to two circles, then the line segment joining the points of tangency on line m has the same length as the line segment joining the points of tangency on line n.

If two circles are tangent, then their centers and their common point of tangency are collinear.

If two circles intersect in two points, then their line of centers is the perpendicular bisector of the chord joining the points of intersection.

Appendix A

Ratios and Proportions

Before we begin our work with similar triangles it may be a good idea to review ratios and proportions. Probably you recall from arithmetic that a ratio is a comparison of two numbers using division. Since fractions can be used to express division, they are often used to express ratios. To compare the number 12 to the number 4, we write $\frac{12}{4}$. We may also use colon notation to express a ratio. The ratio of 12 to 4, using colon notation is 12:4. Generally ratios are reduced to lowest terms, so we reduce the ratio of 12 to 4 to 3 to 1, and we write 3 as the final ratio. Also with ratios, order is important. The ratio of 12 to 4 is not the same as the ratio of 4 to 12. The ratio of 4 to 12 is represented in fraction form with 4 as the numerator and 12 as the denominator.

The ratio of 12 to 4 $\quad \frac{12}{4}$

The ratio of 4 to 12 $\quad \frac{4}{12}$

An expression representing the equality of two ratios is called a proportion. The equation $\frac{6}{8} = \frac{3}{4}$ is a proportion. The numbers in a proportion have special names. Let us consider the proportion $\frac{a}{b} = \frac{c}{d}$ where a, b, c and d represent real numbers:

a is called the first term of the proportion.
b is called the second term of the proportion.
c is called the third term of the proportion.
d is called the fourth term of the proportion.

a and d are called the extremes of the proportion.
b and c are called the means of the proportion.

In the proportion, $\frac{a}{x} = \frac{x}{b}$, notice that the means are equal. The mean, which is x in this case, is called the mean proportional between a and b. In the proportion $\frac{1}{6} = \frac{6}{36}$, 6 is called the mean proportional between 1 and 36.

Let us consider again the proportion $\frac{a}{b} = \frac{c}{d}$. If we multiply both sides of the equation by bd, we obtain the equation ad = bc. Recall that a and d are called the extremes of the proportion, while b and c are called its means. We may say then that for a proportion, the product of the extremes equals the product of the means. In other words, we may cross multiply the terms of a proportion, and the resulting products will always be equal. Applying this rule to the proportion $\frac{3}{4} = \frac{6}{8}$ we obtain the equation $3 \times 8 = 4 \times 6$.

CONCLUSION:

In a proportion, the product of the extremes equals the product of the means.

If we let $\frac{a}{b} = \frac{c}{d}$ represent any proportion, we can show that the terms of a proportion may be interchanged in several ways:

Consider $\frac{a}{b} = \frac{c}{d}$

We obtain $ad = bc$ (Product of extremes equals product of means)

$\frac{ad}{ab} = \frac{bc}{ab}$ (Divide both sides of the equation by ab)

We obtain $\frac{d}{b} = \frac{c}{a}$

CONCLUSION:

We may interchange the extremes of a proportion.

Likewise, if we divide both sides of the equation ad = bc by cd, we obtain the result:

$$\frac{a}{c} = \frac{b}{d}$$

CONCLUSION:

We may interchange the means of a proportion.

Similarly, if we divide both sides of the equation ad = bc by ac, we obtain the following result:

$$\frac{d}{c} = \frac{b}{a}$$

CONCLUSION:

We may interchange both the extremes and the means of a proportion.

Now let us look at some other interesting relationships dealing with proportions. Again we consider the proportion $\frac{a}{b} = \frac{c}{d}$. Adding 1 to both sides of the proportion, we get:

$$\frac{a}{b} + 1 = \frac{c}{d} + 1$$

$$\frac{a}{b} + \frac{b}{b} = \frac{c}{d} + \frac{d}{d} \qquad \left(\frac{b}{b} = \frac{d}{d} = 1\right)$$

$$\frac{a+b}{b} = \frac{c+d}{d}$$

Subtracting 1 from both sides of the proportion yields the result:

$$\frac{a-b}{b} = \frac{c-d}{d}$$

Finally, we show that given an extended ratio, as shown below, the ratio of the sum of the numerators to the sum of the denominators is equal to each of the ratios that form the extended proportion.

$$\frac{a}{b} = \frac{c}{d} = \frac{e}{f} = \frac{g}{h} = k$$

We have $\frac{a}{b} = k$ hence $a = bk$

$\frac{c}{d} = k$ hence $c = dk$

$\frac{e}{f} = k$ hence $e = fk$

$\frac{g}{h} = k$ hence $g = hk$

Adding we get $a + c + e + g = bk + dk + fk + hk$

Factoring out k, $a + c + e + g = k(b + d + f + h)$

Dividing we get $k = \frac{a + c + e + g}{b + d + f + h}$

OUR CONCLUSION:

If $\frac{a}{b} = \frac{c}{d} = \frac{e}{f} = \frac{g}{h} = k$

Then $k = \frac{a + c + e + g}{b + d + f + h}$

EXAMPLE 1:

What proportions and relationships can be derived from the proportion $\frac{3}{4} = \frac{6}{8}$?

Product of Extremes Equals Product Of Means $3 \cdot 8 = 6 \cdot 4$

Interchange Extremes $\frac{8}{4} = \frac{6}{3}$

Interchange Means $\frac{3}{6} = \frac{4}{8}$

Interchange Both Extremes And Means $\frac{8}{6} = \frac{4}{3}$

Add Terms $\frac{3 + 4}{4} = \frac{6 + 8}{8}$

Subtract Terms $\frac{3 - 4}{4} = \frac{6 - 8}{8}$

EXAMPLE 2:

Form a new proportion by adding the terms of the fractions $\frac{6}{8} = \frac{9}{12} = \frac{15}{20} = \frac{3}{4}$

$$\frac{6 + 9 + 15}{8 + 12 + 20} = \frac{3}{4}$$

Exercises Appendix A

1. What are the means of the proportion $\frac{3}{5} = \frac{6}{10}$?
2. What are the extremes of the proportion $\frac{3}{5} = \frac{6}{10}$?
3. What are the means of the proportion $\frac{1}{k} = \frac{r}{s}$?
4. What are the extremes of the proportion $\frac{x}{y} = \frac{7}{8}$?
5. Write a proportion for each of the following:
   a. x is the mean proportional between 5 and 6.
   b. y is the mean proportional between 10 and 15.
   c. c is the mean proportional between d and e.
   d. r is the mean proportional between q and p.
6. Which of the following are valid derivations from the proportion $\frac{7}{9} = \frac{21}{27}$?

   a. $\frac{7}{21} = \frac{9}{27}$ b. $\frac{9}{7} = \frac{27}{21}$ c. $\frac{27}{9} = \frac{21}{7}$

   d. $\frac{7}{27} = \frac{21}{9}$ e. $\frac{9}{21} = \frac{7}{27}$ f. $\frac{7 - 9}{9} = \frac{21 - 27}{27}$

7. Which of the following are valid derivations from the proportion $\frac{a}{x} = \frac{b}{y}$?

a. $\frac{a}{y} = \frac{b}{x}$ b. $\frac{x}{a} = \frac{y}{b}$ c. $\frac{a}{b} = \frac{x}{y}$

d. $\frac{a + x}{x} = \frac{b + y}{y}$ e. $\frac{x}{a + x} = \frac{y}{b + y}$

8. From each of the following relations, derive at least 3 proportions. You can check your answers by cross multiplying (Recall that the product of the extremes equals the product of the means).

a. $\frac{8}{3} = \frac{48}{18}$ b. $\frac{p}{q} = \frac{r}{s}$

9. Use the conclusion, "Product of the extremes equals the product of the means," to solve for x in each proportion:

a. $\frac{x}{8} = \frac{3}{4}$ b. $\frac{5}{9} = \frac{x}{8}$ c. $\frac{x + 2}{3} = \frac{5}{8}$

d. $\frac{x + 2}{x - 3} = \frac{4}{7}$ e. $\frac{x}{x - 2} = \frac{3}{x - 4}$ f. $\frac{x + 8}{x - 4} = \frac{x + 3}{x + 2}$

Answers To Selected Exercises Appendix A

1. 5, 6 2. 3, 10 3. k, r 4. x, 8

5. a. $\frac{5}{x} = \frac{x}{6}$ or $\frac{6}{x} = \frac{x}{5}$
   b. $\frac{10}{y} = \frac{y}{15}$ or $\frac{15}{y} = \frac{y}{10}$
   c. $\frac{d}{c} = \frac{c}{e}$ or $\frac{e}{c} = \frac{c}{d}$
   d. $\frac{q}{r} = \frac{r}{p}$ or $\frac{p}{r} = \frac{r}{q}$
6. a, b, c, f
7. b, c, d, e
9. a. 6
   b. $\frac{40}{9}$
   c. $-\frac{1}{8}$
   d. $-\frac{26}{3}$
   e. 6, 1
   f. $-\frac{28}{11}$

Appendix B

Introduction to Solid Geometric Figures

A **polyhedron** is a solid bounded by polygons. The word "polyhedron" comes from two Greek words: 'poly' meaning 'many' and 'hedron' meaning 'face.' The plural for the word polyhedron is polyhedra. Here are some examples of polyhedra:

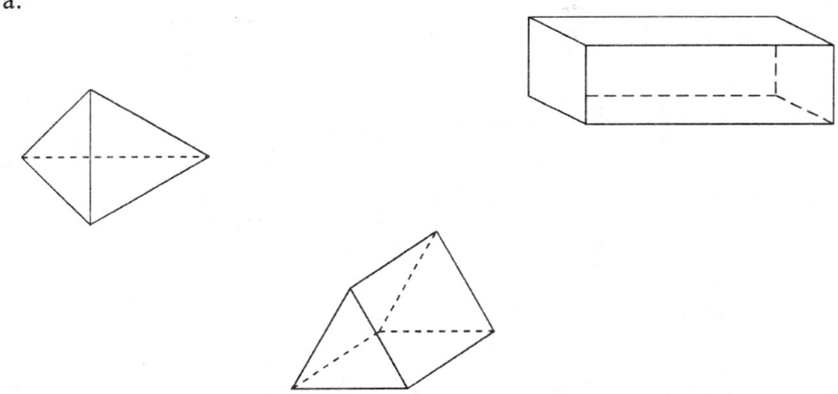

The sides of a polyhedron are called its **faces;** it's corners are called **vertices**; and the intersection of its faces are called **edges**. The polyhedron that you see in the following illustration has 5 faces, which include its base and 4 lateral faces, 5 vertices and 8 edges.

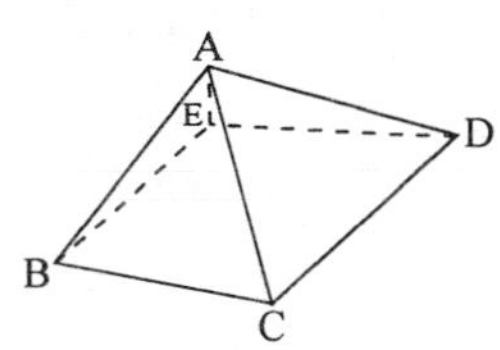

FACES

△ ABC, △ ACD, △ ADE, △ AEB (Lateral Faces)
Quadrilateral BCDE (Base)

VERTICES

Points A, B, C, D and E

EDGES

$\overline{AB}$, $\overline{AC}$, $\overline{AD}$, $\overline{AE}$, $\overline{BE}$, $\overline{BC}$, $\overline{CD}$ and $\overline{DE}$

Regular polyhedra

For a polyhedron to be a **regular polyhedron,** all of its faces must be congruent regular polygons, and the configuration of polygons at each of its vertices must be the same. *There are only five regular polyhedra.* They are shown in the illustration.

The Platonic Solids

Regular Tetrahedron (4 Faces)

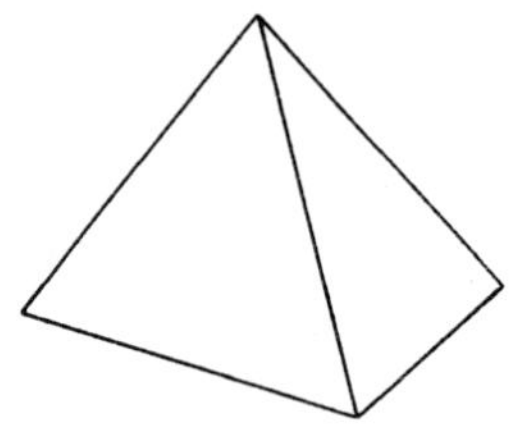

Regular Hexahedron (Cube) (6 Faces)

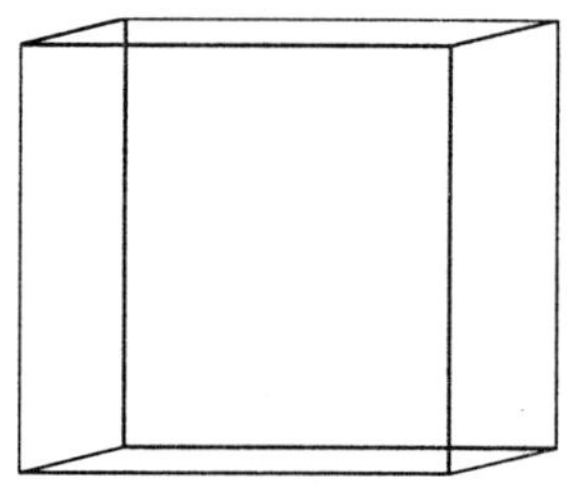

Regular Octahedron (8 Faces)

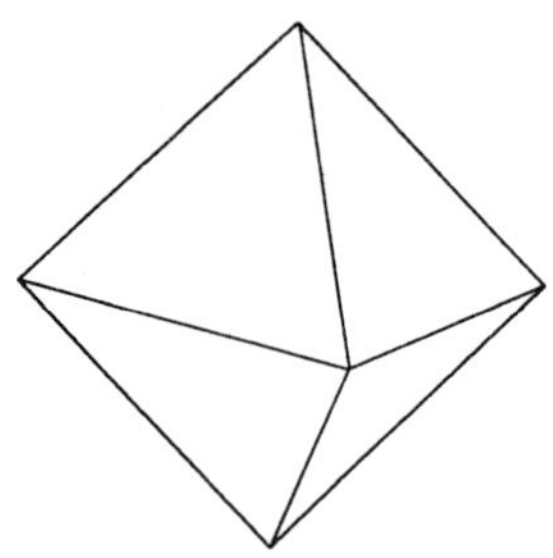

Regular Dodecahedron (12 Faces)

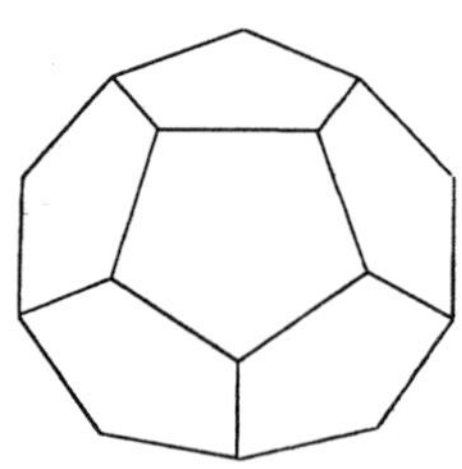

Regular Icosahedron (20 Faces)

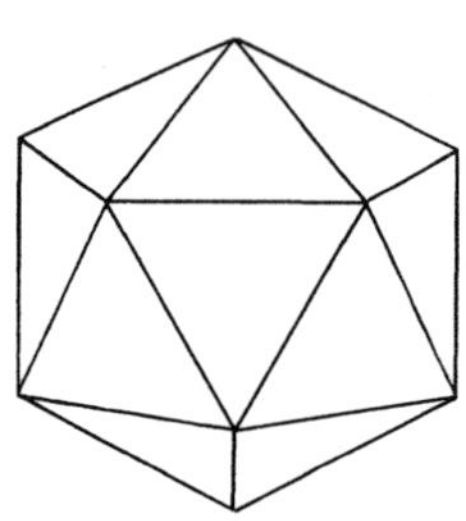

These five polyhedra are known as the Platonic Solids after Plato, the Greek philosopher. Plato believed that an atom of earth had the shape of a cube; an atom of air had the shape of a regular octahedron; an atom of fire had the shape of a regular tetrahedron; and an atom of water had the shape of a regular icosahedron. He believed that the universe itself had the shape of a regular dodecahedron.

A **semiregular polyhedron** is a polyhedron whose faces consist of two or more types of regular polygons. In addition the polyhedron must have identical configurations at all of its vertices. A soccer ball is modeled after a semiregular polyhedron.

It can be proved that *there are only 13 semiregular polyhedra.* These polyhedra are known as the Archimedean Solids. Some of these solids are shown in the illustration.

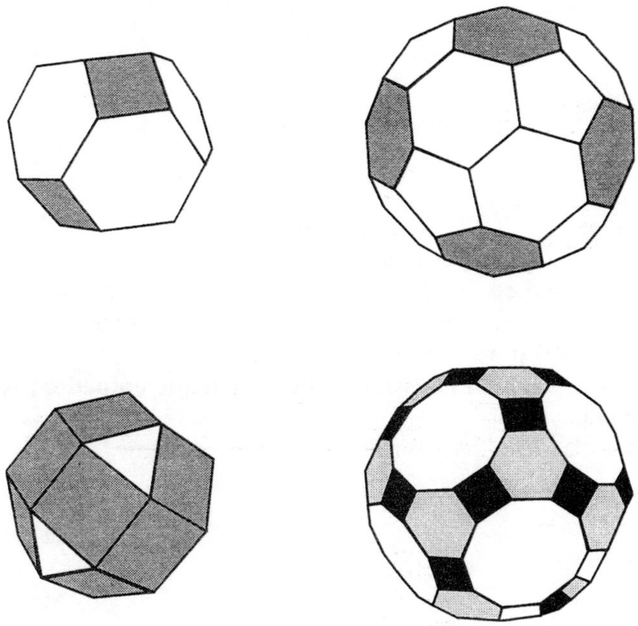

Some Archimedean Solids

Pyramids and Prisms

A **pyramid** is a polyhedron in which one face can be any polygon (This face is identified as the base of the pyramid) but the remaining faces must be triangles. The triangular faces of the pyramid intersect at a point called the **apex** of the pyramid.

If the base of a pyramid is a regular polygon and its triangular faces are congruent isosceles triangles, the pyramid is called a **regular pyramid.**

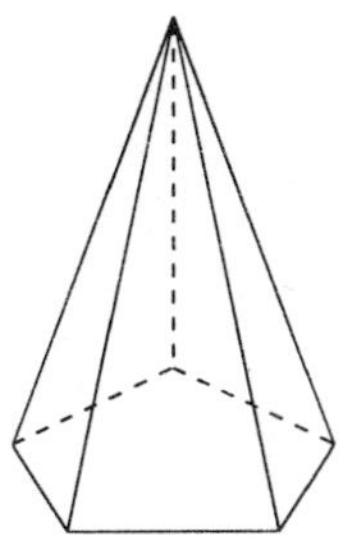

Regular Pyramid

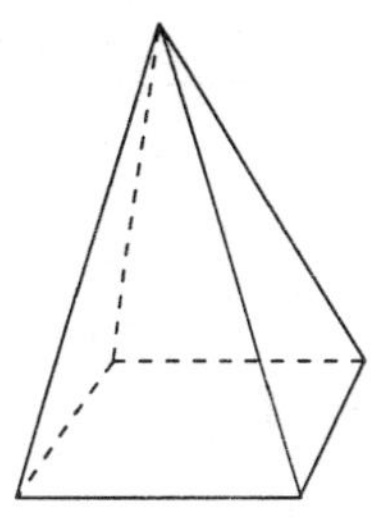

Non-Regular Pyramid

A **prism** is a polyhedron that has a pair of parallel faces that are congruent polygons and parallelograms as its other faces. Each of the parallel faces is called a **base of the prism**. If the parallelograms are rectangles, the prism is called a **right prism.** If the parallelograms are rectangles, and the bases are regular polygons, the prism is called a **right regular prism.**

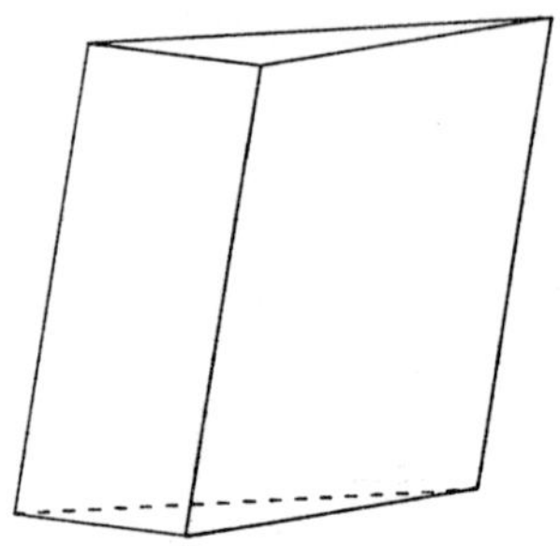

Prism

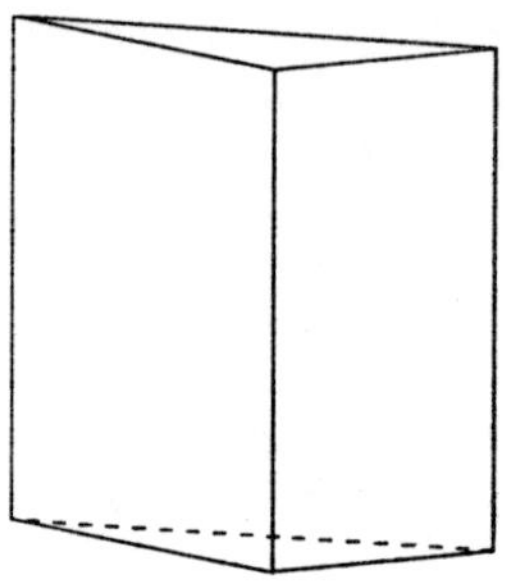

Right Prism

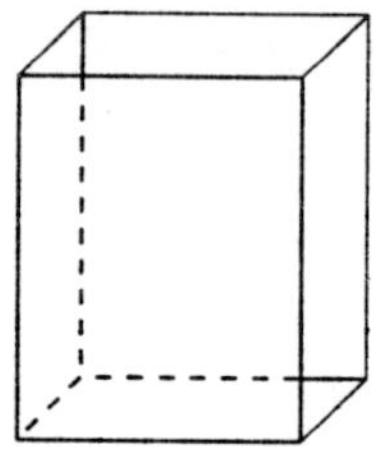

Right Regular Prism

Pyramids and prisms may be classified by the type of polygons that form their bases. A **rectangular prism** has a rectangle for each base. A **triangular prism** has a triangle for each base. Likewise, a **square pyramid** has a square for a base, while a **hexagonal pyramid** has a hexagon for a base.

A rectangular prism also may be called a **parallelepiped or parallelepipedon.** A parallelepiped is simply a prism whose bases are parallelograms.

Some examples of polyhedra classified by their bases are shown below:

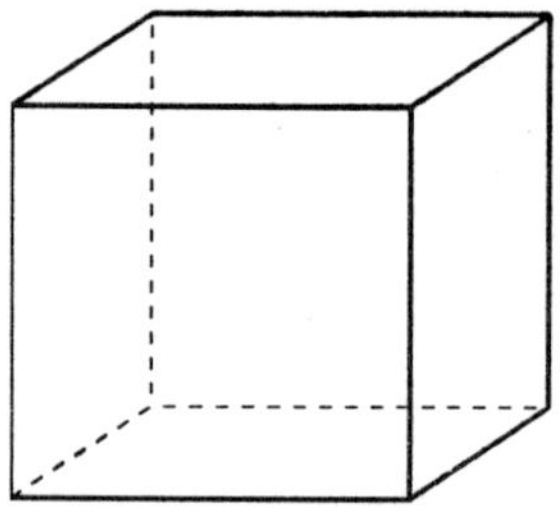

Square Prism or Cube

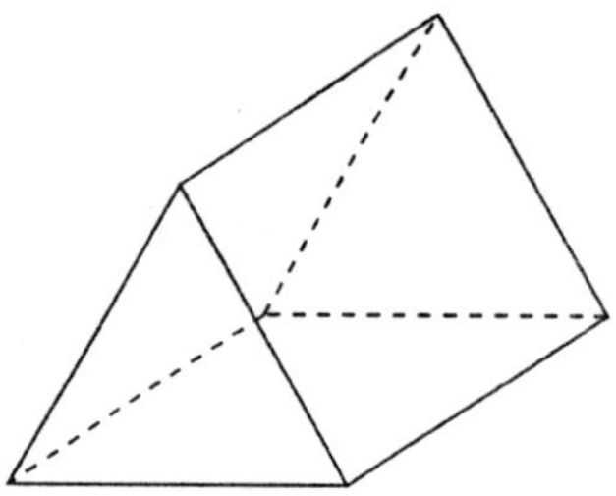

Triangular Prism

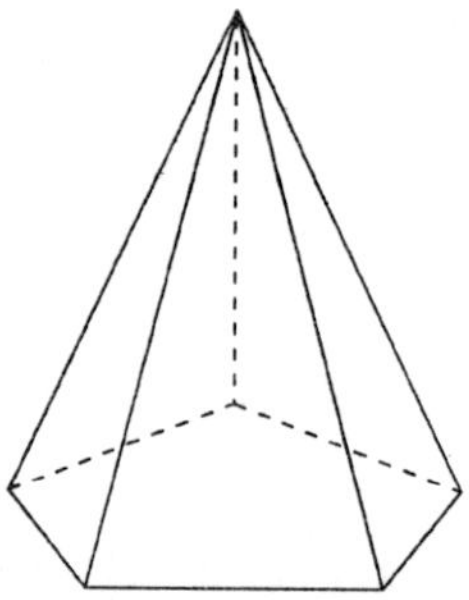

Pentagonal Pyramid

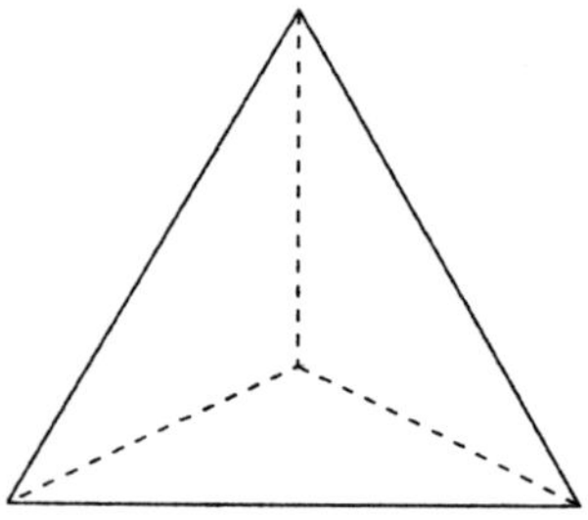

Triangular Pyramid

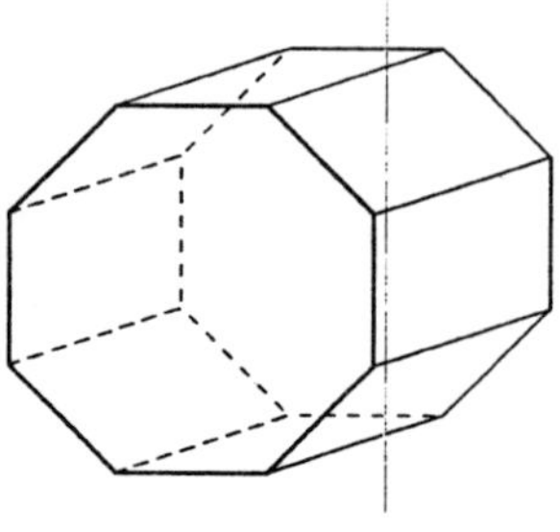

Octogonal Prism

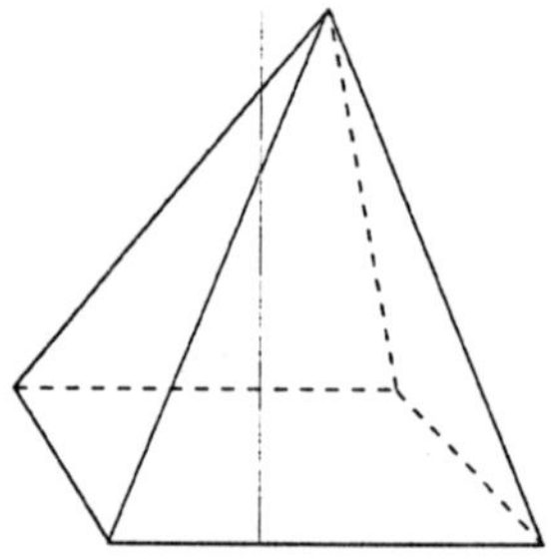

Rectangular Pyramid

Solids with Curved Surfaces

A **sphere** is the surface formed by all points equidistant from a given point in space. The given point is called the **center of the sphere**, while the fixed distance between each point on the sphere and the given point is called the **radius of the sphere.**

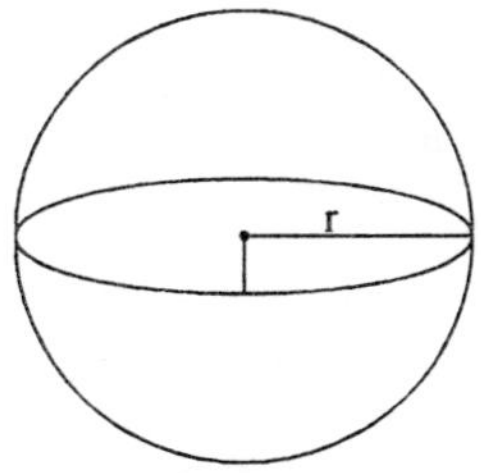

A Sphere

A **circular cylinder** may be thought of as the set of points traced out by a line segment which moves along a circle and which always points in the same direction. We may classify circular cylinders as right or oblique. A **right circular cylinder** is one in which the moving line is vertical when the circle lies in a horizontal plane. The cylinder is, otherwise, called **oblique.**

Right Circular Cylinder

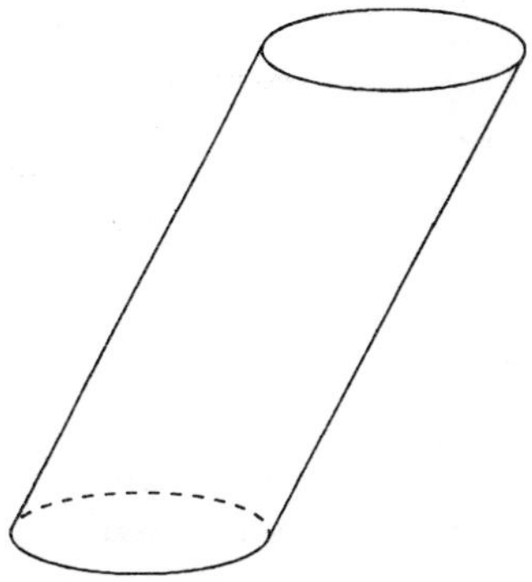

Oblique Circular Cylinder

A **circular cone** is the set of all points traced out by a line segment which moves along a circle as one point on the line stays fixed. A right **circular cone** is one in which the fixed point lies directly above the center of the circle when the circle lies in a horizontal plane. If the fixed point does not lie directly above the center of the circle, the cone is called **oblique.**

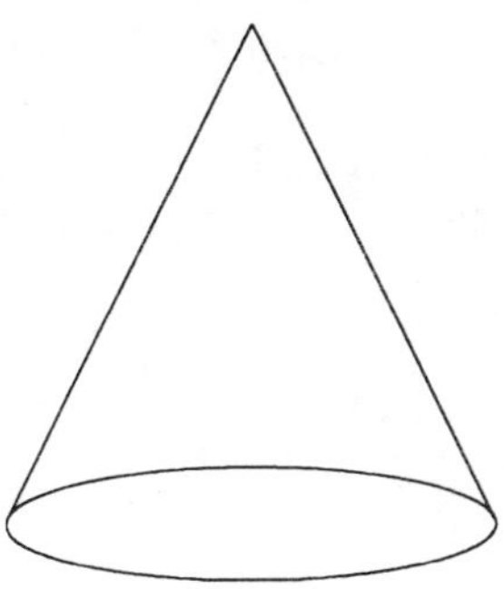

Right Circular Cone

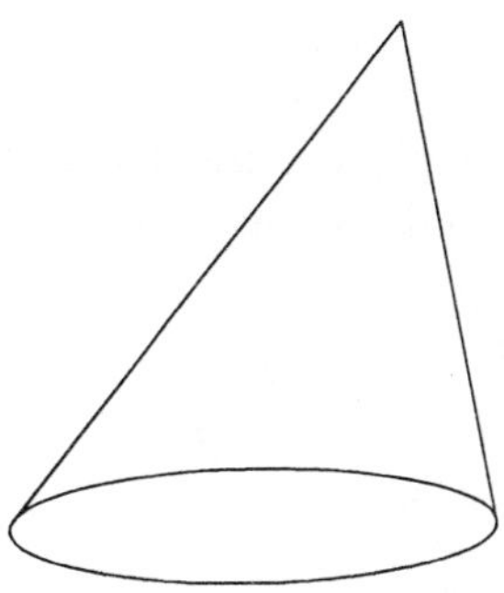

Oblique Circular Cone

Volume of Solid Figures

The **volume of a solid object** is the amount of space which it encloses. Volume is a three-dimensional quantity and is measured in cubic units. We may therefore define the volume of a solid to be the number of cubic units necessary to fill it up. Here are some examples of cubic units used to find volumes.

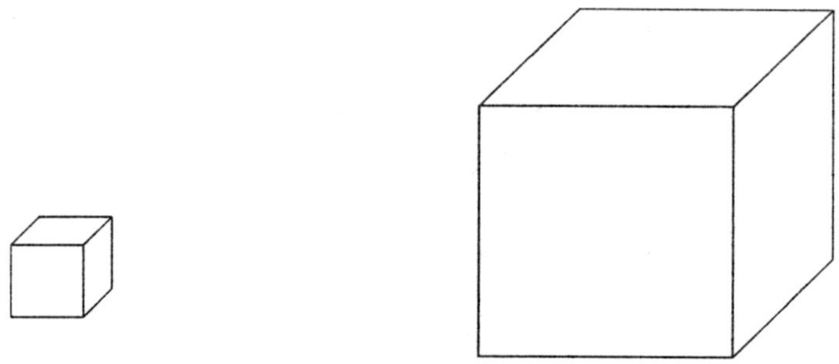

Observe the rectangular prism in the illustration below:

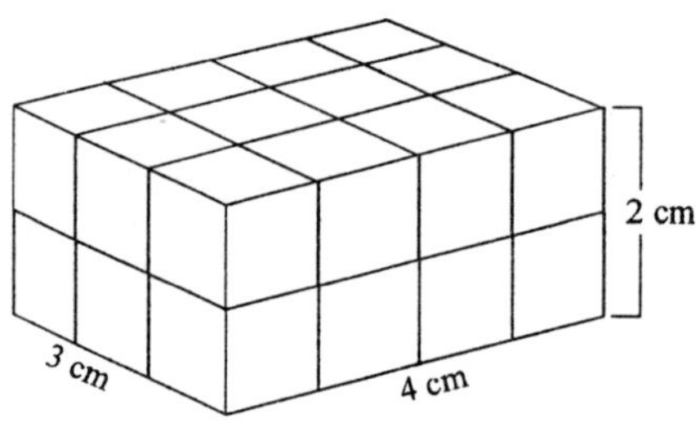

This prism has a length of 4 cm, a width of 3 cm and a height of 2 cm. Notice that there are two layers of cubes and each of these layers contains $(3 \cdot 4)$ or 12 cubes. The volume of the prism is therefore $(2 \cdot 12)$ or 24 cubic centimeters.

In any rectangular prism, we can find the volume by multiplying the length times the width times the height of the prism. Notice that the area of the base of the rectangular prism in the illustration is the product of l and w. If we use an upper case B to represent the area of the base and an h to represent the corresponding height, then the volume V of the prism may be obtained by using the formula $\mathbf{V = Bh}$. Even if the base of a **right prism** is not a rectangle, this formula is still true.

The volume of a **circular cylinder** is computed exactly as it is for a prism. We first determine the area of the base, which is a circle for a cylinder, and multiply this number by the height of the cylinder. A right circular cylinder and an oblique circular cylinder with the same base and the same height have the same volume.

The volume of a **pyramid** is exactly $1/3$ of the volume of a prism having the same base and height, that is, if V represents volume, B represents the area of the base of the pyramid and h represents the height of the pyramid, then $\mathbf{V = 1/3\, Bh}$.

The volume of a **circular cone** is $1/3$ of the volume of a circular cylinder having the same height and base, that is, $\mathbf{V = 1/3\, Bh}$, where V represents the volume of the cone, B represents the area of its base, and h represents its height.

Exercises Appendix B

1. This is a picture of a regular hexagon. Can you represent a cube by drawing three line segments in the figure?

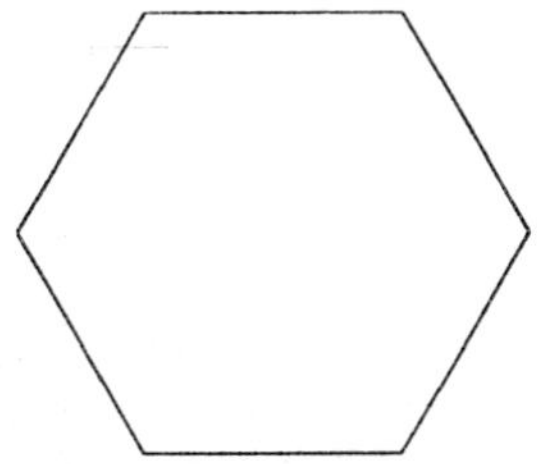

2. The following figure is a polyhedron that has been cut open and laid out as shown. What type of polyhedron is it?

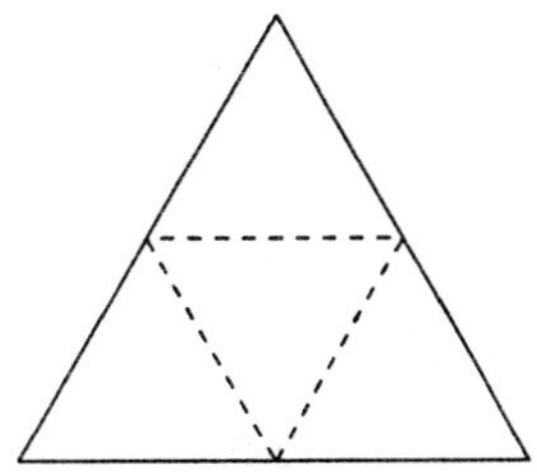

3. The following figure is a polyhedron that has been cut open and laid out as shown. What type of polyhedron is it?

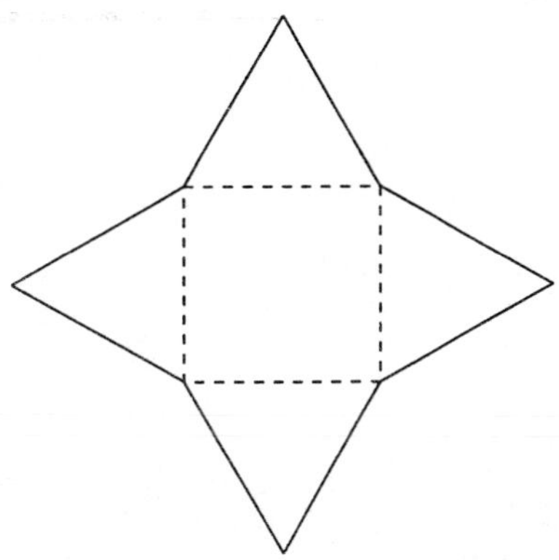

4. Can you identify the polyhedron that can be formed from each of these following patterns?

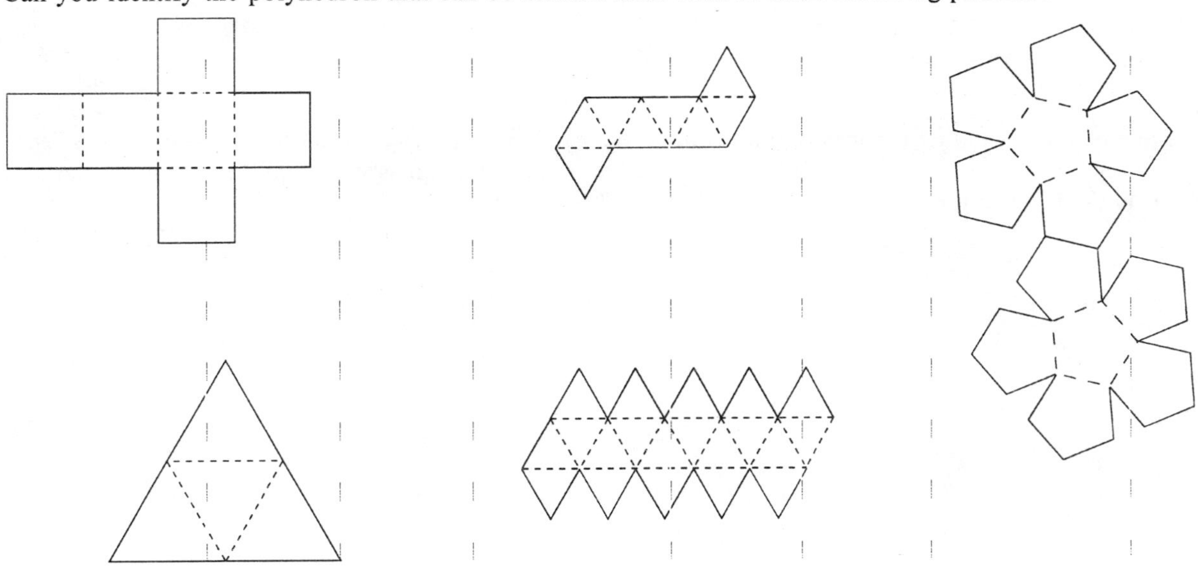

5. We can think of the *angular deficiency at a vertex* of a polyhedron as the number of degrees lost in the circle around that vertex.

The following figure is a square pyramid with equilateral triangles as its faces. If we look at the configuration of polygons at each vertex of the square pyramid, we notice that at each vertex, the number of degrees is less than 360. Consider the polygons surrounding vertex A.

Degrees lost at vertex A
360 – (60 + 60 + 90)
360 – 210
150

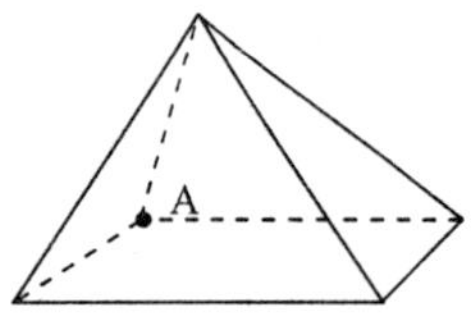

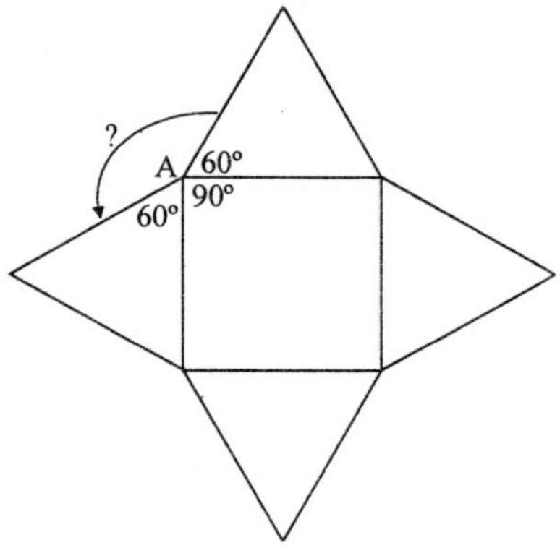

At the apex of the pyramid (the point of intersection of its triangular faces), the deficiency in degrees is (360–240) or 120. Why?

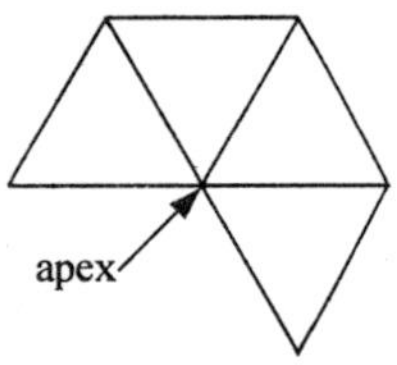

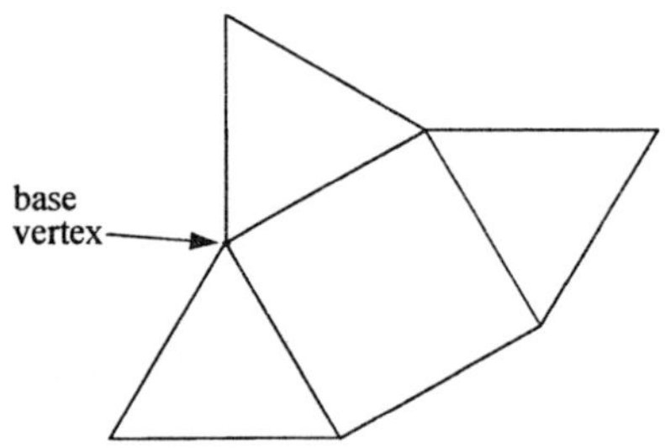

At each of the remaining vertices of the pyramid, the deficiency in degrees is (360–210) or 150. Why? The sum of the angular deficiencies at all of the vertices of the pyramid is called its total angular deficiency. What is the total angular deficiency of this pyramid?

6. a. There must be an angular deficiency at each vertex of any polyhedron. Can you explain why?

   b. Find the angular deficiency at each vertex of the Platonic solids. Use this information to explain why these five solids are the only regular polyhedra?

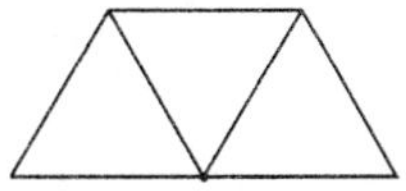

Regular Tetrahedron

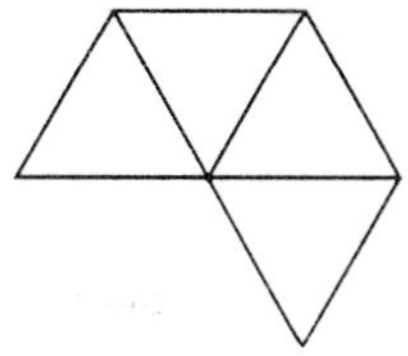

Regular Octahedron

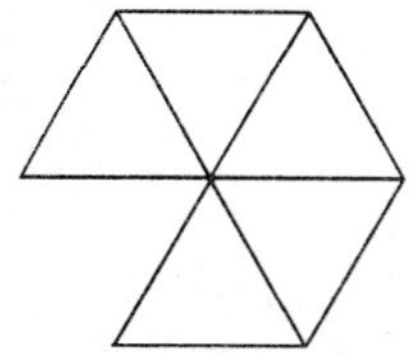

Regular Icosahedron

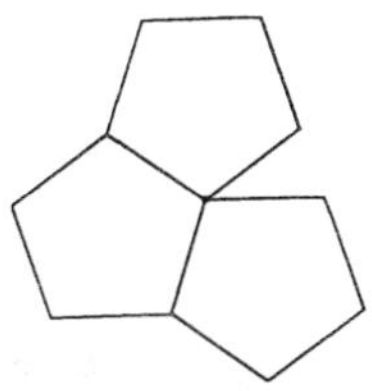

Regular Dodecahedron

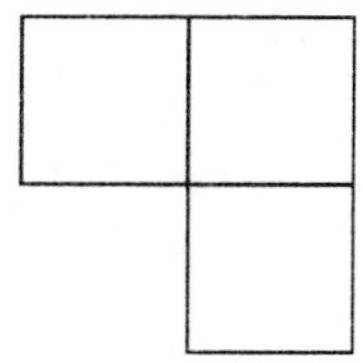

Regular Hexahedron or Cube

7. Find the total angular deficiency for each of the regular polyhedra. What number did you get?

8. Find the total number of faces, vertices and edges for five different polyhedra. Then use this information to complete the following table.

Polyhedron	*Faces*	*Vertices*	*Edges*	*Sum of Vertices & Faces*

For each polyhedron, how is the sum of its vertices and faces related to its number of edges?

9. Find the volume of the following solids:

a.

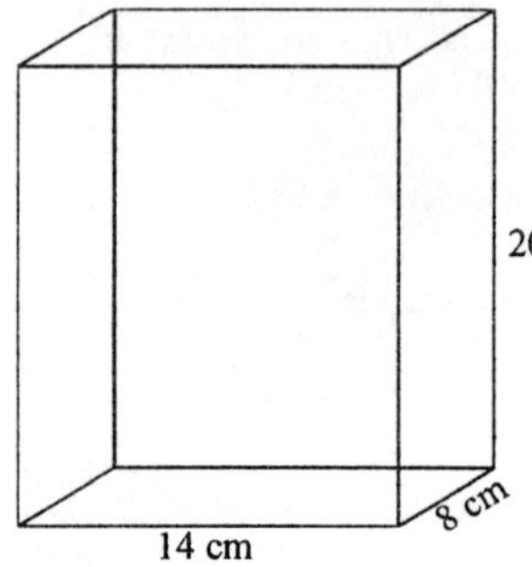

e.

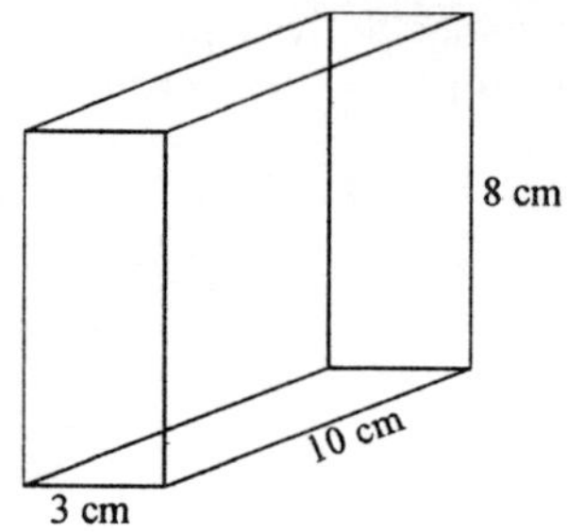

b.

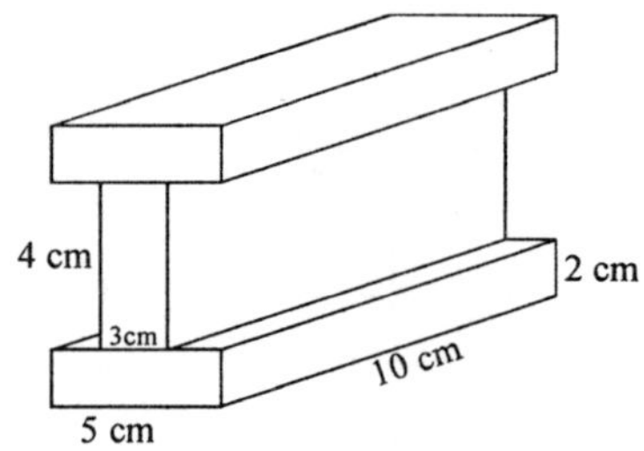

f.

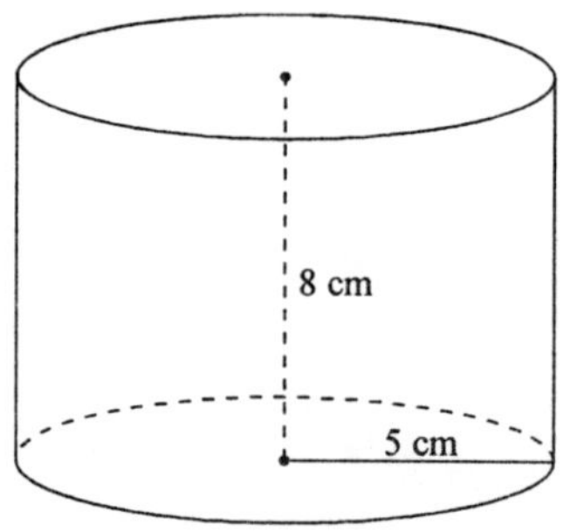

c.

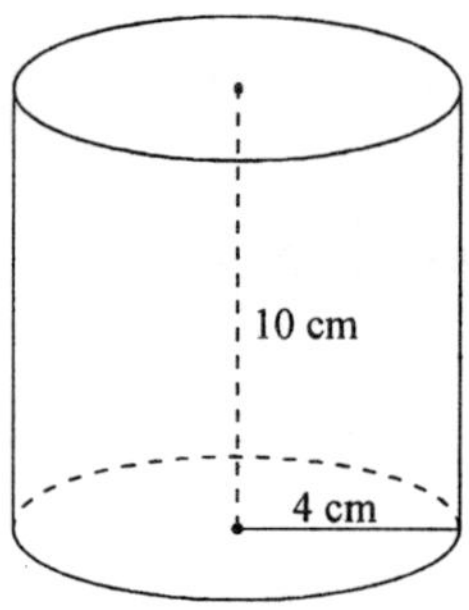

g.

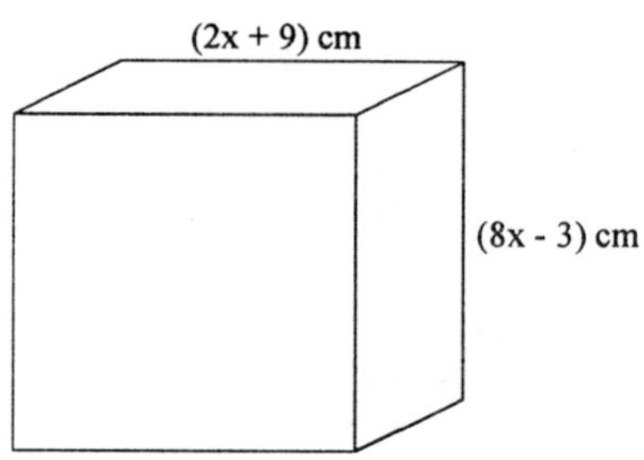

d.

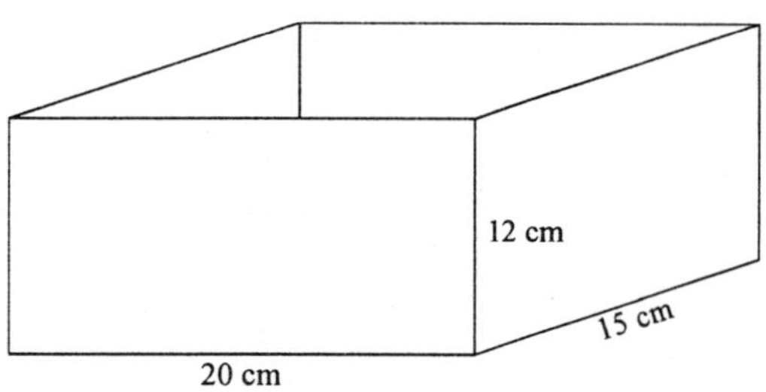

h.

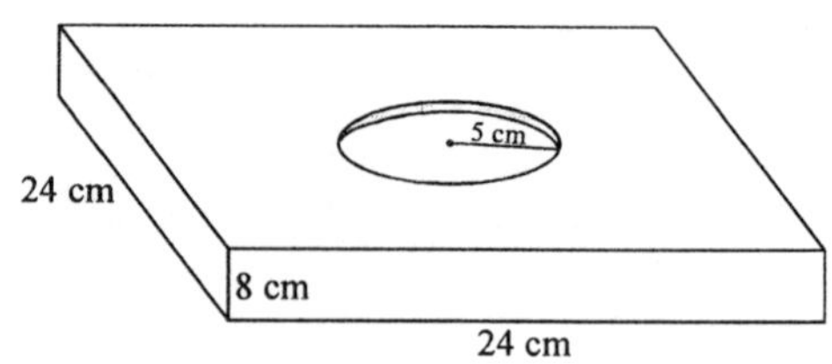

10. Find the volume of each solid in Exercises 4.3.

11. What happens to the volume of a cube when its edge is doubled?

12. What happens to the dimensions of a cube when the volume of the cube is doubled?

13. What happens to the volume of a cube if each of its dimensions is doubled?

Appendix C

Introduction to Transformations

A **transformation** is a one-to-one correspondence between two sets of points in a plane. You should note that such a correspondence may assign a point of the given figure to another point of the figure itself.

We will discuss three types of transformations: *translations, rotations* and *reflections.*

Translations (Slides)

A **translation** can be thought of as the result of sliding, without turning, a geometric figure from one position to another position in a plane. The process of translating a figure in a plane involves a linear shift of the figure in that plane. In the illustration, the quadrilateral ABCD has been translated to a new position indicated by A'B'C'D'. Notice that the line segments joining the corresponding points are all parallel.

AA' , BB' , CC' , and DD' are parallel.

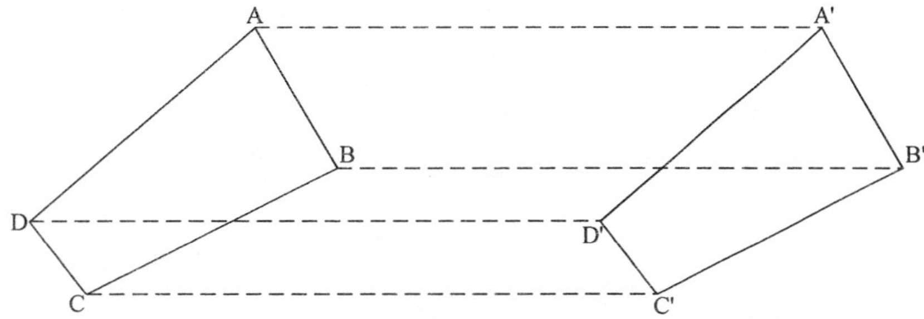

The following transformation is not a translation because it is not **linear**. Notice that a rotation of the quadrilateral has taken place. The line segments joining the corresponding points are not parallel.

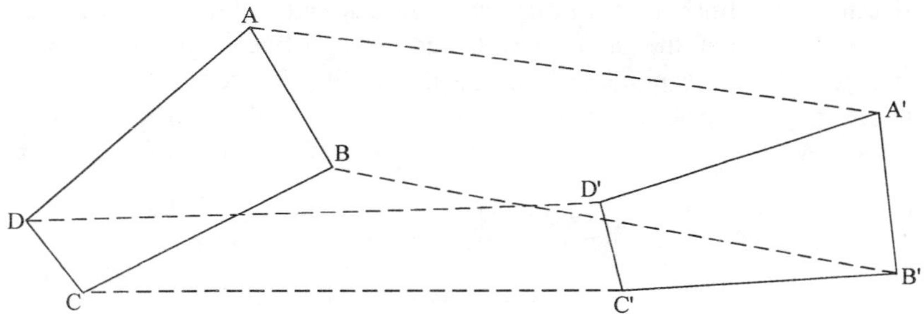

When a translation of a geometric figure takes place, its size and its shape remain unchanged. The measures of angles and the lengths of line segments are the same. *In other words, the original figure and its translation are congruent.*

Rotations (Turns)

The **rotation** of a figure in a plane involves turning the figure about a given point in the plane called the **center of rotation.** Each point of the rotated figure must lie on the arc of concentric circles that have as their center the center of rotation. In the illustration, triangle OSR has been rotated to another position in the plane indicated by triangle OS'R'. Point O is the center of rotation.

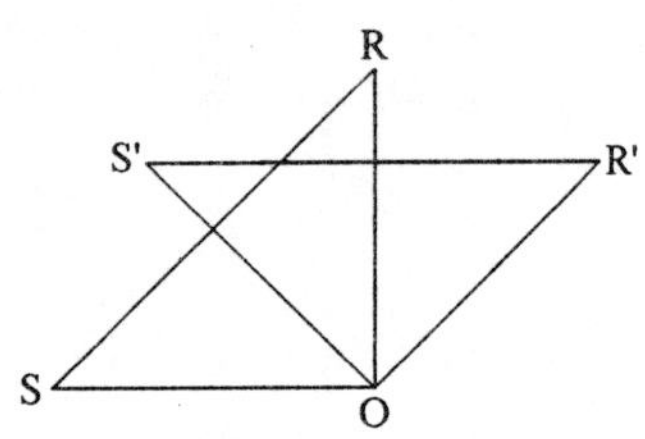

The center of rotation does not have to be a point of the figure itself. Notice in the following illustration that the center of rotation is not a point of the figure.

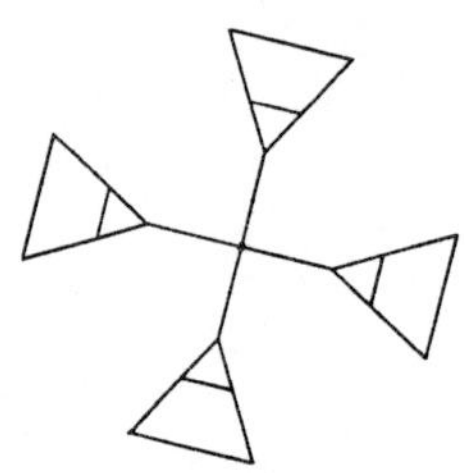

When a rotation of a figure occurs, the size and the shape of the figure remain unchanged.

Reflections (Flips)

We think of a reflection as a flipping of a figure in a plane about a line called the line of reflection or the mirror of reflection. In the illustration, the figure is reflected through line ℓ. Each point of the reflected figure is called a reflection or image of its corresponding point in the original figure.

More formally, we define a point A' to be a reflection of point A through line ℓ iff, ℓ is the perpendicular bisector of the line segment AA'. Note that a point on the line of reflection itself is its own reflection.

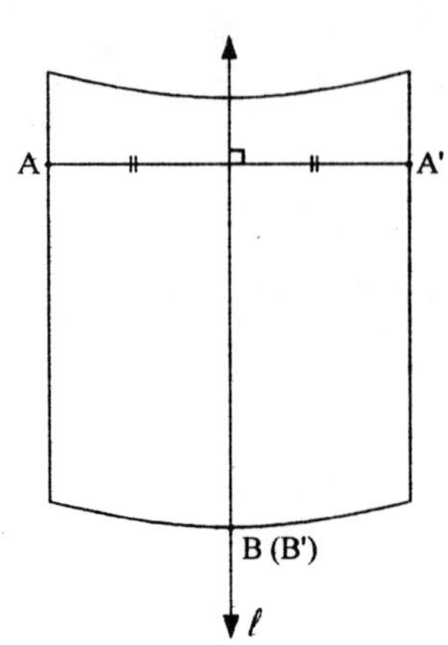

In the illustration that follows, points A', B' and C' are the images of points A, B and C with respect to line ℓ. You may want to try placing a mirror on line ℓ so that it faces toward the left. Doing so, you will notice that the points A, B and C appear in the mirror to be in the location of points A', B' and C'.

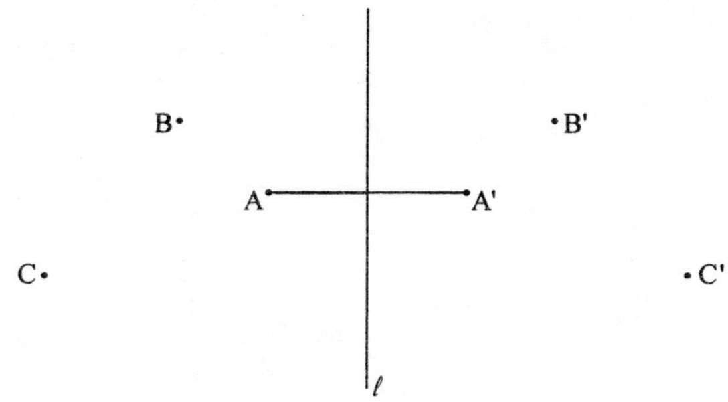

In the next illustration we see that the line of reflection is the perpendicular bisector of each line segment joining each point to its image.

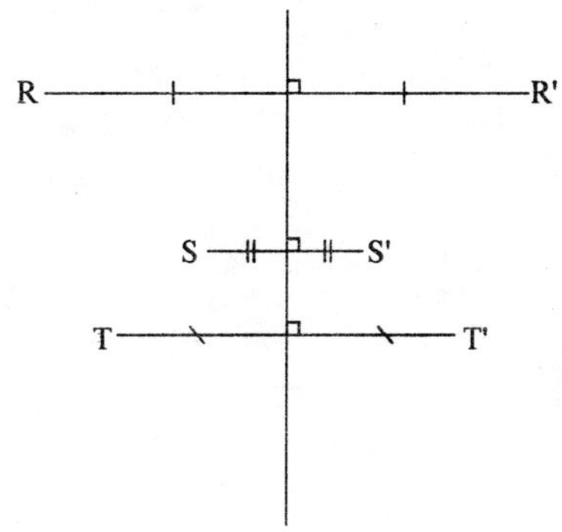

As with translations and rotations, *properties of size and shape are not changed by a reflection.* However, there is one basic property that does change when a figure is reflected through a line. The orientation of the figure is changed. Let us see what is meant by the orientation of a figure.

Consider triangle ABC:

If we trace our pencil along the sides of triangle ABC from A to B then from B to C then back to A, we move in a clockwise direction.

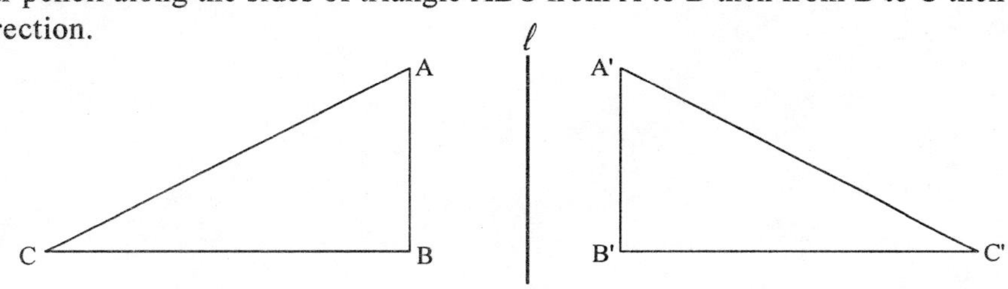

Now, if we trace our pencil along the sides of triangle A'B'C', which is the reflection of triangle ABC through line ℓ, moving from A' to B', B' to C' and then back to A', our movement is in a counterclockwise direction. One triangle is said to have a clockwise orientation; while the other triangle is said to have a counterclockwise orientation.

We conclude that *the orientation of a figure is changed when the figure is reflected through a line.*

We can find the image of a point through a given line by first constructing the perpendicular line segment from the point to the line. We then extend this perpendicular segment and mark off on it a segment having the same length as the perpendicular segment.

Sometimes a translation is called a composite of two successive reflections through parallel lines. In the illustration △ ABC is first reflected through line m. It then is reflected through line n. Note that lines m and n are parallel.

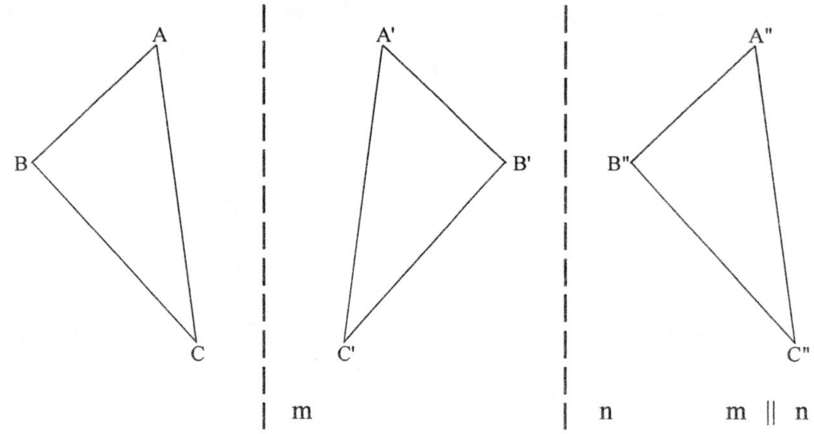

Likewise, a rotation can be called a composite of two successive reflections through intersecting lines. In the next illustration, triangle ABC is reflected first through line m and then through line n.

Notice that m and n are intersecting lines.

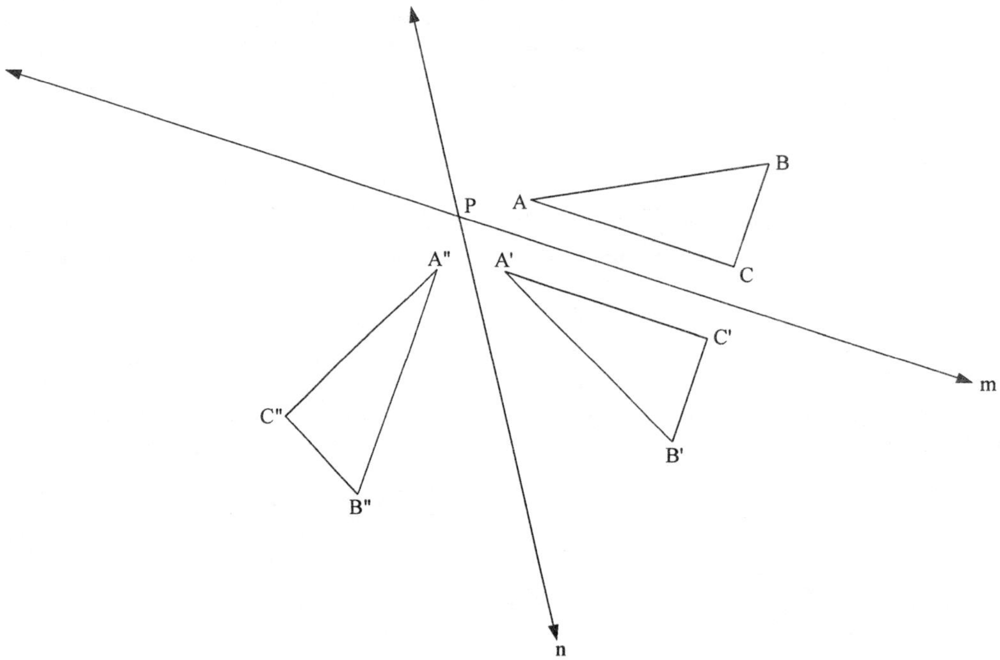

Symmetry

A geometric figure has line **symmetry** if there is a line of reflection such that the image of each point of the figure, through this line, is also a point of the same figure.

This figure has line symmetry.

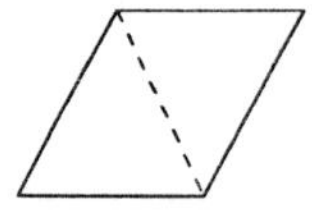

A simple way to determine if a figure has line symmetry is to fold the figure along the supposed line of symmetry and see if the two halves of the figure coincide. If the two halves coincide, the fold line is called the line of symmetry.

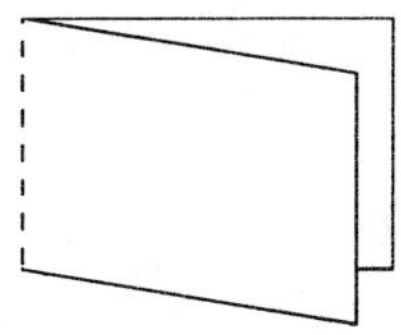

Another way to see if a figure has line symmetry is to place a mirror along the assumed line of symmetry. If the figure and its mirror reflection are identical to the given figure, then the given figure has line symmetry.

A figure may have more than one line of symmetry. The square in the illustration has four lines of symmetry. The circle has infinitely many lines of symmetry, since every line through its center is a line of symmetry. An equilateral triangle has three lines of symmetry.

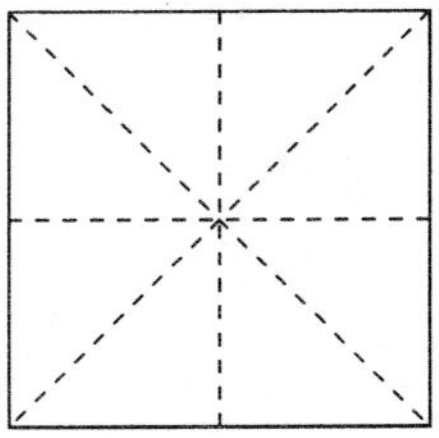
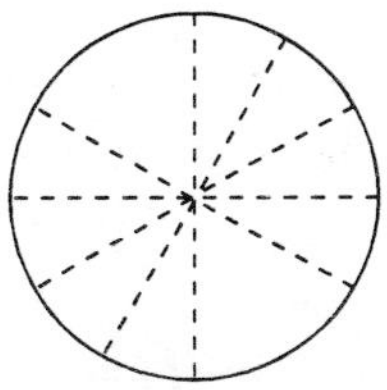
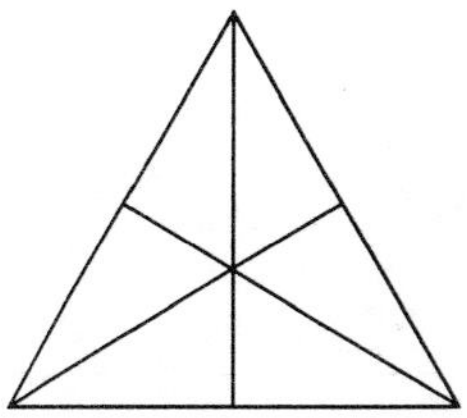

Line symmetry may be referred to as **mirror symmetry, reflective symmetry, axial symmetry or bilateral symmetry.**

Other names for the line of symmetry are:
axis of symmetry, mirror line and line of reflection.

Point Symmetry

A figure has point symmetry if a rotation of 180° (a half turn) makes the figure coincide with itself. A simple way to determine if a figure has point symmetry is to turn the figure upside down and see if it looks the same as it did when it was in its original position. A figure that has point symmetry is unchanged in its appearance by a 180 degree rotation. Another name for point symmetry is **radial symmetry**. These figures have point symmetry.

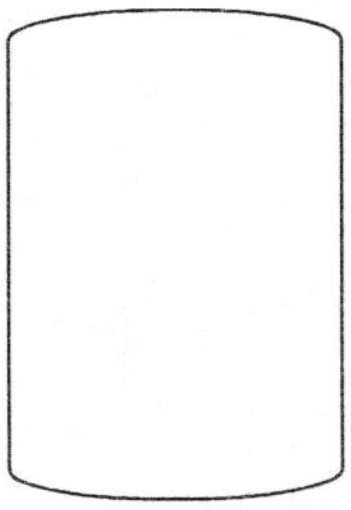
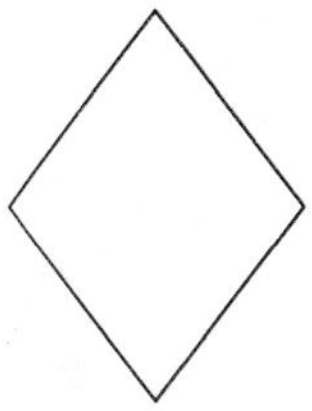

Rotational Symmetry

Point symmetry is a special case of a more general type of symmetry called rotational symmetry. A figure has rotational symmetry if it can be rotated about a point through less than 360° to coincide with itself. An equilateral triangle has rotational symmetry but not point symmetry. Rotations of 120° and 240° return the triangle to its original position.

Exercises Appendix C

1. How many lines of symmetry does each figure have?
   a. a regular triangle
   b. a regular quadrilateral
   c. a regular pentagon
   d. a regular hexagon
   e. a regular octagon
   f. a regular n-sided polygon

2. 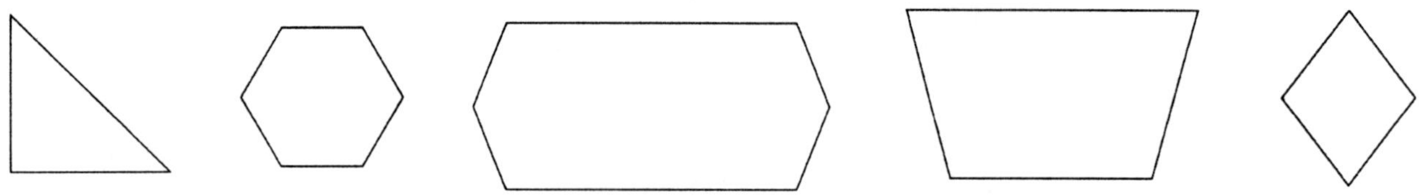
   a. Which of the above figures have exactly one line of symmetry?
   b. Which of the above figures have exactly two lines of symmetry?
   c. Which of the above figures have rotational symmetry?

3. Which letters of the alphabet have vertical lines of symmetry?
   Which letters of the alphabet have horizontal lines of symmetry?
   Which letters of the alphabet have point symmetry?

   A B C D E F H K M N O

4. Sketch all axes of symmetry for the following figures.

5. Look up information on tessellations. How are they related to transformations? Try making your own tessellation.

6. Find the reflection of triangle ABC through line m.

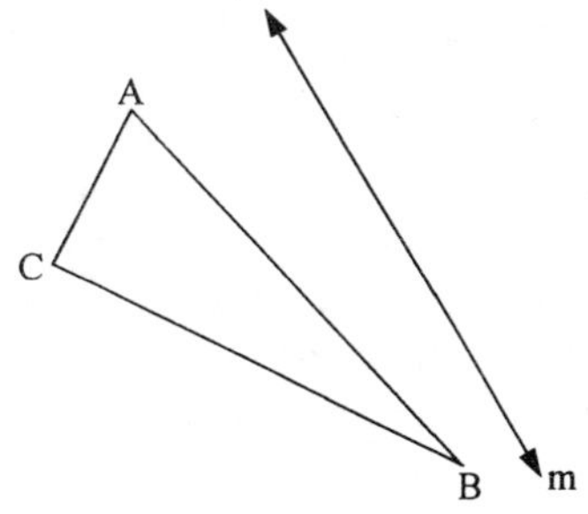

7. In the graph, $\triangle$ A'B'C' is the reflection of $\triangle$ ABC through the y-axis. The coordinates of the vertices of $\triangle$ ABC are as follows:

A(2, 5) B(7, 2) C(2, 2)

a. What are the coordinates of the vertices of $\triangle$ A'B'C'?
b. How are the coordinates of the vertices of $\triangle$ ABC related to the coordinates of $\triangle$ A'B'C'?

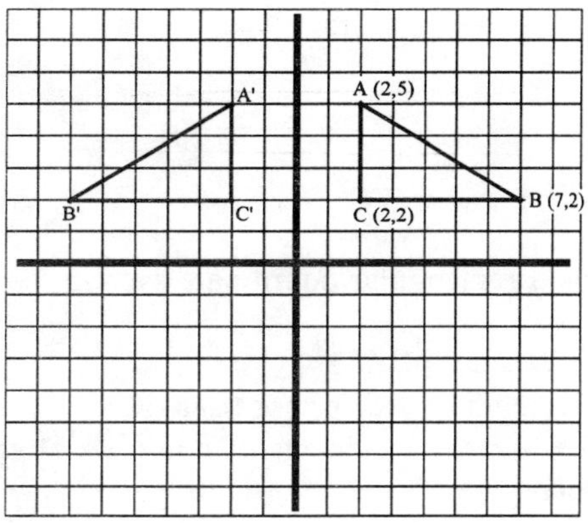

8. Complete the following exercises:

a. Graph the following points on a rectangular coordinate system

A(2, 5) B(7, 2) C(2, 2)

b. Connect these points with line segments to form a triangle.
c. Interchange the coordinates of the points A, B and C. You should now have coordinates for three new points.
d. Graph these points on the same rectangular coordinate system, and draw the line segments connecting these points. The resulting figure will be a triangle.
e. Now, graph the line determined by the equation $y = x$.
f. What can you say about the two triangles with respect to the line?

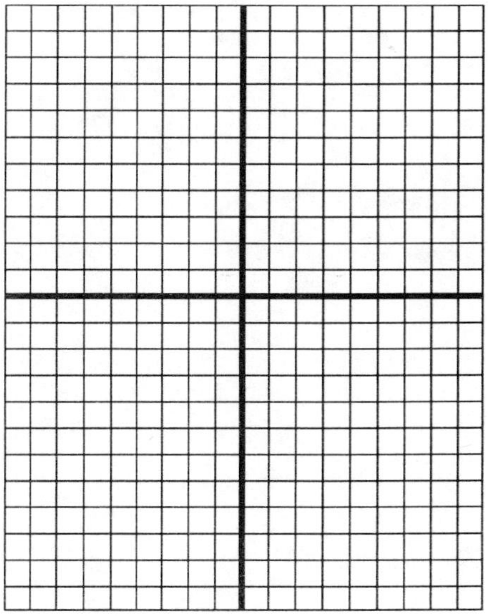

9. Complete the following exercises:

   a. Graph the following points on a rectangular coordinate system

      A(2, 5) B(7, 2) C(2, 2)

   b. Connect these points with line segments to form a triangle.
   c. Change the signs of the y-coordinates of points A, B and C. You should now have coordinates for three new points.
   d. Graph these new points on the same rectangular coordinate system, and draw the line segments connecting these points. The resulting figure will be a triangle.
   e. What can you say about the two triangles with respect to the x-axis?

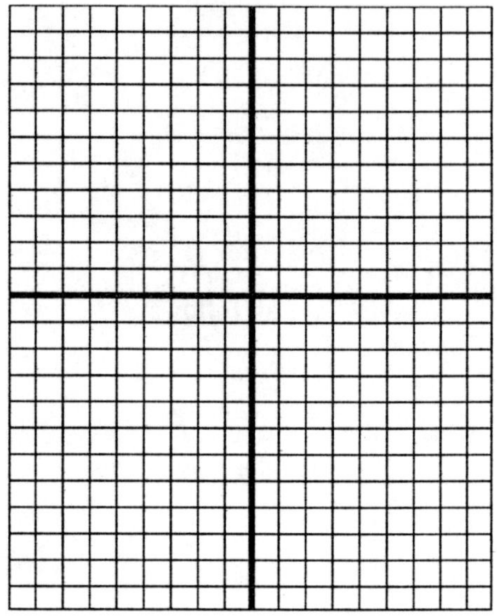

10. Complete the following exercises:

   a. Graph the following points on a rectangular coordinate system

      A(2, 5) B(7, 2) C(2, 2)

   b. Connect these points with line segments to form a triangle.
   c. Change the signs of both the x and y coordinates of points A, B and C. You should now have coordinates for three new points.
   d. Graph these new points on the same rectangular coordinate system, and draw the line segments connecting these points. The resulting figure will be a triangle.
   e. What can you say about the second triangle with respect to the original triangle?

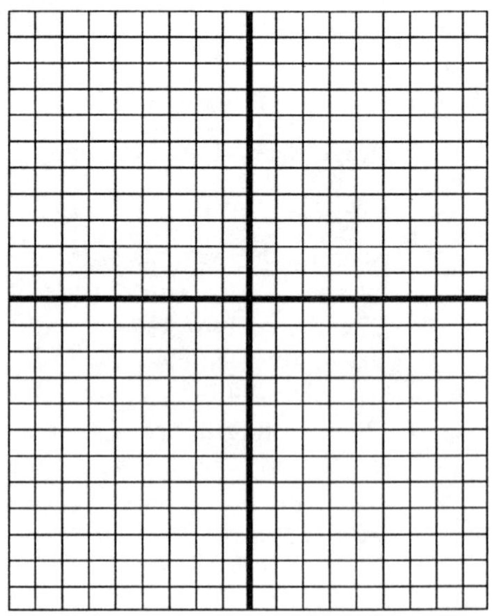

Appendix D

Definitions

Altitude (Height)

altitude of a parallelogram
Any perpendicular line segment joining two parallel sides of the parallelograms; or
The distance between two parallel sides of the parallelogram

altitude of a trapezoid
Any perpendicular line segment joining the two parallel sides of the trapezoid; or
The distance between the two parallel sides of the trapezoid

altitude of a triangle
A perpendicular line segment joining a vertex of a triangle to the line containing the opposite side of the triangle; or
The length of this segment; or
The line that contains this line segment

Angles

angle
Two rays with the same endpoint

sides of an angle
The two rays that form the angle

vertex of an angle
The intersection of the two rays that form the angle

measure of an angle
The positive number associated with the amount of rotation needed to swing one side of an angle about its vertex so that it coincides with the other side of the angle.

degree
A unit used to measure an angle. It is 1/360 of a complete revolution.

minute
A unit used to measure an angle. It is 1/60 of a degree.

second
A unit used to measure an angle. It is 1/3600 of a degree and 1/60 of a minute.

protractor
A tool used to measure angles

acute angle
An angle that has a measure less than 90 degrees

right angle
An angle that has a measure of 90 degrees

obtuse angle
An angle that has a measure greater than 90 degrees but less than 180 degrees

straight angle
An angle that has a measure of 180 degrees

reflex angle
An angle that has a measure between 180 and 360 degrees

adjacent angles
Two angles that have a common side and a common vertex

vertical angles
Two nonadjacent angles formed by two intersecting lines

linear pair of angles
Two adjacent angles with opposite rays as sides

supplementary angles
Two angles whose total measure is 180 degrees

complementary angles
Two angles whose total measure is 90 degrees

exterior angle of a triangle
An angle that forms a linear pair with an interior angle of the triangle

remote interior angles of an exterior angle of a triangle
The two interior angles of the triangle that do not form a linear pair with the exterior angle

Base of a Polygon

base of a triangle
Any side of a triangle (corresponding altitude is the perpendicular line segment joining the selected side to the opposite vertex of the triangle; or the length of this perpendicular segment)
The length of this side

bases of a parallelogram
Two parallel sides of a parallelogram (corresponding altitude is the distance between these parallel sides)
The lengths of these sides

bases of a trapezoid
The two parallel sides of a trapezoid (corresponding altitude is the distance between these parallel sides)
The lengths of the two parallel sides

base of an isosceles triangle
The side opposite the vertex angle of the triangle (the vertex angle is the angle that contains the two sides of the triangle that have equal lengths)

Circle

circle
A set of points in a plane, all points of which are the same distance from a fixed point called the center of the circle

radius of the circle
Any line segment that joins the center of a circle to a point on the circle is called a radius of the circle
The length of such a line segment

interior point of a circle
A point is in the interior of a circle iff the length of the line segment joining the point to the center of the circle is less than or equal to the radius of the circle.

exterior point of a circle
A point is in the exterior of a circle iff the length of the line segment joining the point to the center of the circle is greater than the radius of the circle.

concentric circles
Two or more circles that have the same center but different radii.

chord of a circle
A line segment that joins two points of the circle.

diameter of a circle
A chord that contains the center of the circle
The length of such a chord

secant of a circle
A line that intersects the circle in two points

tangent of a circle
A line that intersects the circle in exactly one point

point of tangency or point of contact
The point where a tangent intersects a circle

common internal tangent of two circles
A tangent to both circles. It intersects the line segment joining the centers of the two circles

common external tangent of two circles
A tangent to both circles. It does not intersect the line segment joining the centers of the two circles

arc of a circle
An arc consists of two points of the circle (called endpoints of the arc) together with all the points of the circle that lie between those two points

semicircle
An arc whose endpoints are also the endpoints of a diameter of the circle

central angle
An angle that has the center of the circle as its vertex

major arc
An arc that has a measure that is more than 180 degrees

minor arc
An arc that has a measure that is less than 180 degrees

measure of a minor arc
The measure of its central angle

measure of a major arc
360 minus the measure of the minor arc with the same endpoints

measure of a semicircle
180 degrees

circumference of a circle
The distance around the circle

pi
The ratio of the circumference of a circle to its diameter

A secant ray
A ray that intersects the circle in two points

A secant segment
A segment that contains two points of the circle

A tangent ray
A ray that intersects a circle in exactly one point and the line that contains it is a tangent of the circle

A tangent segment
A segment that intersects a circle in exactly one point and the line that contains it is a tangent of the circle

measure of a semicircle
½ a revolution or 180 degrees

circumscribed polygon
A polygon that has each of its sides tangent to a circle ("tangent to" means each side touches the circle in one point)

inscribed polygon
A polygon that has each of its vertices lying on a circle.

inscribed angle of a circle
An angle that has its vertex on a circle and each of its sides contains a chord of the circle

A common tangent of two or more circles
A line tangent to each of the circles

tangent circles
Circles that have the same tangent line at the same point of contact

line of centers
The line containing the centers of two circles

Conditional Statements

conditional statement
A statement that can be written in an if-then form

Venn Diagrams
Circles used to represent sets

converse
The statement obtained by interchanging the hypothesis and the conclusion of a conditional statement

inverse
The statement obtained by negating the hypothesis and the conclusion of a conditional statement

contrapositive
The statement obtained by interchanging and negating the hypothesis and the conclusion of a conditional statement

logically equivalent
When one statement is true, the other statement is true and when one statement is false, the other statement is false.

if and only if (iff)
Two statements: the statement as written and its converse

Constructions

geometric constructions
Drawings done with only two tools: a straightedge and a compass

straightedge
A tool for drawing straight lines; unlike a ruler, it can not be used for measuring

compass
A tool for drawing circles and arcs

concurrent lines or line segments
Three or more lines or line segments intersecting in a common point.

Distances

distance between two points
The length of the line segment joining the two points

distance from a point to a line not containing the point
The length of the perpendicular line segment joining the point to the line

distance between two parallel lines
The length of any perpendicular line segment joining the two lines

Deductive System

deductive system
A system of study that uses undefined terms, definitions, postulates and theorems. Basically, such a system involves the use of relatively simple statements to derive more complex statements through a process called deductive logic.

definition
An agreement that a word or group of words may be used in place of another word or group of words

postulate
A statement that is accepted as being true without proof

axiom
A statement that is accepted as being true without proof

theorem
A statement that must be proved before it is accepted as true

deductive proof
A group of statements that compose a deductive argument

deductive argument.
An argument for which a conclusion is inescapable once a major premise is accepted and an instance of the hypothesis of the major premise called the minor premise is accepted

Diagonal

Diagonal of a polygon
A line segment that joins two nonadjacent vertices of the polygon

Figure/Geometric Figure

Figure/Geometric Figure
A point or a set of points

Legs

legs of a right triangle
The sides opposite the acute angles of the triangle

legs of an isosceles triangle
Two sides of the triangle that have equal lengths

Length

Length of a line segment
The distance between its two endpoints

Lines

concurrent lines
Two or more lines that intersect in the same point

intersecting lines
Two lines that have a common point

perpendicular lines
Two lines that form right angles when they intersect

parallel lines
Two lines in the same plane that do not intersect

skew lines
Two lines that do not intersect and do not lie in the same plane

Line Segments (Segments)

line segment
Two points of a line together with all the points on the line that lie between these two points

bisector of a line segment
A point, ray, line or line segment that contains the midpoint of the segment or separates the segment into two segments of equal lengths

length of a line segment
The distance between its two endpoints

perpendicular line segments
Two intersecting line segments that are contained in perpendicular lines

perpendicular bisector of a line segment
A line, ray or line segment that is perpendicular to the line segment and also bisects it

Measure/Measurements

perimeter of a polygon
The sum of the lengths of the sides of the polygon

circumference of a circle
The distance around the circle

area
The number of square units necessary to cover a surface

measure of an angle
The positive number associated with the amount of rotation needed to swing one side of an angle about its vertex so that it coincides with the other side of the angle

measure of a minor arc
The measure of its central angle

measure of a major arc
360 minus the measure of the minor arc with the same endpoints

measure of a semicircle
180 degrees

Midpoint

midpoint of a line segment
A point on the segment that separates it into two line segments of equal lengths

Plane

plane figures or figures
A point or a set of points that lie in a plane

Points

collinear points
Points that lie on the same line

coplanar points
Points that lie in the same plane

coordinate of a point
A number assigned to a point on a numberline

Polygons

polygon
A closed geometric plane figure formed by joining line segments end to end

pentagon
A polygon with exactly five sides

hexagon
A polygon with exactly six sides

heptagon
A polygon with exactly seven sides

octagon
A polygon with exactly eight sides

sides of a polygon
The line segments that form the polygon

vertex of a polygon
The intersection of two sides of the polygon

interior angle of the polygon
An angle that contains two sides of the polygon and whose interior contains all the other sides of the polygon

exterior angle of the polygon
An angle that forms a linear pair with an interior angle of the polygon

convex polygon
A polygon each of whose interior angles has a measure less than 180 degrees

concave polygon
A polygon that has an interior angle with a measure greater than 180 degrees

regular polygon
A polygon that is both equiangular and equilateral

equianglar polygon
A polygon that has all angles of equal measure

equilateral polygon
A polygon that has all sides of equal length

Quadrilaterals

quadrilateral
A polygon with exactly four sides

consecutive angles of a quadrilateral or parallelogram
A pair of angles of the quadrilateral whose vertices are the endpoints of a side of the quadrilateral

parallelogram
A quadrilateral that has two pairs of parallel sides (both pairs of opposite sides are parallel)

rhombus
A quadrilateral with four sides of equal length

rectangle
A quadrilateral with four angles of equal measure

square
A quadrilateral that is equiangular and equilateral

trapezium
A quadrilateral with no equal sides

trapezoid
A quadrilateral with exactly one pair of parallel sides

Rays

ray
A point on a line together with all points on the line that lie on one side of that point

opposite rays
Two rays that lie on the same line and have a common endpoint

perpendicular rays
Two intersecting rays contained in perpendicular lines
Two intersecting rays that form one or more right angles

Space

space
The set that contains all points

Transversals

transversal
A line that intersects two lines in two distinct points

corresponding angles
Pairs of angles formed by the intersection of a transversal with two lines in such a way that all interior points of the angles lie on the same side of the transversal and on the same side of the two lines

Same side interior angles
Pairs of angles formed by the intersection of a transversal with two lines in such a way that all interior points of the angles lie on the same side of the transversal and all interior points in the intersection of the two angles lie between the two lines

Alternate interior angles
Pairs of nonadjacent angles formed by the intersection of a transversal with two lines in such a way that all points in the interiors of these angles lie on different sides of the transversal and some interior points of each angle lie between the two lines

Trapezoid

trapezoid
A quadrilateral with exactly one pair of parallel sides

isosceles trapezoid
A trapezoid whose legs have equal lengths

legs of a trapezoid
The nonparallel sides of the trapezoid

bases of a trapezoid
The parallel sides of the trapezoid

base angles of a trapezoid
A pair of angles opposite a base of the trapezoid

Triangles

triangle
A polygon that has three sides

congruent triangles
Two triangles that have three pairs of corresponding sides with equal lengths and three pairs of corresponding angles with equal measures

similar triangles
Two triangles that have three pairs of corresponding angles with equal measures and that has corresponding sides proportional

scalene triangle
A triangle that has no sides with equal lengths

isosceles triangle
A triangle that has at least two sides with equal lengths

equilateral triangle
A triangle that has three sides with equal lengths

equiangular triangle
A triangle whose three angles have equal measures

right triangle
A triangle that contains a right angle

obtuse triangle
A triangle that contains an obtuse angle

acute triangle
A triangle that has three acute angles

median of a triangle
A line segment joining a vertex of a triangle to the midpoint of the opposite side of the triangle

centroid of a triangle
The point of intersection of the medians of a triangle

circumcenter of a triangle
The point of intersection of the perpendicular bisectors of the sides of a triangle

incenter of a triangle
The point of intersection of the bisectors of the angles of a triangle

altitude of a triangle (height)
A perpendicular line or line segment joining a vertex of the triangle to the line containing the side of the triangle opposite that vertex
The length of such a segment

orthocenter
The point of intersection of the altitudes of a triangle

Appendix E

Postulates

General Postulates (Axioms)

Section 6.1

1. A number may be substituted for its equal in any expression.
   If $a = b$ then a may be replaced by b, and b may be replaced by a in any expression.
   Substitution Postulate of Equality

2. Any number equals itself. $a = a$
   Reflexive Postulate of Equality

3. The members of an equation may be interchanged.
   If $a = b$ then $b = a$
   Symmetric Postulate of Equality

4. Numbers equal to the same number are equal to each other.
   If $a = b$ and $b = c$ then $a = c$
   Transitive Postulate of Equality

5. If equal numbers are added to equal numbers, the sums are equal.
   If $a = b$ and $c = d$ then $a + c = b + d$
   Addition Postulate of Equality

6. If equal numbers are subtracted from equal numbers, the differences are equal.
   If $a = b$ and $c = d$ then $a - c = b - d$
   Subtraction Postulate of Equality

7. If equal numbers are multiplied by equal numbers, the products are equal.
   If $a = b$ and $c = d$ then $ac = bd$
   Multiplication Postulate of Equality

8. If equal numbers are divided by nonzero equal numbers, the quotients are equal.
   If $a = b$ and $c = d$ where c and d are not zero, then $\frac{a}{c} = \frac{b}{d}$
   Division Postulate of Equality

9. Like powers of equal numbers are equal.
   If $a = b$ then $a^n = b^n$
   Powers Postulate of Equality

10. Like roots of equal numbers are equal.
    If $a = b$ then $\sqrt{a} = \sqrt{b}$
    Roots Postulate of Equality

11. If $a + b = c$, then $c > a$ and $c > b$.
    Whole Greater Than Part Postulate

12. If a and b are real numbers, then exactly one of the following statements must be true:
    $a = b$ or $a > b$ or $a < b$.
    Trichotomy Postulate of Inequality

13. If $a > b$ and $b > c$ then $a > c$.
Transitive Postulate of Inequality

14. If $a > b$ then $a + c > b + c$.
Addition Postulate of Inequality

15. If $a > b$ and $c > 0$ then $ac > bc$.
If $a > b$ and $c < 0$ then $ac < bc$.
Multiplication Postulate of Inequality

Section 7.1

16. If $a + b = c$, then $c > a$ and $c > b$.
Whole Greater Than Part Postulate

17. If a and b are real numbers, then exactly one of the following statements must be true: $a = b$ or $a > b$ or $a < b$.
Trichotomy Postulate

18. If $a > b$ and $b > c$ then $a > c$.
Transitive Postulate of Inequality

19. If $a > b$ then $a + c > b + c$.
Addition Postulate of Inequality

Geometric Postulates

Section 6.1

1. If points A, B and C are collinear with B between A and C, then $AB + BC = AC$.
Whole Equals Sum of Parts Postulate

2. If $\overrightarrow{BD}$ is between $\overrightarrow{BA}$ and $\overrightarrow{BC}$, and the rays are coplanar, then $\angle ABD + \angle DBC = \angle ABC$.
Whole Equals Sum of Parts Postulate

3. If two angles form a linear pair, then they are supplementary.
Linear Pair Postulate

Section 6.2

4. Two triangles are congruent if three sides of one triangle have the same lengths, respectively, as three sides of the other triangle.
Side Side Side Postulate (SSS Postulate)

5. Two triangles are congruent if two sides and the included angle of one triangle have the same lengths and measure respectively, as two sides and the included angle of the other triangle.
Side Angle Side Postulate (SAS Postulate)

6. Two triangles are congruent if two angles and the included side of one triangle have the same measures and length, respectively, as two angles and the included side of the other triangle.
Angle Side Angle Postulate (ASA Postulate)

Section 6.3

7. An angle has exactly one bisector, that is, a ray with the vertex of the angle as its endpoint which separates the angle into two angles of equal measure.
The Angle Bisector Postulate

Section 7.1

8. If two parallel lines are cut by a transversal, then corresponding angles have equal measures.
The Transversal Postulate

9. Given a line and a point not on the line, there is exactly one line through the point that is parallel to the given line.
The Parallel Postulate and Theorem

10. Given a line and a point not on the line, there can be no more than one line through the point that is parallel to the given line.
The Parallel Postulate

Given a line and a point not on the line, there can be no more than one line through the point that is perpendicular to the given line.
In a plane, given a line and a point on the line, there is exactly one line through the point that is perpendicular to the given line.
The Perpendicular Postulates

11. If A and B are two points, then there is exactly one line that contains points A and B OR
Two points determine a line.
The Unique Line Postulate

12. If $\overrightarrow{AB}$ is a ray then there is a unique point P on ray $\overrightarrow{AB}$ such that AP equals any specified length.
The Unique Point Postulate

A line contains an infinite number of points.
The Line Postulate

Section 9.1

13. If the measures of two angles of one triangle equal, respectively, the measures of two angles of another triangle, then the two triangles are similar.
Angle Angle Postulate (AA Postulate)

14. If a region is the union of two or more regions that have no interior points in common, then its area is the sum of their areas.
Area Addition Postulate

Section 10.1

15. If A and C are endpoints of an arc and B is another point on the arc, then $m\widehat{AB} + m\widehat{BC} = m\widehat{AC}$.
Arc Addition Postulate Or Whole Equals Sum Of Parts Postulate For Arcs

Appendix F

Theorems

Section 6.1

1. Supplements of equal angles are equal.
2. Supplements of the same angle are equal.
3. Complements of equal angles are equal.
4. Complements of the same angle are equal.
5. Vertical angles are equal.
6. All right angles are equal.
7. If two angles form a linear pair and are equal, then each is a right angle.
8. The bisector of an angle divides the angle into two angles, each ½ as large as the original angle.
9. The bisector of a line segment divides the segment into two segments, each ½ as large as the original line segment.

Section 6.2

10. Two triangles congruent to a third triangle are congruent to each other.

Section 6.3

11. If two sides of a triangle have equal lengths, then the angles opposite those sides have equal measures, or
    The base angles of an isosceles triangle are equal.
12. The bisector of the vertex angle of an isosceles triangle separates the triangle into two congruent triangles.
13. The bisector of the vertex angle of an isosceles triangle is the perpendicular bisector of the base of the triangle.
14. If two angles of a triangle have equal measures then the sides opposite those angles have equal lengths.
15. If a triangle is equilateral, then it is equiangular.
16. If a triangle is equiangular, then it is equilateral.
17. *ALTERNATE INTERIOR ANGLES THEOREM*
    If two parallel lines are cut by a transversal, then alternate interior angles have equal measures.
18. *SAME SIDE INTERIOR ANGLES THEOREM*
    If two parallel lines are cut by a transversal, then same side interior angles are supplementary.
19. The sum of the measures of the angles of a triangle is 180.
20. *EXTERIOR ANGLE THEOREM*
    The measure of an exterior angle of a triangle equals the sum of the measures of its two remote interior angles.
21. *REMOTE INTERIOR ANGLES THEOREM*
    The measure of an exterior angle of a triangle is greater than the measure of each of its remote interior angles.
22. If the measures of two angles of a triangle equal, respectively, the measures of two angles of another triangle, then the two remaining angles have equal measures.

23. In a right triangle, the two angles that are not right angles are complementary.

24. In a right triangle, the two angles that are not right angles are acute.

25. In a right triangle, the acute angles are complementary.

26. Each angle of an equiangular triangle has a measure of 60.

27. *AAS THEOREM*
If the measures of two angles and the length of the side opposite one of these angles equal, respectively, the measures of two angles and the length of the corresponding side in another triangle, then the two triangles are congruent.

28. *HL THEOREM*
If the lengths of the hypotenuse and a leg of one triangle equal, respectively, the lengths of the hypotenuse and a leg of another triangle, then the triangles are congruent.

29. *HA THEOREM*
If the length of the hypotenuse and the measure of an acute angle of one triangle equal, respectively, the length of the hypotenuse and the measure of an acute angle of another triangle, then the two triangles are congruent.

30. *GREATER SIDE THEOREM*
Given two sides of a triangle and the angles opposite those sides, the angle that has the greater measure is opposite the side that has the greater length.

31. *GREATER ANGLE THEOREM*
Given two angles of a triangle and the sides opposite those angles, the side with the greater length is opposite the angle with the greater measure.

32. *TRIANGULAR INEQUALITY THEOREM*
The sum of the lengths of two sides of a triangle is always greater than the length of the remaining side of the triangle.

33. *ADDITION THEOREM OF INEQUALITY*
If $a > b$ and $c > d$, then $a + c > b + d$.

Section 7.1 Exercises

34. Each angle of an equiangular triangle has a measure of 60.

35. HA THEOREM
If the length of the hypotenuse and the measure of an acute angle of one triangle equal, respectively, the length of the hypotenuse and the measure of an acute angle of another triangle, then the two triangles are congruent.

Section 7.2

36. *CORRESPONDING ANGLES/PARALLEL LINES THEOREM*
If two lines are cut by a transversal so that corresponding angles have equal measures, then the two lines are parallel.

37. *ALTERNATE INTERIOR ANGLE/PARALLEL LINE THEOREM*
If two lines are cut by a transversal so that alternate interior angles have equal measures, then the lines are parallel.

38. *SAME SIDE INTERIOR ANGLES/PARALLEL LINES THEOREM*
If two lines are cut by a transversal so that same side interior angles are supplementary, then the lines are parallel.

39. In a plane, if two lines are perpendicular to the same line, then the two lines are parallel.

Section 7.2 Exercises

40. Each point on the perpendicular bisector of a line segment is equidistant from the endpoints of the line segment.

Section 8.1

41. The sum of the measures of the angles of a quadrilateral is 360.
42. A diagonal of a parallelogram separates it into two congruent triangles.
43. The diagonals of a parallelogram bisect each other.
44. The opposite sides of a parallelogram have equal lengths.
45. The opposite angles of a parallelogram have equal measures.
46. The consecutive angles of a parallelogram are supplementary.

Section 8.2

47. A quadrilateral is a parallelogram if both pairs of opposite sides have equal lengths.
48. A quadrilateral is a parallelogram if two sides have equal lengths and are parallel.
49. A quadrilateral is a parallelogram if both pairs of opposite angles have equal measures.
50. A quadrilateral is a parallelogram if its diagonals bisect each other.

Section 8.3

51. A rhombus is a parallelogram.
52. The diagonals of a rhombus are perpendicular.
53. If the diagonals of a parallelogram are perpendicular, then the parallelogram is a rhombus.
54. A rectangle is a parallelogram.
55. The angles of a rectangle are right angles.
56. The diagonals of a rectangle have equal measures.
57. If the diagonals of a parallelogram have equal measures, then the parallelogram is a rectangle.
58. A square is a parallelogram.
59. A square is a rhombus.
60. A square is a rectangle.

Chapter 8 Exercises

61. Parallel lines are always the same distance apart.
62. The base angles of a trapezoid have equal measures.
63. The diagonals of an isosceles trapezoid are equal in length.

Section 9.1

AREA THEOREMS ARE LISTED AS THEOREMS 101–105

64. In a proportion, the product of the extremes equals the product of the means.
65. We may interchange the extremes of a proportion.
66. We may interchange the means of a proportion.

67. We may interchange both the extremes and the means of a proportion.

68. If $\frac{a}{b} = \frac{c}{d}$ then $\frac{a + b}{b} = \frac{c + d}{d}$ and $\frac{a - b}{b} = \frac{c - d}{d}$.

69. If $\frac{a}{b} = \frac{c}{d} = \frac{e}{f} = \frac{g}{h} = k$ then $k = \frac{a + c + e + g}{b + d + f + h}$.

70. *ALTITUDE TO THE HYPOTENUSE THEOREM.*
The altitude to the hypotenuse of a right triangle separates the triangle into two triangles that are similar to the original triangle and to each other.

71. The altitude to the hypotenuse of a right triangle separates the hypotenuse of the triangle into two segments (line segments) for which the following is true:

   a. Each leg of the triangle is a mean proportional between the hypotenuse and the segment of the hypotenuse next to that leg.

   b. The altitude is the mean proportional between the two segments of the hypotenuse.

72. *THE PYTHAGOREAN THEOREM*
In a right triangle, the square of the hypotenuse equals the sum of the squares of its legs.

73. *CONVERSE OF THE PYTHAGOREAN THEOREM.*
If the sum of the squares of two sides of a triangle equals the square of the remaining side of the triangle, then the triangle is a right triangle.

74. *30-60 RIGHT TRIANGLE THEOREM*
In a right triangle with an angle of 30 degrees, the length of the hypotenuse is twice the length of the side opposite the 30 degree angle. The length of the other side of the triangle, that is, the side opposite the 60 degree angle, is the product of $\sqrt{3}$ and the length of the side opposite the 30 degree angle.

75. *45-45 RIGHT TRIANGLE THEOREM.*
In a right triangle with an angle of 45 degrees, the length of the hypotenuse is the product of $\sqrt{2}$ and the length of a leg of the triangle.

Section 9.2

76. *TRIANGLE MIDPOINT THEOREM*
The line segment joining the midpoints of two sides of a triangle is parallel to the remaining side and one-half as long.

77. A line parallel to one side of a triangle cuts off a triangle that is similar to the given triangle.

78. A line parallel to one side of a triangle cuts the other two sides of the triangle proportionally.

79. The bisector of one angle of a triangle divides the opposite side of the triangle in the same ratio as the other two sides.

Section 10.1

80. Radii of the same circle are equal.
81. A tangent of a circle is perpendicular to the radius at the point of tangency.

Section 10.2

82. The measure of an inscribed angle is ½ the measure of the arc that it intercepts.
83. The measure of an angle with its vertex on the circle and whose sides are secant and tangent rays of the circle is ½ the measure of the arc that it intercepts.

Section 10.3

84. The measure of an angle whose vertex is an exterior point of the circle and whose sides are secant or tangent rays is ½ the absolute value of the difference in the measures of the arcs that it intercepts.

85. The measure of an angle whose vertex is in the interior of the circle and whose sides are formed by two intersecting secants is ½ the sum of the measures of the intercepted arcs.

86. Angles inscribed in the same arc have equal measures.

Section 10.4

87. If two chords intersect in the interior of a circle, the product of the lengths of the segments of one chord equals the product of the lengths of the segments of the other chord.

88. If a tangent segment and a secant segment intersect in an exterior point of the circle, the tangent segment is the mean proportional between the secant segment and the external segment of the secant segment.

89. If two secant segments intersect in an exterior point of the circle, the product of one of the segments and its external segment equals the product of the other secant segment and its external segment.

90. An angle inscribed in a semicircle is a right angle.

91. The perpendicular from the center of a circle to a chord of the circle, bisects the chord.

92. The two segments formed by two tangents to a circle from an external point of the circle have equal lengths.

Section 10.5

93. If m and n are two external tangents to two circles, then the line segment joining the points of tangency on line m has the same length as the line segment joining the points of tangency on line n.

94. If two circles are tangent, then their centers and their common point of tangency are collinear.

95. If two circles intersect in two points, then their line of centers is the perpendicular bisector of the chord joining the points of intersection.

Chapter 10 Exercises

96. Two chords of a circle have equal lengths iff their intercepted arcs have equal measures.

97. The opposite angles of an inscribed quadrilateral are supplementary.

Introduced in Section 4.1

98. The perimeter P of a rectangle with length L and width W is $P = 2L + 2W$.

99. The perimeter P of a square with side of length S is $P = 4S$.

100. The circumference C of a circle with radius r and diameter d is $C = 2\pi r$ or $C = \pi d$.

101. The area A of a rectangle with width W and length L is $P = LW$.

102. The area A of a square with side S is $A = S^2$.

103 The area A of a parallelogram with height h and base b is $A = bh$.

104. The area A of a trapezoid with height h and bases b_1 and b_2 is $A = \frac{1}{2}h(b_1 + b_2)$.

Appendix G

105. The area of a triangle with sides of lengths a, b and c is equal to $\sqrt{(s-a)(s-b)(s-c)s}$ where s is ½ the perimeter of the triangle.

Appendix G

Heron's Formula

This formula for finding the area of a triangle using the lengths of its sides is credited to Heron of Alexandria (circa 250 AD) although some historians believe that it was first proved by Archimedes. The proof requires some rather challenging algebraic manipulations.

THEOREM:

The area of a triangle with sides of lengths a, b and c is equal to $\sqrt{s(s-a)(s-b)(s-c)}$ where s is $\frac{1}{2}$ the perimeter of the triangle.

GIVEN:

Triangle ABC with sides of lengths a, b and c.

s is the semiperimeter; $s = \frac{1}{2}(a + b + c)$.

PROVE:

$A = \sqrt{s(s-a)(s-b)(s-c)}$ where A is the area of $\triangle$ ABC and

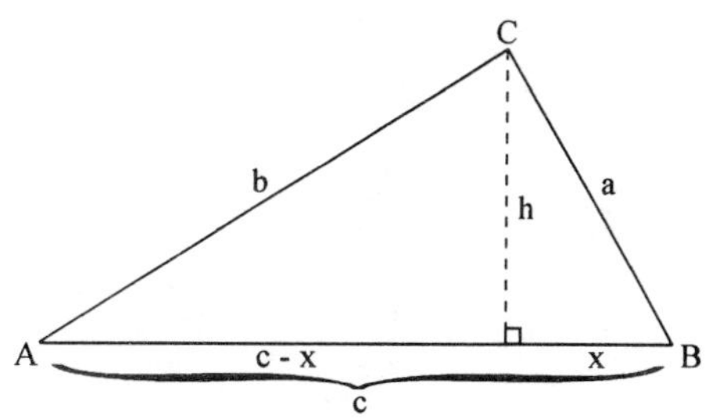

Similar proofs can be done for altitudes in other positions.

STEP 1

Triangle ABC with sides of lengths a, b and c.
s is the semiperimeter; $s = \frac{1}{2}(a + b + c)$.

STEP 2

Construct the altitude from C to $\overline{AB}$.

STEP 3

$$s = \frac{a+b+c}{2}$$

$a + b + c = 2s$	$a + b + c = 2s$	$a + b + c = 2s$
subtract 2a	subtract 2b	subtract 2c
$b + c - a = 2s - 2a$	$a + c - b = 2s - 2b$	$a + b - c = 2s - 2c$
$b + c - a = 2(s - a)$	$a + c - b = 2(s - b)$	$a + b - c = 2(s - c)$

We will use these results later.

STEP 4

$h^2 + x^2 = a^2$ Why?

$h^2 = a^2 - x^2$

STEP 5

$h^2 + (c - x)^2 = b^2$ Why?

$h^2 = b^2 - (c - x)^2$

$h^2 = b^2 - c^2 + 2cx - x^2$

We now use the **Transitive Postulate of Equality** to equate the results obtained in step 5 and step 4.

STEP 6

$a^2 - x^2 = b^2 - c^2 + 2cx - x^2$

Solving this equation for x, we obtain $x = \frac{a^2 - b^2 + c^2}{2c}$.

STEP 7

Now, back to step 4 where we have $h^2 = a^2 - x^2$.

Factoring, we obtain $h^2 = (a - x)(a + x)$.

We now replace each x in this equation with the result obtained in step 6.

STEP 8

$$h^2 = \left[a - \frac{a^2 - b^2 + c^2}{2c}\right]\left[a + \frac{a^2 - b^2 + c^2}{2c}\right]$$

$$h^2 = a^2 - \frac{(a^2 - b^2 + c^2)^2}{4c^2}$$

$4c^2h^2 = 4c^2a^2 - (a^2 - b^2 + c^2)^2$ Multiplying by $4c^2$

$4c^2h^2 = (2ac)^2 - (a^2 - b^2 + c^2)^2$ Rewriting $4c^2a^2$

STEP 9

$4c^2h^2 = [(2ac) - (a^2 - b^2 + c^2)]\,[(2ac) + (a^2 - b^2 + c^2)]$ Factoring

$4c^2h^2 = [2ac - a^2 + b^2 - c^2]\,[2ac + a^2 - b^2 + c^2]$

$4c^2h^2 = [b^2 - (a^2 - 2ac + c^2)]\,[(a + c)^2 - b^2]$

$4c^2h^2 = [b^2 - (a - c)^2]\,[(a + c)^2 - b^2]$

STEP 10

$4c^2h^2 = [b - (a - c)]\,[b + (a - c)]\,[a + c - b]\,[a + c + b]$ Factoring again

STEP 11

$4c^2h^2 = [b - a + c)]\,[b + a - c)]\,[a + c - b]\,[a + c + b]$ Removing Parentheses

Now we substitute the results obtained in step 3 into this final equation.

$\mathbf{b + c - a = 2(s-a)}$ $\mathbf{a + c - b = 2(s-b)}$ $\mathbf{a + b - c = 2(s-c)}$

STEP 12

$4c^2h^2 = 2(s - a)\,2(s - c)\,2(s - b)\,2s$ Note that $(a + b + c)$ is twice the semiperimeter.

$4c^2h^2 = 16(s - a)\,(s - b)\,(s - c)s$

$c^2h^2 = 4(s - a)\,(s - b)\,(s - c)s$

STEP 13

$ch = \sqrt{4\,(s - a)\,(s - b)\,(s - c)\,s}$ Extracting roots

$ch = 2\sqrt{(s - a)\,(s - b)\,(s - c)\,s}$

STEP 14

$\frac{1}{2}$ ch = $\sqrt{(s - a)(s - b)(s - c)s}$ Multiplying by $\frac{1}{2}$

But the area A of $\triangle$ ABC is the product of $\frac{1}{2}$ its base and height.

STEP 15

A = $\frac{1}{2}$ ch

Substituting the result obtained in step 14 into this equation, we obtain the conclusion:

STEP 16

A = $\sqrt{(s - a)(s - b)(s - c)s}$

And We Can Say With Pride, **QED.**

Appendix H

Symbols Used in the Text

$\overleftrightarrow{AB}$	line through points A and B
$\overline{AB}$	line segment with endpoints A and B
$\overrightarrow{AB}$	ray through point B with endpoint A
$\overset{\frown}{AB}$	arc with endpoints A and B
AB	length of line segment AB
$m\overset{\frown}{AB}$	measure of arc AB
$\measuredangle A$	angle with vertex A
$\angle A$	measure of angle A
$\triangle$	triangle
$\perp$	perpendicular
$\cong$	is congruent to
$\parallel$	parallel
$\sim$	is similar to
$\longrightarrow$	implies
$\longleftrightarrow$	one-to-one correspondence
π	pi
7°	7 degrees
$7'$	7 minutes
$7''$	7 seconds
$\therefore$	therefore

right angle

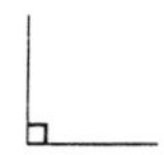

equal lengths

equal measures

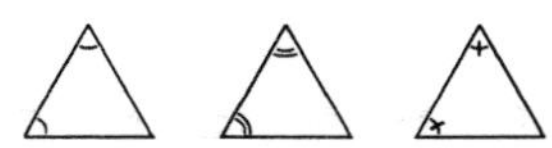

parallel lines

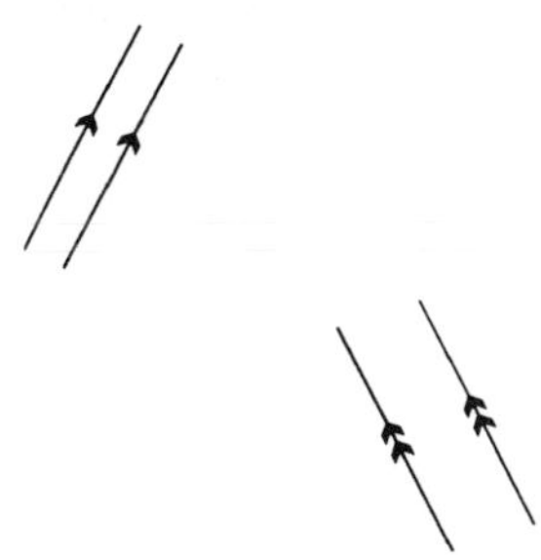

Appendix I

Formulas

Area and Perimeter Formulas for Plane Figures

b represents the base; h represents the corresponding height;
r represents the radius.

$P = 2L + 2W$	Perimeter of a rectangle
$P = 4S$	Perimeter of a square
$C = 2\pi r$	Circumference of a Circle
$C = \pi d$	Circumference of a Circle
$A = LW$	Area of a Rectangle
$A = S^2$	Area of a Square
$A = bh$	Area of A Parallelogram
$A = \frac{1}{2}bh$	Area of a Triangle
$A = \pi r^2$	Area of a Circle
$A = \frac{1}{2}h(b_1 + b_2)$	Area of a Trapezoid

Surface Area of Solids (A_s)

$A_s = 6e^2$	Cube (edge e)
$A_s = 2lw + 2lh + 2wh$	Rectangular Prism
$A_s = 2\pi rh + 2\pi r^2$	Right Circular Cylinder
$A_s = 4\pi r^2$	Sphere

Volume of Solids (B represents the area of the Base)

$V = Bh$ Rectangular Prism

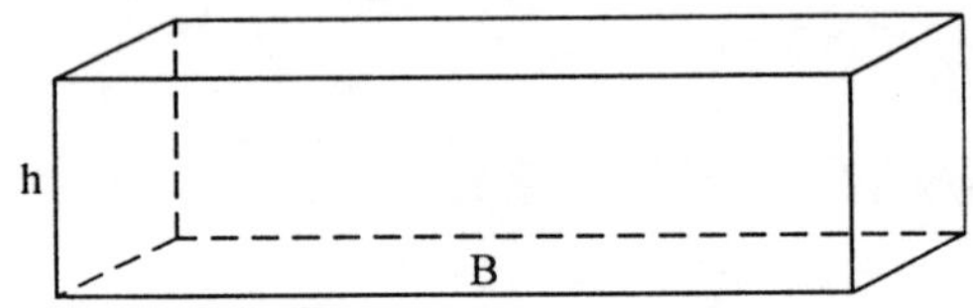

$V = \frac{1}{3}Bh$ Pyramid

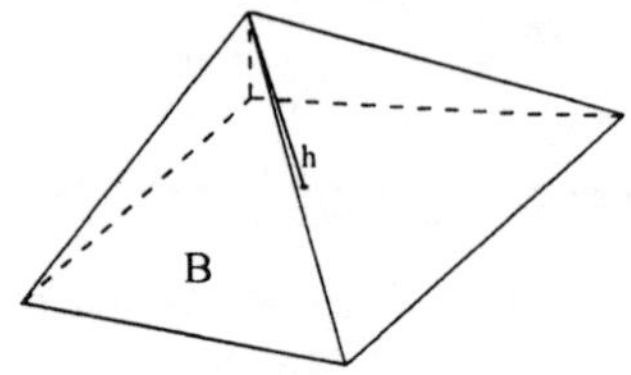

$V = Bh$ Triangular Prism (Any Prism)

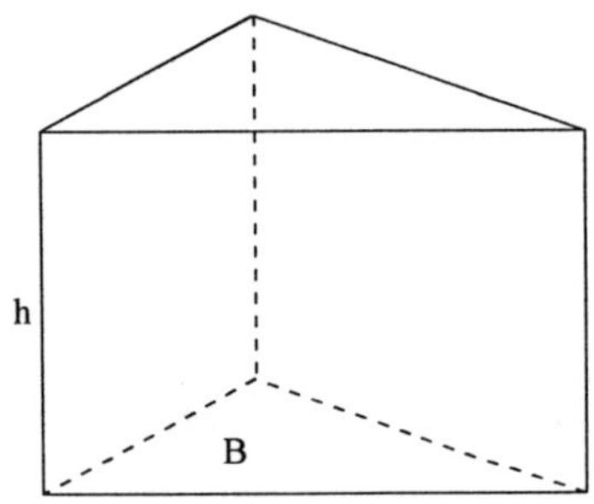

$V = \frac{1}{3}Bh$ Cone

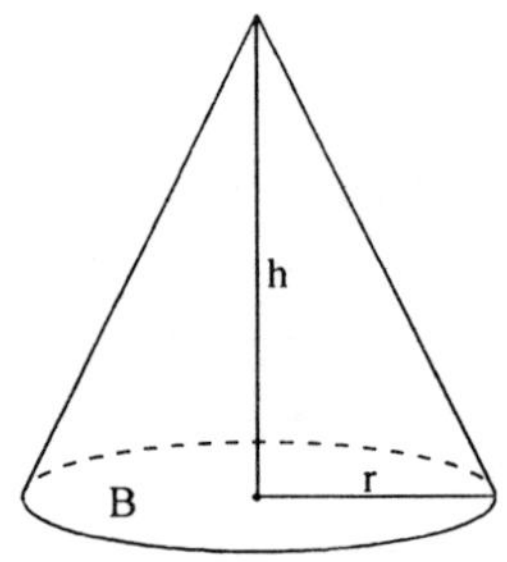

$V = Bh$ Cylinder

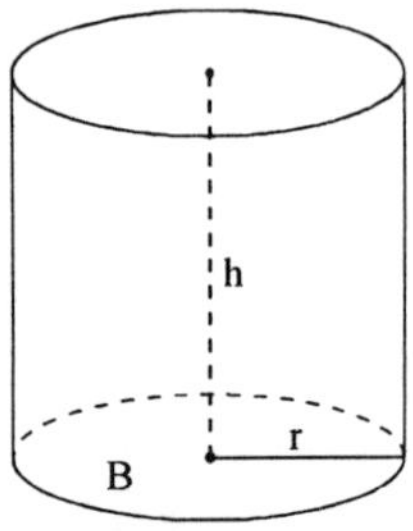

$V = \frac{4}{3}\pi r^3$ Sphere

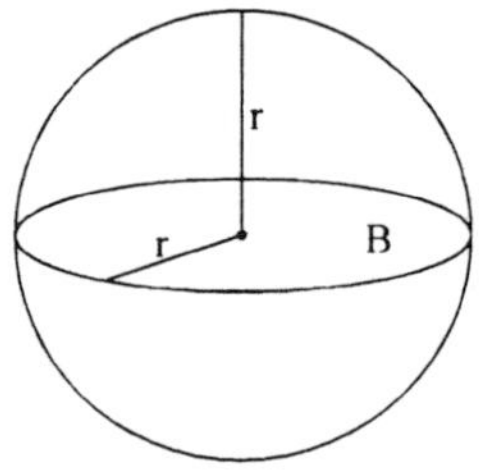

Appendix J

Answers to Selected Exercises

Exercises 1.1

1. $\overleftrightarrow{AB}$ or $\overleftrightarrow{BA}$, $\overleftrightarrow{BC}$ or $\overleftrightarrow{CB}$, $\overleftrightarrow{AC}$ or $\overleftrightarrow{CA}$;

3. $\overrightarrow{CA}$, $\overrightarrow{CB}$

5. a. ray-one endpoint and indefinite length/line segment-two endpoints and definite length
   b. line-no endpoint and indefinite length/line segment-two endpoints and definite length
   c. line segment-a set of points/length of a line segment—a number
7. a. $\overleftrightarrow{HE}$ or $\overleftrightarrow{HO}$ or $\overleftrightarrow{OE}$ (order of letters not important)
   b. $\overline{HE}$ or $\overline{EH}$
   c. $\overrightarrow{OE}$
   d. HE or EH
9. P
11. a. 3
    b. 3
    c. 5
    d. 2
    e. 5
    f. 5
    g. 2
    h. 2
    i. 4
    j. $|a + 3|$ or $|-3 - a|$ or $a + 3$

13. an infinite number of points

15. $\overline{AB}$
17. a. (3 , 5)
    b. (0 , 0)
    c. (0 , 2)
    d. (–3 , 4)
19. a. line segment-no thickness
    b. plane-no thickness, indefinite width and length
    c. point-no dimensions
    d. plane-no thickness, indefinite width and length

Exercises 1.2

1. $\angle$ Q , $\angle$ 2 , $\angle$ PQR or $\angle$ RQP

3. $\angle$ DPA , $\angle$ DPC , $\angle$ CPB or $\angle$ BPA

5. a. $\angle$ DCA or $\angle$ BCA
   b. $\angle$ CAE
   c. $\angle$ DCB
7. $\angle A = 90$, $\angle B = 50$, $\angle C = 115$, $\angle D = 70$
9. a. 150.21
   b. 28.7522
   c. 120.5033
   d. 15.95
   e. 42°48'
   f. 15'
   g. 0.75

h. 9.6'
i. 36'
j. 1988.52'
k. $1.468\overline{3}$
l. 205200"
m. 2520'
n. 37°24'54"

11. a. 90°
b. 60°
c. 15°
d. 7.5°
e. 96°

13. ∠ R = 27, ∠ P = 63

15a. ∠ A = 32, ∠ B = 15

17. ∠ N = 51, ∠ M = 106

19. $14 < x < 32$

Exercises 1.3

1. 55

3. adjacent

5. 13

7. 44

9. ∠ 2 = 180 – m, ∠ 3 = m, ∠ 4 = 180 – m

11. ∠ TAR = 55, complementary

13. a. r
b. ∡ 3 and ∡ 7
c. ∡ 3 and ∡ 8
d. ∡ 6 and ∡ 8
e. ∡ 5 and ∡ 6 or ∠ 7 and ∠ 8
f. ∡ 5 and ∡ 3 or ∠ 3 and ∠ 6

15. a. 25, 65
b. 50, 130
c. 90, 90
d. 56, 84
e. 27, 48
f. 32, 148

17. AB

19. ∠ A = 90, ∠ APB = 18

20. a. y = 125 , x = 55 , z = 55
b. x = 144
c. cannot determine
d. a = 153 , b = 27 , c = 153 , e = 27, f = 153 , g = 153, h = 27
e. x = 52
f. y = 107
g. y = 125
h. x = 158
i. x = 30
j. x = 15
k. cannot determine
l. x = 60, y = 60
m. x = 50, y = 60
n. x = 28, y = 96
o. x = 42

Exercises 2.1

1. Impossible to draw a right equilateral triangle and an obtuse equilateral triangle
2. b , d , g
3. square
5. a. 60
   b. 90
   c. 108
   d. 120
   e. $128\frac{4}{7}$
   f. 135
   g. 144
   h. $\frac{180(n-2)}{n}$
7. legs $\overline{AN}$ and $\overline{AP}$; base $\overline{NP}$; base angles $\angle$ N and $\angle$ P; vertex angle $\angle$ A
9. Sides $\overline{AE}$, $\overline{ED}$, $\overline{DC}$, $\overline{CB}$, $\overline{BA}$; vertices A, B, C, D, E ; angles $\angle$ A, $\angle$ B, $\angle$ C, $\angle$ D, $\angle$ E ; diagonals $\overline{AD}$, $\overline{AC}$, $\overline{BE}$, $\overline{BD}$, $\overline{CE}$
10. a. 30-60-90
    b. 36-54-90
    c. 35-40-105
    d. 157 all sides
    e. each leg has length $15\frac{1}{2}$
    f. 45-45-90
    g. 70-80-30
17. a. parallelogram
    b. rectangle
    c. square
18. a and b

Exercises 2.2

1. $\angle$ P = $\angle$ N, $\angle$ Q = $\angle$ A, $\angle$ R = $\angle$ D ; PQ = NA, QR = AD, PR = ND
2. $\triangle$ WYZ $\cong$ $\triangle$ CAB
3. b ,c ,e , f ,g , h
5. no
7. a , b , d
9. b , d
10. $\triangle$ ADB $\cong$ $\triangle$ ADC
11. 8
12. 6
13. 42
14. 11
15. x = 8 , y = 4
16. x = 4 , y = 1
17. ASA
18. $\triangle$ DBC
19. a. $\triangle$ ABC and $\triangle$ EFD
    b. none

Exercises 2.3

2. Show all pairs of corresponding sides proportional
   Show that two pairs of corresponding angles have equal measures
3. AA
4. yes
5. no
6. yes
7. yes
8. no
9. $\angle$ A = $\angle$ A and $\angle$ AMN = $\angle$ APE, $\triangle$ APE ~ $\triangle$ AMN
10. $\angle$ C = $\angle$ C and $\angle$ CAB = $\angle$ CEA, $\triangle$ CAB ~ $\triangle$ CEA, $\frac{CA}{CE} = \frac{CB}{CA} = \frac{AB}{EA}$
11. $\angle$ B = $\angle$ B, $\angle$ CAB = $\angle$ BEA, $\triangle$ CAB ~ $\triangle$ AEB, $\frac{CA}{AE} = \frac{CB}{AB} = \frac{AB}{EB}$
13. corresponding sides are not proportional, that is, $\frac{30}{12}$ is not equal to $\frac{21}{7}$ and $\frac{15}{5}$

14. yes, AA

15. $\frac{c}{e} = \frac{a}{d} = \frac{b}{f}$

16. $\frac{a}{c} = \frac{d}{e}$

21. a. $x = 5.25$, $y = 7.5$
    b. $x = 125$, $y = 95$
    c. $x = 7.5$, $y = 6$
    d. $x = 7\frac{5}{7}$ and $y = 3\frac{6}{7}$
    e. $x = 9$, $y = 15$
    f. $x = 12$
    g. $x = 4.5$
    h. $y = 21$
    i. $x = 14\frac{2}{5}$ and $y = 19\frac{1}{5}$
    j. cannot determine
    k. $x = 5\frac{1}{2}$ and $y = 24$
    l. $x = 10$
    m. $x = 20$, $y = 16$
    n. $x = 10$
    o. cannot determine
    p. $x = 2\sqrt{105}$

22. a. $93\frac{1}{3}$
    b. 12 feet

Exercises 3.1

1. infinite number

2. infinite number

5. a. chord
   b. center
   c. radius
   d. secant
   e. tangent
   f. diameter
   g. point of tangency or point of contact

6. a. $\overset{\frown}{BC}$ or $\overset{\frown}{AC}$
   b. $\overset{\frown}{ABC}$ or $\overset{\frown}{BAC}$
   c. $\overset{\frown}{AB}$ or $\overset{\frown}{ACB}$

7. a. 70°
   b. 290°
   c. 180°
   d. 10
   e. 30°
   f. 150
   g. 140
   h. 5

8. a. T
   b. T
   c. T
   d. T
   e. F

9. AP = BP (radii of the same circle have equal lengths). In an isosceles triangle the base angles have equal measures.

10. 19

11. a. $OB < r$
    b. $OC > r$
    c. $OA = r$

16. 35

17. a. 88
    b. 88
    c. 72
    d. 160
    e. 100
    f. 54
    g. 108
    h. 90

18. a. $52°$
b. $85°$
c. $x°$
d. $(3x + 2y)°$

19. a. $26°$
b. $42.5°$
c. $(\frac{1}{2} x)°$
d. $\frac{1}{2}(3x + 2y)°$

20. $x = \frac{40}{31}$, $y = -\frac{16}{31}$

Exercises 4.1

1. a. 60
b. 6
c. $6\sqrt{2}$
d. $6\sqrt{7}$

2. a. 64
b. $\frac{9}{25}$
c. 7
d. 75

3. a. 18
b. $\frac{7}{4}$
c. 6
d. $15\sqrt{10}$
e. $x^2 - 9$

4. a. 12
b. $\frac{7}{8}$
c. $\frac{15}{2}x^2 + \frac{59}{2}x + 28$
d. 3
e. $\frac{75}{2}\sqrt{6}$

5. a. 84
b. $108\sqrt{3}$

6. a. 24
b. 25
c. can't do now
d. 25
e. 64
f. 84
g. can't do now
h. 60
i. 120

7. 16
8. 4
9. Legs are 3 and 5.
10. $6\sqrt{2}$
11. 4
12. One altitude is twice the other one.
13. a. Area is doubled. b. Area is doubled. c. Area is quadrupled.
14. a. Area is quadrupled. b. Area is 9 times as much.

15. a. 625π
b. 3π
c. 324π
d. $x^2\pi + 8x\pi + 16\pi$

16. a. 48
b. 36
c. 92
d. $12x$

17. 18π

21. a. 4π
b. 39π
c. 16π
d. $25\pi - 50$

22. $A = \frac{1}{2}ap$

Exercises 4.2

1. a. 20
   b. 39
   c. 15
   d. $\sqrt{39}$
   e. $\sqrt{2}$
   f. $8\sqrt{2}$
   g. $6\sqrt{2}$
   h. 10
   i. 2
2. a, c, d, e
3. 60
4. $5\sqrt{14}$
5. Lengths from right to left:
   $\sqrt{2}, \sqrt{3}, 2, \sqrt{5}, \sqrt{6}, \sqrt{7}, \sqrt{8}, (2\sqrt{2})$
   2 is the only rational length. The others are irrational.
7. 9

Exercises 4.3

1. a. 600cm^2
   b. 360cm^2
   c. 1168cm^2
   d. 192cm^2
   e. $(6x^2 - 2y^2)\ \text{cm}^2$
   f. 376cm^2
2. a. 1056cm^2
   b. $170\pi\ \text{cm}^2$
   c. $384\pi\ \text{cm}^2$
   d. $(24x^2 + 72x + 54)\text{cm}^2$
   e. $152\pi\ \text{cm}^2$
   f. 1140cm^2

Exercises 5.1

1. Both are used to make conclusions/inductive (specific to general)/deductive(general to specific).
2. Observed pattern may not continue as conjectured/also, perception and interpretation of data may not be valid.
3. The generalization may not be true/if it is false, conclusions based upon it may be false.
6. Used to formulate conjectures.

Exercises 5.2

1. Undefined terms, definitions, postulates and theorems
3. To avoid circular definitions
4. We prove theorems/we accept postulates.
5. a. commutative property of addition
   b. distributive property
   c. multiplicative identity property
6. It is accepted as true/this does not mean true.

Exercises 5.3

1. a. Two lines form a linear pair (hypothesis).
   They are supplementary (conclusion).
   b. They are vertical angles (hypothesis).
   They are equal (conclusion).
   c. It is a square (hypothesis).
   It is a rectangle (conclusion).

d. It is a line (hypothesis).
   It has no endpoint (conclusion); or
   It has an endpoint (hypothesis)
   It is not a line (conclusion).
e. They are the base angles of an isosceles triangle (hypothesis).
   They are equal (conclusion).
f. They form a right angle (hypothesis).
   The two lines are perpendicular (conclusion).
g. It is a triangle (hypothesis).
   It does not have two right angles (conclusion).
   It has two right angles (hypothesis).
   It is not a triangle (conclusion).
h. A quadrilateral is equiangular (hypothesis).
   It is a rectangle (conclusion).
i. It is equilateral (hypothesis).
   A quadrilateral is a rhombus (conclusion).
j. Two triangles are congruent (hypothesis).
   They are similar (conclusion).

2. a. If two angles are supplementary, then they form a linear pair (converse).
   If two angles do not form a linear pair, then they are not supplementary (inverse).
   If two angles are not supplementary, then they do not form a linear pair (contrapositive).
   b. If a triangle is equilateral, then it is equiangular (converse).
   If a triangle is not equiangular, then it is not equilateral (inverse).
   If a triangle is not equilateral, then it is not equiangular (contrapositive).
   c. If an angle is acute, then its measure is less than 90 (converse).
   If an angle does not have a measure that is less than 90, then it is not acute (inverse).
   If an angle is not acute, then it does not have a measure that is less than 90 (contrapositive).
   d. If it is not a polygon, then it is a circle (converse).
   If it is not a circle, then it is a polygon (inverse).
   If it is a polygon, then it is not a circle (contrapositive).
3. a. If it is a fruit, then it is a banana (converse/false).
   If it is not a banana, then it is not a fruit (inverse/false).
   If it is not a fruit, then it is not a banana (contrapositive/true).
   b. If a figure is a polygon, then it is a rectangle (converse/false).
   If a figure is not a rectangle, then it is not a polygon (inverse/false).
   If a figure is not a polygon, then it is not a rectangle (contrapositive/true).
   c. If a number is divisible by 2, then it is even (converse/true).
   If a number is not even, then it is not divisible by 2 (inverse/true).
   If a number is not divisible by 2, then it is not even (contrapositive/true).
   d. If a triangle is equilateral, then it is isosceles (converse/true).
   If a triangle is not isosceles, then it is not equilateral (inverse/true).
   If a triangle is not equilateral, then it is not isosceles (contrapositive/false).
4. if and only if
5. A polygon is regular if it is equiangular and equilateral.
   A polygon is equiangular and equilateral if it is regular.
6. a. false
   b. true
   c. true
   d. false
   e. true
   f. true
   g. false

Exercises 5.4

1. no deduction/Angles do not form a vertical pair/no deduction/Angles have equal measures
2. not valid/valid/not valid
3. adbec
4. dacb
5. acdbe
6. cab

Exercises 6.1

1. 1. given
   2. Reflexive Postulate of Equality
   3. Addition Postulate of Equality
   4. Whole Equals Sum of Parts Postulate
   5. Whole Equals Sum of Parts postulate
   6. Substitution Postulate of Equality
3. 1. Given
   2. Whole Equals Sum of Parts Postulate
   3. Whole Equals Sum of Parts Postulate
   4. Substitution Postulate of Equality
   5. Reflexive Postulate of Equality
   6. Subtraction Postulate of Equality
6. 1. Given
   2. Definition of Linear Pair
   3. Linear Pair Postulate
   4. Theorem: Supplements of equal angles are equal

Exercises 6.2

16. a. yes
    b. yes
    c. no

Exercises 7.1

3. $\overline{RN}$, $\overline{MR}$, $\overline{MN}$
4. b
5. c
6. $\overline{BC}$
7. $\overline{AD}$
8. a and c
9. a. $x > 3$
   b. $x > 1$
   c. $x > 6$
   d. $x > 9$
   e. $x > \frac{5}{3}$
10. a. $41\frac{2}{3}$, 63, $75\frac{1}{3}$
    b. 27, 63, 90
13. 1. Given
    3. Whole Equals Sum of Parts Postulate
15. 1. Given
    3. Whole Equals Sum of Parts Postulate
17. 1. Remote Interior Angle Theorem
    2. Remote Interior Angle Theorem
19. 1. Given
    2. Remote Interior Angle Theorem
    3. Transitive Postulate of Inequality

Exercises 7.2

2. The corresponding angles have equal measures.
3. The coplanar lines are perpendicular to the same line.
4. The alternate interior angles have equal measures.

Exercises Chapter 8

4. a. T
   b. F
   c. T
   d. F
   e. F
   f. F
   g. T
   h. T
   i. F
   j. F
   k. F
5. a. 30, 30, 150, 150
   b. 45, 45, 135, 135
   c. 72, 72, 108, 108
   d. 90, 90, 90, 90
6. a. 30, 30, 150, 150
   b. 60, 60, 120, 120
   c. 82, 82, 98, 98
   d. 50, 50, 130, 130
7. a. $\angle D = \angle B = 80$; $\angle A = \angle C = 100$
   b. $\angle D = \angle B = 18.5$; $\angle A = \angle C = 161.5$
   c. $\angle D = \angle B = 73$; $\angle A = \angle C = 107$
   d. $\angle D = \angle B = 109$; $\angle A = \angle C = 71$
8. a. $\angle A = \angle C = 40$; $\angle D = \angle B = 140$
   b. $\angle A = \angle C = 149$; $\angle D = \angle B = 31$
   c. $\angle A = \angle C = 110$; $\angle D = \angle B = 70$
   d. $\angle A = \angle C = 73$; $\angle D = \angle B = 107$
18. Slopes of opposite sides are equal.
19. Two sides are contained in vertical lines and are therefore parallel and the other two sides are contained in horizontal lines and are therefore parallel.

Exercises 9.1

6. a. $AB = 12$, $AC = 6\sqrt{3}$
   b. $AB = 26$, $AC = 13\sqrt{3}$
   c. $AB = 30$, $AC = 15\sqrt{3}$
   d. $AB = 2$, $AC = \sqrt{3}$
   e. $AB = 4\sqrt{6}$, $AC = 6\sqrt{2}$
7. a. $BC = 5$, $AC = 5\sqrt{3}$
   b. $BC = 4$, $AC = 4\sqrt{3}$
   c. $BC = 2\sqrt{2}$, $AC = 2\sqrt{6}$
   d. $BC = \frac{\sqrt{15}}{2}$, $AC = \frac{3\sqrt{5}}{2}$
8. a. $BC = \sqrt{5}$, $AB = 2\sqrt{5}$
   b. $BC = 1$, $AB = 2$
   c. $BC = \frac{10\sqrt{3}}{3}$, $AB = \frac{20\sqrt{3}}{3}$
   d. $BC = 4\sqrt{3}$, $AB = 8\sqrt{3}$
9. a. $AB = \sqrt{2}$, $BC = 1$
   b. $AB = 12\sqrt{2}$, $BC = 12$
   c. $AB = \sqrt{6}$, $BC = \sqrt{3}$
   d. $AB = 4\sqrt{3}$, $BC = 2\sqrt{6}$
   e. $AB = 2\sqrt{5}$, $BC = \sqrt{10}$
10. a. $AC = BC = \sqrt{5}$
    b. $AC = BC = \sqrt{3}$
    c. $AC = BC = 6\sqrt{2}$
    d. $AC = BC = 2\sqrt{2}$
11. a. 50
    b. $25\sqrt{3}$
    c. $25\sqrt{3}$
12. a. 450
    b. $225\sqrt{3}$
    c. $225\sqrt{3}$
13. a. 9
    b. $3\sqrt{3}$
    c. $9\sqrt{3}$
14. a. 100
    b. $\left(\frac{100\sqrt{3}}{3}\right)$
    c. $100\sqrt{3}$
15. a. 25
    b. 36
    c. 192
    d. 72

15. e. 50
16. a. 49
    b. 112
    c. 22.5

17. a. $32 + 32\sqrt{3}$
    b. $12 + 4\sqrt{2} + 4\sqrt{3}$
    c. $18 + 18\sqrt{3}$
    d. $8 + 8\sqrt{3}$

18. a. $10, 10\sqrt{3}, 50\sqrt{3}$
    b. $12, 12\sqrt{3}, 72\sqrt{3}$
    c. $8, 8\sqrt{3}, 32\sqrt{3}$
    d. $6, 6\sqrt{3}, 18\sqrt{3}$
    e. $x, x\sqrt{3}, \left(\sqrt{\frac{3}{2}}\right) x^2$

19. a. $(\frac{1}{2})xy$
    b. diagonals

20. a. 24
    b. $24\sqrt{2}$
    c. $24\sqrt{3}$
    d. $24\sqrt{3}$

22. a. $36\sqrt{3}$
    b. $16\sqrt{3}$
    c. $4\sqrt{3}$
    d. $\left(\sqrt{\frac{3}{4}}\right)(a^2)$

23. a. 10 miles
    b. 10 miles

24. $90\sqrt{2}$

25. a. 10
    b. 10
    c. $8\sqrt{2}$
    d. 192
    e. 4
    f. $\sqrt{30}$
    g. $7\frac{9}{17}$
    h. $2\sqrt{105}$
    i. 2.2
    j. 3
    k. $x = 6.4, y = 4.8$
    l. $3\sqrt{10}$

Exercises 9.2

1. a. 12
   b. 7.5
   c. 15
   d. $4\frac{6}{7}$
   e. $5\sqrt{2} + 10$

2. a. 8
   b. 8
   c. $8\frac{4}{5}$
   d. 12
   e. $8 + 4\sqrt{3}$
   f. 10

7. a. 12
   b. 6.4
   c. 32
   d. 60

8. a. $1\frac{1}{3}$
   b. 3
   c. 15
   d. 12
   e. 5

Exercises Chapter 10

1. 58
2. 74
3. 71
4. 5
5. 50
6. 250
7. 204
8. 27.5
9. 120
10. 214
11. a. 30
    b. 16
    c. 34
    d. 130
12. a. 22.5
    b. 2
    c. 84
    d. 20
13. a. 6.8
    b. 4.5
14. 90
15. 90
16. a. 120
    b. 6
    c. $6\sqrt{3}$
    d. 60
17. 90
18. a. 200
    b. 20
    c. 140
    d. 40
    e. 40
    f. 60
    g. 120
    h. 100
19. a. 100
    b. 130
    c. 100
    d. 100
    e. 200
    f. 10
    g. 20
26. a. same center
    b. 23
    c. 7
27. a. same center
    b. 26
    c. 14